# Essential Biochemistry
# for Medicine

# Essential Biochemistry for Medicine

**Dr Mitchell Fry**

*Biological Sciences, University of Leeds*

(W)WILEY-BLACKWELL

A John Wiley & Sons, Ltd., Publication

This edition first published 2010
© 2010 John Wiley & Sons Ltd.

Wiley-Blackwell is an imprint of John Wiley & Sons, formed by the merger of Wiley's global Scientific, Technical and Medical business with Blackwell Publishing.

*Registered office*: John Wiley & Sons Ltd, The Atrium, Southern Gate, Chichester, West Sussex, PO19 8SQ, UK

*Other Editorial Offices*:
9600 Garsington Road, Oxford, OX4 2DQ, UK

111 River Street, Hoboken, NJ 07030-5774, USA

For details of our global editorial offices, for customer services and for information about how to apply for permission to reuse the copyright material in this book please see our website at www.wiley.com/ wiley-blackwell

*Library of Congress Cataloging-in-Publication Data*

Fry, Mitchell.
Essential biochemistry for medicine / Mitchell Fry.
    p. ; cm.
  Includes index.
  ISBN 978-0-470-74328-7 (cloth)
  1. Clinical biochemistry. 2. Biochemistry. I. Title.
  [DNLM: 1. Biochemical Phenomena. QU 34 F947e 2011]
    RB112.5.F79 2011
    616.07–dc22
                                        2010018059

A catalogue record for this book is available from the British Library

ISBN: 978-0-470-74328-7

Typeset in 9/11 Times by Laserwords Private Limited, Chennai, India

Printed in Singapore by Markono Print Media Pte Ltd.

First Impression 2010

# Contents

# Preface

To the uninitiated, biochemistry is a complex and intricate subject, but importantly it is a subject that underpins the biosciences, including medicine. As a university lecturer, and by training a biochemist, I have taught my subject to both 'my own' students, and to those on allied degree schemes and pre-clinical medicine. Of course, the lines so conveniently drawn (for teaching purposes) between the different bio-disciplines are very artificial; there is far more commonality than difference between these subjects. As a biochemist I am pleased to see the subject have such eminence, and rightly so, but at the same time it should not be delivered as a *fate accompli*, but rather as an aid to understand and clarify, a foundation to build upon and allow explanation. When I set out to write this book, it was not my intention to write a 'biochemistry' text, nor a 'medical' text, but rather something that provided a more complete picture. This is not meant to be a reference work, but rather a companion, and hopefully one that accurately reflects the type, depth and amount of biochemistry that is appropriate for medical and biomedical undergraduate students alike.

*Essential Biochemistry for Medicine* should provide a useful and helpful supplement to lectures and workshops, a biochemical–physiological–medical continuum, full of numerous medical examples, additional factual material and FOCUS sections on some favourite medical topics. I have tried to keep the book simply presented but packed with information, and it contains a full index to aid quick navigation. Indeed, it may be the only biochemistry book you need.

**Mitch Fry**

# CHAPTER 1

# Nutritional requirements

Food consists of water, macronutrients (carbohydrates, fats and proteins) and micronutrients (vitamins, minerals).

The amount of energy contained in food is typically measured in calories; a dietary calorie (C) is actually a thousand calories (kcal) (*a calorie is defined as the amount of heat energy that is required to increase the temperature of 1 gram of water by 1 degree Celsius*). Carbohydrates (a hydrated energy source) and proteins produce about 4 kcal per gram, while fat (an anhydrous energy source) produces about 9 kcal of heat per gram.

## 1.1 Carbohydrates and sugars

Carbohydrates are mostly used for energy; limited amounts can be stored in the liver and muscles in the form of glycogen. They vary widely in their complexity, and in the speed with which they are digested and metabolised. Sugars are a class of carbohydrates. Sugar monosaccharides include glucose, fructose and galactose. Disaccharides, composed of two monosaccharide units, include sucrose (common table sugar, glucose and fructose), lactose (found mostly in milk), glucose and galactose (Figure 1.1).

Polysaccharides are polymers of monosaccharides. Starch is a polysaccharide composed of amylose, an essentially linear polysaccharide, and amylopectin, a highly branched polysaccharide; both are polymers of D-Glucose.

Amylose (Figure 1.2) consists typically of 200–20 000 glucose units, which form a helix as a result of the bond angles between the units; the linkages between glucose molecules are referred to as 1–4 (between carbon 1 and carbon 4 of adjacent glucose molecules; see Figure 1.1 for numbering of ring structure).

Amylopectin differs from amylose in being highly branched. Short side chains of about 30 glucose units are attached with 1–6 linkages approximately every 20–30 glucose units along the chain.

*Essential Biochemistry for Medicine*   Dr Mitchell Fry
© 2010 John Wiley & Sons, Ltd

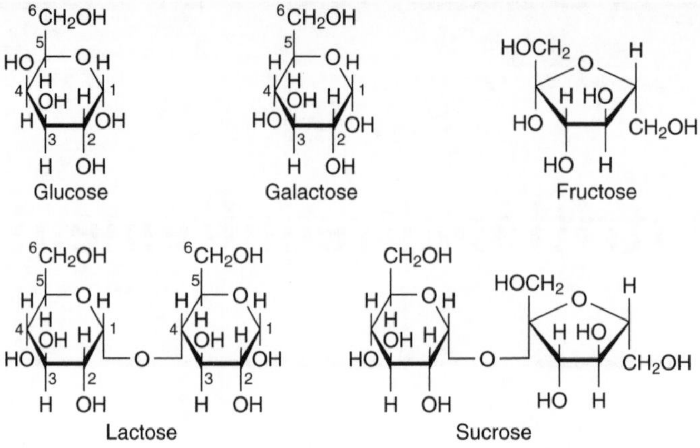

**Figure 1.1** Simple sugar structures.

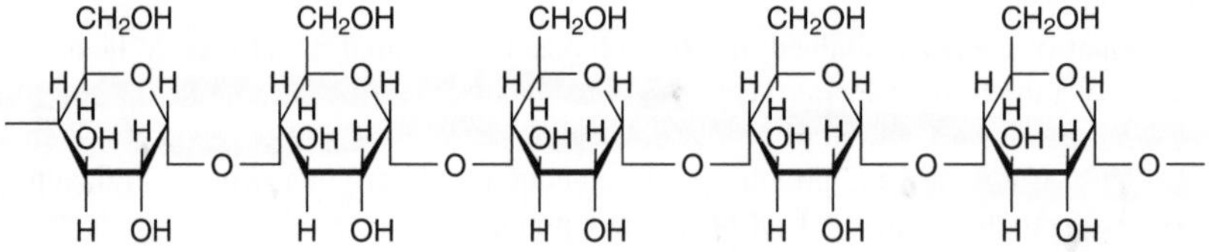

**Figure 1.2** Amylose.

## 1.2 Glycogen

Glycogen is similar in structure to amylopectin, but branches more frequently (Figure 1.3). Starch and glycogen polysaccharides provide structures that are used for energy storage, in plants and animals respectively.

Fibre is a polymer carbohydrate. Most fibre is derived from the cell walls of plants and is indigestible, for example cellulose.

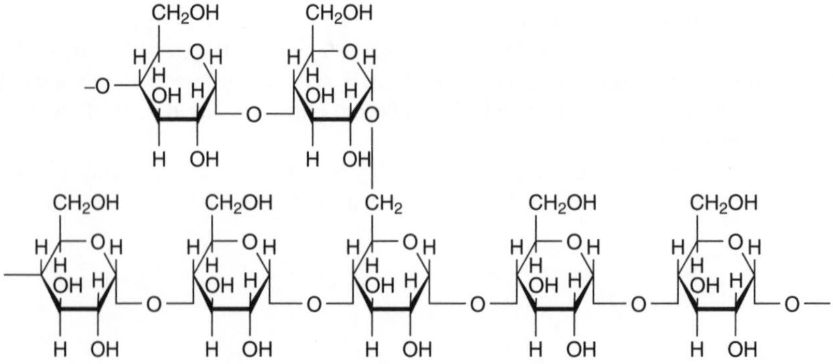

**Figure 1.3** Glycogen.

**Table 1.1**  Glycaemic indices of some common foods.

| Classification | GI range | Common examples |
|---|---|---|
| Low GI | 55 or less | Most fruit and vegetables (except potatoes, watermelon), grainy breads, pasta, legumes/pulses, milk, products that are low in carbohydrates (e.g. fish, eggs, meat, nuts, oils), apples. |
| Medium GI | 56–69 | Whole wheat products, brown rice, basmati rice, sweet potato, table sugar, ice cream. |
| High GI | 70–99 | Corn flakes, baked potato, watermelon, boiled white rice, croissant, white bread. |
| | 100 | Pure glucose. |

## 1.3  Glycaemic index

The ability of the body to digest different carbohydrates can be described by the glycaemic index (GI) (Table 1.1).

Low GI foods release glucose more slowly and steadily; high GI foods cause a more rapid rise in blood glucose levels. The latter are suitable for energy recovery after endurance exercise or for a person with diabetes experiencing hypoglycaemia. Only foods containing carbohydrates have a glycaemic index. Fats and proteins have little or no direct effect on blood sugar.

## 1.4  Lipids

Lipids (fats) provide energy and constitute a major energy store, as well as being an important body mass builder. Up to 20% of a healthy male's total weight comprises fat; this can be as much as 25% in females. Fat is a normal and healthy constituent of the body, cushioning internal organs from shock and providing heat insulation. As an energy source, fat contains over twice the energy per gram as does carbohydrate. Carbohydrates (in the form of glucose) are typically used to provide rapid energy, while fat is burned during sustained exercise. Fat is the primary fuel of choice during slow aerobic exercise, while glucose is used during fast aerobic or anaerobic exercise.

Lipids include fats and oils; oils tend to be liquid at room temperature, fats tend to be solid.

A fat molecule consists of one molecule of glycerol, bonded by dehydration synthesis (the loss of water) to three fatty acid molecules (this is a triacylglycerol, Figure 1.4). Fatty acids are

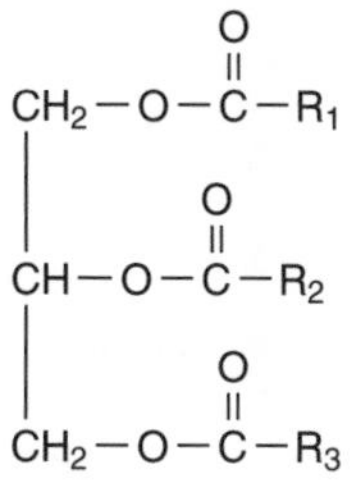

$$CH_2-O-\overset{\overset{\textstyle O}{\|}}{C}-R_1$$

$$CH-O-\overset{\overset{\textstyle O}{\|}}{C}-R_2$$

$$CH_2-O-\overset{\overset{\textstyle O}{\|}}{C}-R_3$$

**Figure 1.4**  A triacylglycerol molecule. The glycerol backbone is bonded to three fatty acids ($R_1$, $R_2$ and $R_3$).

long-chain hydrocarbon molecules which contain a carboxylic acid group (COOH) on one end. During dehydration synthesis, three fatty acid molecules each bond to one of the three –OH groups of the glycerol.

Phospholipids (Figure 1.5) are an important class of lipid; they are the fundamental building blocks of cellular membranes and a major constituent of surfactant, the film that occupies

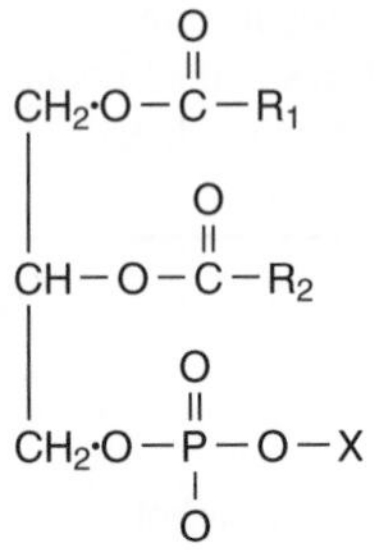

$$
\begin{array}{c}
\phantom{CH_2 \cdot O - } \overset{\displaystyle O}{\overset{\|}{\phantom{C}}} \\[-2pt]
CH_2 \cdot O - C - R_1 \\[4pt]
\phantom{CH_2 \cdot O - } \overset{\displaystyle O}{\overset{\|}{\phantom{C}}} \\[-2pt]
CH - O - C - R_2 \\[4pt]
\phantom{CH_2 \cdot O - } \overset{\displaystyle O}{\overset{\|}{\phantom{P}}} \\[-2pt]
CH_2 \cdot O - P - O - X \\[2pt]
\phantom{CH_2 \cdot O - P - }\underset{\displaystyle O}{\phantom{O}}
\end{array}
$$

**Figure 1.5** A phospholipid molecule. In a phospholipid molecule, one fatty acid is replaced with a phosphate group, to which is attached (X) a nitrogen-containing molecule, for example choline, ethanolamine, serine or inositol (giving the phospholipid phosphatidylcholine, phosphatidylethanolamine, phosphatidylserine or phosphatidylinositol, respectively).

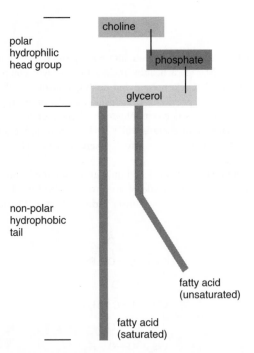

**Figure 1.6** An amphipathic phospholipid (phosphatidylcholine) molecule. The amphipathic phospholipid molecule contains a polar head group and non-polar tail; this is crucial to the ability of such molecules to self assemble in water to form lipid membranes.

the air/liquid interfaces in the lung. Consisting of a polar (charged) head group and a pair of non-polar fatty acid tails, they are amphipathic molecules (Figure 1.6); 'amphipathic' describes the tendency of these molecules to assemble at interfaces between polar and non-polar phases.

Lipids may be saturated or unsaturated (or polyunsaturated), depending on whether their fatty acids contain carbon–carbon double bonds (Figure 1.7). A cis conformation around the double bond causes a 'kink' in the fatty acid chain, preventing adjacent chains from closely aligning, and therefore increasing the fluidity (lower melting point).

Common sources of saturated fats are beef, veal, lamb, pork and dairy products made from whole milk, as well as coconut and palm oil. Common sources of mono-unsaturated fats are olive oil and peanut oil, while poly-unsaturated fats are found in sunflower and sesame oils.

Food manufacturers frequently employ trans-fats (Figure 1.8) to improve taste, consistency and shelf-life. These may further be hydrogenated. Artificially hydrogenating (adding hydrogen atoms to) vegetable oil makes it more solid. Margarine and shortening contain hydrogenated fats.

Although some amount of saturated fat in the diet is beneficial, general dietary advice is to avoid saturated fats. Trans-fats have been implicated in raising total cholesterol, lowering HDL ('good cholesterol') and raising LDL ('bad cholesterol').

Certain fatty acids must be included in the food we consume (essential fatty acids). Two essential fatty acids are the polyunsaturated omega-3 (linolenic acid) and omega-6 fatty acids (linoleic acid) (Figure 1.9). Modern diets often contain an overabundance of the omega-6 fatty acids and a deficiency in omega-3 fatty acids. Fish oil is a good source of omega-3 fatty acids, as is flaxseed oil.

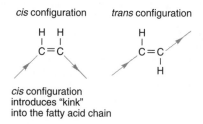

**Figure 1.7**   Cis and trans unsaturated fatty acids.

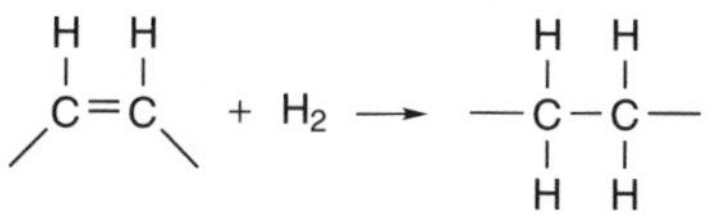

**Figure 1.8**   Hydrogenation of the double (unsaturated) carbon bond.

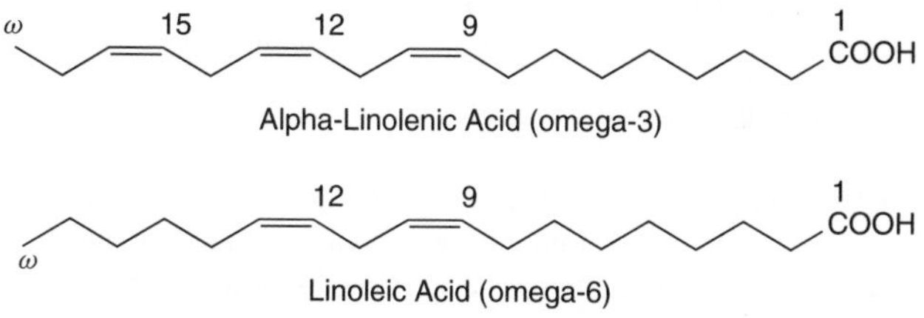

**Figure 1.9**   Linolenic and linoleic acids.

# 1.5 Proteins and amino acids

Proteins and amino acids can also be used for energy (during starvation or extreme exercise), but are primarily used to provide tissue mass. Amino acids that cannot be synthesised by the body are referred to as essential amino acids (Table 1.2).

All tissues have some capability for synthesis of non-essential amino acids, through the inter-conversion (transamination) of amino acids and their keto-acid carbon skeletons (Figure 1.10).

Glucogenic amino acids are those that can give rise to tricarboxylic acid (TCA) cycle inter-mediates, such as pyruvate, $\alpha$-ketoglutarate and oxaloacetate, which can then be used for the net synthesis of glucose through the gluconeogenesis pathway. Lysine and leucine are the only amino acids that are solely ketogenic; these give rise to acetyl-CoA or acetoacetyl-CoA, both of which can enter the TCA cycle, but neither of which can bring about *net* glucose production. In times of dietary surplus, the potentially toxic nitrogen of amino acids is eliminated via transamination, deamination and urea formation.

Unlike fat and carbohydrate, nitrogen has no designated storage depots in the body. Since the half-life of many proteins is short (of the order of hours), insufficient dietary quantities of even one amino acid can quickly limit the synthesis and lower the body levels of many essential

**Table 1.2**    Essential vs. non-essential amino acids.

| Essential | Arginine[a] | Methionine[a] | Phenylalanine[a] | Histidine | Isoleucine | Leucine | Lysine |
|---|---|---|---|---|---|---|---|
| | Threonine | Typtophan | Valine | | | | |
| Non-essential | Alanine | Asparagine | Aspartate | Cysteine | Glutamate | Glutamine | Glycine |
| | Proline | Serine | Tyrosine | | | | |

[a] Arginine and methionine are synthesised *in vivo*, but not in sufficient amounts, while phenylalanine is required in higher amounts to form tyrosine.

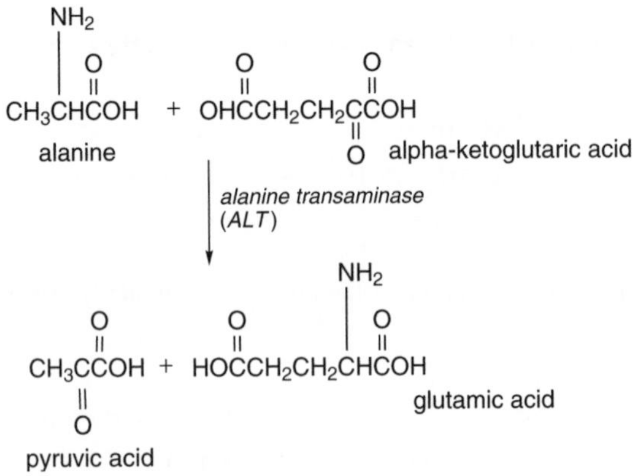

**Figure 1.10**   The transamination of alanine to glutamic acid. In the transamination of alanine, the amino group is transferred to $\alpha$-ketoglutarate, producing a 'new' amino acid, glutamic acid. The corresponding $\alpha$-keto acid (of alanine) is also formed (pyruvic acid).

proteins. Young children, adults recovering from major illness and pregnant women are often in positive nitrogen balance; intake of nitrogen exceeds loss as net protein synthesis proceeds.

## 1.6 Biological value

The biological value of dietary proteins is related to the extent to which they provide all the necessary amino acids. Proteins of animal origin generally have a high biological value, whereas plant proteins may be deficient in lysine, methionine and tryptophan, and are generally less digestible than animal proteins.

## 1.7 Other energy sources

Ketone bodies, produced mainly in the mitochondria of liver cells from acetyl-CoA, provide much of the energy to heart tissue, and during starvation to the brain. They include acetone, acetoacetate and $\beta$-hydroxybutyrate (Figure 1.11). The levels of acetone are much lower than those of the other two ketone bodies; it cannot be converted back to acetyl-CoA and so is excreted in the urine or breathed out.

Acetyl-CoA results from the breakdown of carbohydrates, lipids and certain amino acids. Normally, the acetyl group of acetyl-CoA enters the citric acid cycle to generate energy in the form of ATP, but it can also form ketone bodies; this happens if acetyl-CoA levels are high and the TCA cycle capacity is exceeded (the limiting factor is the availability of oxaloacetate). The creation of ketone bodies is also known as ketogenesis; acetoacetate and $\beta$-hydroxybutyrate are acidic, and if levels of ketone bodies are too high then the pH of the blood falls, resulting in a condition known as ketoacidosis (ketosis). This happens in untreated type I diabetes (diabetic ketosis) and also in alcoholics after heavy drinking and subsequent starvation (alcoholic ketosis).

Ethanol (ethyl alcohol) that is consumed passes to the liver, where it is first converted into acetate, then into ketone bodies. In alcoholic ketosis, alcohol causes dehydration and indirectly blocks the first step of gluconeogenesis; the inhibition of gluconeogenesis by ethanol is caused by the alcohol dehydrogenase reaction, which decreases the [free $NAD^+$]/[free NADH] ratio. This lowers the concentration of pyruvate, which is the immediate cause of the inhibition of gluconeogenesis from lactate. A low pyruvate concentration reduces the rate of the pyruvate carboxylase reaction, one of the rate-limiting reactions of gluconeogenesis. The body is unable to synthesise enough glucose to meet its needs, thus creating an energy crisis resulting in fatty acid metabolism and ketone body formation.

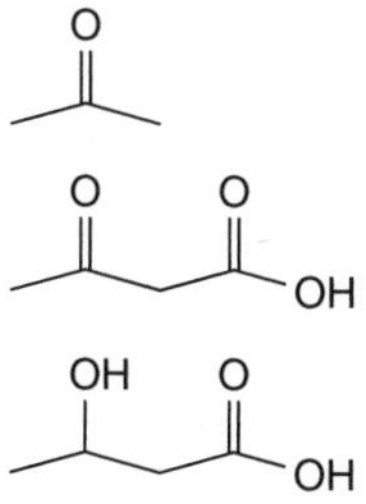

Acetone – $(CH_3)_2CO$

Acetoacetic acid – $CH_3C(O)CH_2CO_2H$

$\beta$-hydroxybutyric acid – $CH_3C(OH)CH_2CO_2H$

**Figure 1.11**  Ketone bodies.

## How the liver breaks down alcohol

(i) Alcohol dehydrogenase catalyses the oxidation of alcohol to acetaldehyde.

(ii) Acetaldehyde dehydrogenase oxidises the acetaldehyde to acetyl-CoA. In these reactions, coenzyme $NAD^+$ is reduced to NADH; therefore, when alcohol is metabolised, $NAD^+$ concentrations are reduced while NADH increases.

During alcohol metabolism, $NAD^+$ becomes unavailable to the many other vital body processes for which it is needed, including glycolysis, the TCA cycle and the mitochondrial respiratory chain. Without $NAD^+$, the energy pathway is blocked and alternative routes are taken, with serious physical consequences:

The accumulation of hydrogen ions shifts the body's balance towards acid.

The depletion of $NAD^+$ (and the change to the NADH to $NAD^+$ ratio) slows the TCA cycle, resulting in a build-up of pyruvate and acetyl-CoA. Excess acetyl-CoA increases fatty acid synthesis and fat deposits in the liver (fatty liver). An accumulation of fat in the liver can be observed after just a single night of heavy drinking.

---

Glycerol is an important by-product of fat metabolism. Fatty acids, from triacylglycerols, are metabolised in the liver and peripheral tissue via $\beta$-oxidation into acetyl-CoA; the remaining glycerol is an important source of glucose (in gluconeogenesis) in the liver. Glycerol enters the reaction sequence:

*Glycerol $\rightarrow$ glycerol 3-phosphate $\rightarrow$ dihydroxyacetone $\rightarrow$ glyceraldehyde 3-phosphate*

Glycerol kinase (which converts glycerol to glycerol 3-phosphate) is a liver enzyme; in this way glycerol enters the gluconeogenesis pathway.

## 1.8 Vitamins

Vitamins are small-molecular-weight organic substances that are necessary for essential biochemical reactions, growth, vitality and the normal functioning of the body. They must be supplied in the diet or in the form of dietary supplements. Central to the definition of a vitamin is that a lack of it will produce a specific deficiency syndrome, and supplying it will cure that deficiency.

- **Fat-soluble vitamins.** Fat-soluble vitamins include vitamins A, D, E and K; they are absorbed, transported, metabolised and stored along with fat.

- **Water-soluble vitamins.** Water-soluble vitamins include vitamin C, and those of the B-complex group: biotin, folate, niacin, pantothenic acid, riboflavin, thiamine, vitamin $B_6$ and vitamin $B_{12}$. They function mainly as coenzymes and prosthetic groups.

The proper role of vitamin supplementation is in the treatment of deficiency in patients who have inadequate intake or absorption, or an increased requirement. In pregnancy, supplementation of certain vitamins may be recommended. The role of vitamins in the body is summarised in Table 1.3.

**Table 1.3** Vitamins and their roles.

| Vitamin | Role in | Natural sources | RDA[a] | Toxicity? |
|---|---|---|---|---|
| Vitamin A – retinol/caro- tene (other carotenoids) Fat soluble | Skin and mucous membranes; growth of strong bones, skin, hair, teeth and gums; resistance to respiratory infections; counteracts night blindness and aids in the treatment of many eye disorders; antioxidant properties. | Green leafy vegetables, carrots, red peppers, sweet potatoes, yellow fruits, apricots, fish-liver oil and eggs. | 1 mg retinol equiva- lent | Yes. Hypervitaminosis A leads to loss of appetite, dry itchy skin often with peeling, intense headaches and enlarged liver. |
| Vitamin B1 – thiamine Water soluble | Prevention of beriberi, a disease of the nervous system; cell growth; normal carbohydrate metabolism; nervous system functioning; stress conditions; anxiety and trauma. | Dried yeast, spirulina, whole wheat, oats, peanuts, soybeans, green vegetables and milk. | 1.5 mg | No, easily excreted from the body, not stored. |
| Vitamin B2 – riboflavin | Cell growth; general metabolism; vision and eye fatigue; skin, nails and hair; mouth sores. | Milk, yeast, cheese, leafy green vegetables, mushrooms, fish and eggs. | 1.7 mg | No |
| Vitamin B3 – niacin, nicotinic acid | Reducing blood pressure; lowering cholesterol levels; preventing pellagra. | Lean meat, whole wheat, tuna, anchovy, yeast, eggs, peanuts and avocados. | 19 mg | No, but contra indicatory for individuals with peptic ulcers or diabetes. |
| Vitamin B5 – pantothenic acid | Utilisation of other nutrients; growth and development of the nervous system; metabolism of fat and sugars. | Fish, whole grains, wheat germ, green vegetables and brewer's yeast. | 7 mg | Few, but very large doses have been known to produce lack of co-ordination in movement and impairment of sensation. |
| Vitamin B6 – pyridoxine | Amino acid metabolism; absorption of $B_{12}$; production of antibodies and red blood cells. | Bananas, brewer's yeast, wheat germ, green and red peppers, nuts, molasses and eggs. | 2 mg | Daily doses of over 500 mg should be avoided. Doses over 2 g can lead to neurological disorders. |

*(continued overleaf)*

**Table 1.3**   (*continued*)

| Vitamin | Role in | Natural sources | RDA$^a$ | Toxicity? |
|---|---|---|---|---|
| Vitamin B12 (cyancobali-min, cobalamin) | Protein and fatty acid metabolism; production of red blood cells; maintenance of nervous system, concentration and memory. | Clams, oysters, beef, eggs and dairy products. Not found in many plant products; strict vegetarians may need to consider B$_{12}$ supplements. | 2 mg | No |
| Biotin – member of the B-complex family Water soluble | Metabolism of fats; synthesis of ascorbic acid; healthy skin; hair balding and greying. | Soya beans, brown rice, nuts, fruit, brewer's yeast and milk. It can be synthesised by intestinal bacteria. | 100 µg | No |
| Folic acid – folate – cofactor of the B-complex vitamins | Cell growth; nucleic acid and protein synthesis; formation of red blood cells and protein metabolism; protects against intestinal parasites; megaloblastic anaemia is caused by folate deficiency. | Raw leafy vegetables, fruit, carrots, avocados, beans and whole wheat. | 200 µg | None, up to 5 mg a day |
| Vitamin C – ascorbic acid Water soluble | Cell growth; bones, gums and teeth; bacterial resistance; antioxidant activity; absorption of iron. | Citrus fruits, hot chilli peppers, broccoli, tomatoes, green leafy vegetables and sweet potatoes. | 60 mg | Vitamin C is non-toxic but not recommended for individuals with peptic ulcers. |
| Vitamin D – calciferol Fat soluble | Calcium and phosphate metabolism; assimilation of vitamin A; treatment of conjunctivitis and rickets. | Sardines, herring and dairy products. Produced by interaction with sunlight and oils (cholesterol) in the skin. | 10 µg | Yes, at high doses. |

(*continued overleaf*)

**Table 1.3** (*continued*)

| Vitamin | Role in | Natural sources | RDA[a] | Toxicity? |
|---|---|---|---|---|
| Vitamin E – tocopherol Fat Soluble | Enhancement of vitamin A function; healing of scar tissue; prevention of anaemia. High alcohol intake may increase oxidation of alpha-tocopherol; increased demand in premature infants and patients with malabsorption. | Wheat germ, whole grains, vegetable oils, soya beans, nuts, apples, apricots and green vegetables. | 1 mg $\alpha$-tocopherol equivalent | No, but some effects at very high doses. |
| Vitamin K – menadione Fat soluble | Regulation of blood clotting | Leafy green vegetables, cauliflower, soybean oil, kelp, cereal grain products, fruits and yoghurt. | 80 mg | No, but supplementation with synthetic vitamin K, exceeding 500 mg, is not recommended. |

[a] RDA – recommended daily allowance, USA.

## 1.9 Minerals

Minerals include sodium, potassium, calcium, phosphorus, magnesium, manganese, sulphur, cobalt and chlorine; trace minerals include iron, zinc, copper, selenium, iodine, fluorine and chromium. Their roles may be generalised within the areas of providing structure in the formation of bones and teeth, maintenance of normal heart rhythm, muscle contractility, neural conductivity, acid–base balance and the regulation of cellular metabolism through their activity/structural associations with enzymes and hormones. The daily requirements of minerals can be obtained from a well-balanced diet.

## 1.5  Minerals

# CHAPTER 2
# Metabolism and energy

Metabolism comprises enzyme-catalysed chemical reactions and pathways that supply the energy and raw materials required for the body to function, grow and reproduce (catabolic = breakdown, anabolic = build-up). Under 'ideal' conditions these systems must operate efficiently, but they must also be able to respond to unexpected shortages and demands, for example fighting, natural disasters, pregnancy, lactation, famine, injury and disease. Enzymes are crucial to metabolism because they allow organisms to drive desirable but energetically unfavourable reactions (usually anabolic) by coupling them to favourable ones (usually catabolic). Enzymes also allow for the regulation of metabolic pathways.

## 2.1   A metabolic strategy

Central to energy metabolism is the maintenance of a blood glucose concentration of about 5 mM, which is essential for normal cerebral function. Confusion and coma can result if blood glucose falls below 3 mM, while serious vascular damage may occur if it exceeds 8 mM for significant periods (see Table 2.1, Section 2.20).

After a meal, glucose concentrations in the portal venous blood can easily reach 20 mM. Much of this excess will be removed by the liver. Stimulation of insulin release results in the uptake of glucose by the peripheral tissues (muscle and adipose tissue). Surplus glucose is stored locally in tissues as glycogen, but mostly it is converted into fats.

A blood glucose of 5 mM is sufficient for just a few minutes of normal activity. This level of glucose is actively defended by the liver, which removes glucose when too high, and replenishes it when too low. Both the supply and the demand for glucose may vary more than 20-fold over a 24 hour period; both can change suddenly and sometimes without warning. The liver can both uptake and secrete glucose; it is one of the few tissues in the body to permit bi-directional glucose transport (enterocytes and kidney are others). Internal hepatic glucose concentrations are similar to those in the bloodstream. Most tissues present a major barrier to glucose entry at the plasma membrane, and glucose is only allowed to enter the cells during periods of intense metabolic activity and in response to circulating insulin. Unlike the liver, most tissues have no export pathway for glucose; their glycogen reserves are strictly for internal use.

*Essential Biochemistry for Medicine*   Dr Mitchell Fry
© 2010 John Wiley & Sons, Ltd

**Table 2.1**   The overall response to blood glucose levels

| Blood (glucose) (mM) | Physiological assessment | Short-term response | Long-term response |
|---|---|---|---|
| >6 | Too high | Convert to glycogen (glycogenesis) | Make triglyceride (lipogenesis) |
| ~5 | Normal | No action | No action |
| <4 | Too low | Convert glycogen to glucose (glycogenolysis) | Make glucose (gluconeogenesis) |

Liver glycogen is the short-term glucose buffer. Long-term surplus glucose is converted into fats via lipogenesis. Long-term shortages are made good via gluconeogenesis from non-carbohydrate precursors.

**Table 2.2**   Energy and water content of fats and carbohydrates

|  | Fats | Carbohydrates |
|---|---|---|
| Energy content (kcal/g) | 9 | 4 |
| Water content (%) | 0 | ~80 |

Although glucose is the essential cerebral energy supply, most tissues preferentially use fat as an energy source. Fats affect a number of metabolic controls that suppress the oxidation of carbohydrates. Most aerobic tissues, such as cardiac muscle, 'prefer' fats; this is reinforced by insulin signalling, and in the absence of insulin most tissues are essentially impermeable to glucose.

Carbohydrate stores are wet and bulky and their energy density is low. They are useful for emergencies and short-term requirements, but are not a cost-effective fuel for longer-term requirements (see Table 2.2).

Carbohydrate stores supply about three hours of average waking activity. The strategy is therefore to conserve limited carbohydrate stores (for emergency use), while fuelling basal metabolic activity with fats.

Liver glycogen provides a short-term source of carbohydrate for emergency use. It is mobilised by adrenalin and glucagon, signalling via calcium ions and cyclic adenosine monophosphate (cAMP), but the total reserve is only sufficient for a few hours of use. Fat, in adipocytes, provides the major energy store in humans, although muscle proteins are also degraded when food intake is inadequate. Most amino acids (except leucine and lysine) are glucogenic, meaning that their carbon skeletons can be converted (at least partially) into glucose via tricarboxylic acid cycle (TCA) cycle intermediates. Fatty acids cannot be converted to glucose, but triacylglycerol droplets comprise 6% by weight of glycerol; glycerol is converted to glyceraldehyde 3-phosphate, which can enter gluconeogenesis or glycolysis. This glycerol component is crucial for survival.

An average-weight human has energy reserves totalling about 500 MJ (about 120 Mcal). Required daily energy intakes are about 12 MJ per day for males, 9.2 MJ for females, so total energy stores would last about 40 days (I J = 4.184 cal).

## 2.2   Carbohydrate metabolism (catabolic)

The catabolic oxidation of glucose, to provide cellular energy, occurs principally through three 'linked' catabolic pathways:

- glycolysis

- tricarboxylic acid cycle (TCA cycle)

- mitochondrial electron transfer/oxidative phosphorylation.

## 2.3  Glycolysis

Sugars such as glucose and fructose are catabolised to pyruvate. This is a central pathway for the catabolism of carbohydrates in which the six-carbon sugars are split to three-carbon compounds, with subsequent net production of cellular energy in the form of ATP (Figure 2.1). Glycolysis can proceed under both anaerobic (without oxygen) and aerobic conditions. This pathway is almost universal to living organisms.

Pyruvate is an intermediate in several metabolic pathways; mostly it is converted to acetyl-CoA to feed into the TCA cycle (Figure 2.2).

Under anaerobic conditions, pyruvate is converted to lactate by the enzyme lactate dehydrogenase, thereby re-oxidising NADH to $NAD^+$ for re-use in glycolysis. Through the Cori cycle, lactate produced in the skeletal muscles can be delivered to the liver and used to regenerate glucose, through gluconeogenesis.

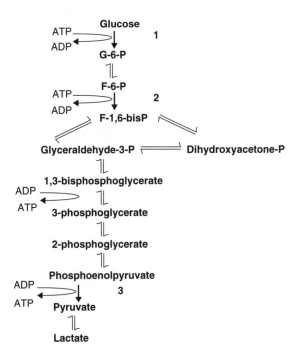

**Figure 2.1**  The glycolytic pathway. The first and third reactions of glycolysis require the input of energy (ATP); the energy investment stage. The 6-carbon sugar is then cleaved to two 3-carbon sugars. In the energy generation stage, ATP is formed. As a consequence, for each molecule of glucose (6-carbon) oxidised, two molecules of ATP are invested and four molecules of ATP are generated, giving a net production of two molecules of ATP. Reactions 1–3 are exergonic and essentially irreversible.

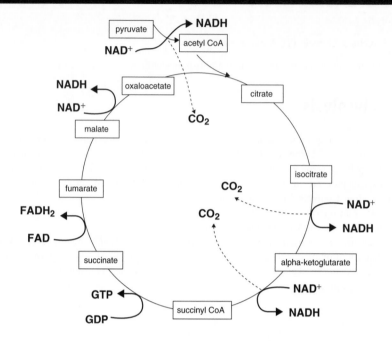

**Figure 2.2** The TCA cycle.

The Cori cycle refers to the metabolic pathway in which lactate, produced by anaerobic glycolysis in the muscle, moves to the liver and is converted to glucose, through gluconeogenesis; glucose can then return to supply the muscle.

Liver

Gluconeogenesis

Glucose

6 ATP

2 Pyruvate

2 Lactate

Blood

Glucose

Muscle

Glucose

2 ATP — Glycolysis

2 Pyruvate

2 Lactate

2 Lactate

Alternatively, in skeletal muscle, pyruvate can be transaminated to alanine (which affords a route for nitrogen transport from muscle to liver); in the liver alanine is used to regenerate pyruvate, which can then be diverted into gluconeogenesis. This process is referred to as the glucose–alanine cycle.

## 2.4 Tricarboxylic acid cycle (TCA cycle – Krebs cycle – citric acid cycle)

Under aerobic conditions the TCA cycle allows for the complete oxidation of pyruvate, generating $CO_2$ as a waste product (which we exhale). Although some ATP is generated directly in the TCA cycle, the most important product is NADH; NADH is subsequently oxidised by the mitochondrial electron transfer chain, which, through oxidative phosphorylation, is the major provider of the cell's ATP requirement.

Each 'turn' of the TCA cycle generates three molecules of NADH. In glycolysis the oxidation of glucose produces two molecules of pyruvate; therefore, the complete oxidation of glucose requires two turns of the TCA cycle, to generate six molecules of NADH. Each molecule of NADH is oxidised by the mitochondrial respiratory chain to generate approximately three molecules of ATP, and so from the NADH alone, approximately 18 molecules of ATP can be formed per molecule of glucose.

## 2.5 Oxidative phosphorylation

NADH, succinate and other substrates are oxidised by the mitochondrial electron transfer chain. The 'chain' consists of a number of redox components, each capable of accepting and donating electrons. Substrates are oxidised (they lose an electron), and the electron passes through the redox chain, directionally from a low to a high redox potential, eventually being added to oxygen and reducing it to water. As electrons pass through the different redox components, low to higher potential, they release energy. Certain of the mitochondrial respiratory complexes (complexes I, III and IV) can use this energy to pump protons (hydrogen ions) across the inner mitochondrial membrane, thereby generating a proton gradient. It is the translocation of protons (driven by the proton gradient) back across the inner mitochondrial membrane, through the ATP synthase, that generates ATP from ADP (Figure 2.3). Complex II (succinate dehydrogenase) does not pump protons.

## 2.6 Brown adipose tissue and heat generation

The uncoupling and collapse of the proton gradient releases the energy of that gradient as heat, rather than using it to generate ATP. This process is a normal physiological function of 'brown' adipose tissue, so called because of the high density of mitochondria in the individual adipose cells. The mitochondria in brown fat contain a protein called thermogenin (also called uncoupling protein 1). Thermogenin acts as a channel in the inner mitochondrial membrane to control the permeability of the membrane to protons.

Newborn babies contain brown fat in their necks and upper backs that serves the function of nonshivering thermogenesis. Muscle contractions that take place in the process of shivering use ATP and produce heat, but nonshivering thermogenesis is a hormonal stimulus for heat generation without the associated muscle contractions of shivering.

The process of thermogenesis in brown fat is initiated by the release of free fatty acids from the triacylglycerol stored in the adipose cells (Figure 2.4). When noradrenaline is released in response to cold sensation it binds to $\beta$-adrenergic receptors on the surface of brown adipocytes, triggering the activation of adenyl cyclase. Activated adenyl cyclase leads to increased production of cAMP and the concomitant activation of cAMP-dependent protein kinase A (PKA), resulting in phosphorylation and activation of hormone-sensitive lipase. The released fatty acids bind to thermogenin, triggering an uncoupling of the proton gradient and the release of the energy of the gradient as heat.

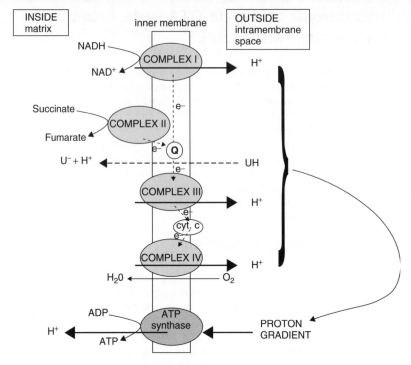

**Figure 2.3** The mitochondrial respiratory chain. The mitochondrial electron transfer chain is localised within the inner mitochondrial membrane. Through the oxidation of NADH or succinate, electrons enter the chain, passing from one redox component to another, handing down a redox potential (low redox potential to a higher redox potential), eventually being added to oxygen at complex IV to form water. At complexes I, III and IV, the energy released by the electron is used to pump protons, from the inside matrix, across the inner membrane, to the intramembrane space (the space between the inner and outer mitochondrial membranes). The proton gradient so formed is 'released' by passing protons back across the inner membrane, through the ATP synthase; this brings about the synthesis of ATP from ADP, known as oxidative phosphorylation. The mitochondrion is said to be 'coupled'; that is, electron transfer is coupled to oxidative phosphorylation. Mitochondria may be 'uncoupled'; compounds that act as uncouplers (UH) work by shuttling protons across the inner membrane, bypassing the ATP synthase and collapsing the proton gradient, and therefore uncoupling electron transport from oxidative phosphorylation.

## 2.7   Glycogenolysis

Glycogenolysis (breakdown of glycogen) is a hormonal response to adrenaline/epinephrine (threat or stress), or glucagon (triggered by low glucose levels). Both hormones cause the conversion of inactive glycogen phosphorylase $b$ to the active glycogen phosphorylase $a$.

Glycogen (n residues) + Pi <-----> Glycogen (n-1 residues) + G-1-P

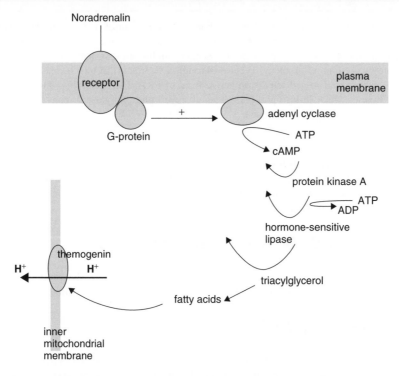

**Figure 2.4**   Activation of thermogenin in brown fat mitochondria. Noradrenalin interacts with its $\beta$-adrenergic receptor, transmitting a signal through the receptor and activating a G-protein, which in turn activates adenyl cyclase. The increase in cAMP activates protein kinase A, which phosphorylates (and activates) the hormone-sensitive lipase. Fatty acids, from the lipolysis of triacyglycerol, bind to thermogenin, which is then able to transport protons across the inner mitochondrial membrane, effectively uncoupling the mitochondria and releasing the energy derived from electron transfer as heat.

Glycogen phosphorylase *a* cleaves the bond at the 1 position by substitution of a phosphoryl group. It breaks down the glucose polymer at $\alpha$-1-4 linkages until only 5-linked glucoses are left on the branch. A second step involves a debranching enzyme to remove 5-linked glucose. In a final step, phosphoglucomutase converts G-1-P (glucose-1-phosphate) to G-6-P (glucose-6-phosphate). The key regulatory enzyme in this process is the glycogen phosphorylase, which is activated by phosphorylation and inhibited by dephosphorylation.

Liver (hepatic) cells will either consume the glucose-6-phosphate in glycolysis or remove the phosphate group (using the enzyme glucose-6-phosphatase) and release the free glucose into the bloodstream for uptake by other cells. Muscle cells do not possess glucose-6-phosphatase and hence will not release glucose, but will rather use the glucose-6-phosphate 'internally' in glycolysis. Liver glycogen is a short-term glucose buffer, muscle glycogen a short-term energy supply.

Glycogen phosphorylase is regulated by both allosteric control and by phosphorylation (covalent modification). Hormones such as adrenaline/epinephrine and glucagon regulate glycogen phosphorylase using second messenger amplification systems that are linked to G-proteins. Activation of adenyl cyclase (through a G-protein) in turn increases levels of cAMP, which bind to and release an active form of PKA. PKA phosphorylates phosphorylase kinase, which in turn phosphorylates glycogen phosphorylase *b*, converting it into the active glycogen phosphorylase *a*.

The goal of glycolysis, glycogenolysis and the TCA cycle, is to conserve energy as ATP from the catabolism of carbohydrates. If cells have sufficient supplies of ATP then these pathways and cycles are inhibited; under such conditions the liver will convert a variety of excess molecules into glucose and/or glycogen.

## 2.8　Carbohydrate metabolism (anabolic)

In order to 'defend' blood glucose levels, the body needs to generate glucose, through either its synthesis (gluconeogenesis) or its release from glycogen stores (see Section 2.7).

## 2.9　Gluconeogenesis

Gluconeogenesis is the generation of glucose from organic molecules such as pyruvate, lactate, glycerol and amino acids (primarily alanine and glutamine). Gluconeogenesis takes place in the liver, and to a lesser extent in the kidneys. The process occurs during periods of starvation or intense exercise. It is an energetically unfavourable pathway that requires the coupling of exergonic and endergonic reactions. While most steps in gluconeogenesis are the reverse of those found in glycolysis, the three regulated and strongly exergonic reactions of glycolysis (1–3 in Figure 2.1) are replaced with more energetically favourable reactions (4–7 in Figure 2.5). The rate of gluconeogenesis is ultimately controlled through the control of the key enzyme fructose-1,6-bisphosphatase. However, both acetyl-CoA and citrate activate pyruvate carboxylase and fructose-1,6-bisphosphatase, and also inhibit the activity of pyruvate kinase (the corresponding negative free energy reaction in glycolysis), so promoting gluconeogenesis.

## 2.10　Glycogenesis

Glycogenesis is the formation of glycogen from glucose (Figure 2.6). Glycogen synthesis depends on the demand for glucose and ATP (energy); if both are present in relatively high amounts then an excess of insulin will promote glucose conversion into glycogen for storage in liver and muscle cells.

Glucose is converted into glucose-6-phosphate by the action of glucokinase (GK) (liver) or hexokinase (muscle), which is then converted to glucose-1-phosphate by the action of phosphoglucomutase, passing through the intermediate glucose-1,6-phosphate. Glucose-1-phosphate is converted to UDP-glucose by the action of uridyl transferase (also called UDP-glucose pyrophosphorylase). Glucose molecules are collected in a chain by glycogen synthase, which

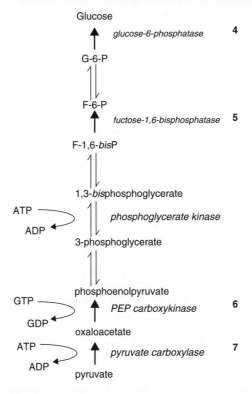

**Figure 2.5** Gluconeogenesis is the 'reversal' of glycolysis, attained through the use of four unique enzymes: glucose-6-phosphatase (4), fructose-1,6-bisphosphatase (5), PEP carboxykinase (6) and pyruvate carboxylase (7). Although phosphoglycerate kinase is shared with glycolysis, in gluconeogenesis this reaction requires the input of ATP.

must act on a pre-existing glycogen primer or glycogenin (a small protein that forms the primer). Chain branches in the growing glucose polymer are made by branching enzyme (also known as amylo-$\alpha$(1 : 4)-$\alpha$(1 : 6) transglycosylase), which transfers the end of the chain on to an earlier part via $\alpha$-1 : 6 glucosidic bonds, forming branches which further grow by addition of more $\alpha$-1 : 4 glucosidic units.

Insulin promotes glycogen synthesis by stimulating hexokinase/GK, inhibiting the glucose-6-phosphatase (of gluconeogenesis) and stimulating glycogen synthase (see also Figure 2.10).

---

Glycogen synthase is a tetrameric enzyme consisting of four identical subunits. Its activity is regulated by phosphorylation of serine residues in the subunit proteins. Phosphorylation of glycogen synthase reduces its activity towards UDP-glucose; in the non-phosphorylated state (synthase $a$, active), glycogen synthase does not require glucose-6-phosphate as an allosteric activator, but when phosphorylated (synthase $b$, inactive), it does.

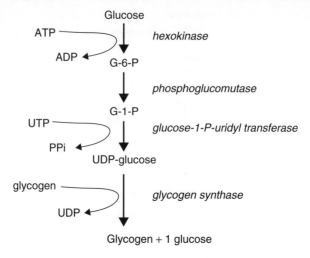

**Figure 2.6** Glycogenesis. The addition of glucose to glycogen depends upon the presence of a pre-existing glycogen primer; glucose monomers are arranged and added to the primer by glycogen synthase, a key regulatory enzyme that is subject to control by covalent phosphorylation.

## 2.11 Fatty acid catabolism

Fats are catabolised by hydrolysis to free fatty acids and glycerol. The free fatty acid is metabolised in the liver and peripheral tissue via $\beta$-oxidation into acetyl CoA; acetyl-CoA then enters the TCA cycle. Glycerol is used by the liver for triacylglycerol synthesis or for gluconeogenesis (following its conversion to 3-phosphoglycerate).

Fatty acids are the preferred energy source for the heart and an important energy source for skeletal muscle during prolonged exertion. During fasting the bulk of the body's energy needs must be supplied by fat catabolism.

Fatty acids must be activated in the cytoplasm in order to enter the mitochondrion (where the $\beta$-oxidation pathway occurs (Figure 2.7)). Activation is catalysed by fatty acyl-CoA ligase (also called acyl-CoA synthetase or thiokinase). The net result of this activation process is the consumption of 2 molar equivalents of ATP.

$$\text{Fatty acid} + \text{ATP} + \text{CoA} \rightarrow \text{Acyl-CoA} + \text{PP} + \text{AMP}$$

Transport of the fatty acyl-CoA into the mitochondrion is accomplished via an acyl-carnitine intermediate; inside the mitochondrion the fatty acyl-CoA molecule is regenerated.

The process of fatty acid oxidation is termed $\beta$-oxidation since it occurs through the sequential removal of 2-carbon units by oxidation at the $\beta$-carbon position of the fatty acyl-CoA molecule. Each round of $\beta$-oxidation produces one molecule of NADH, one molecule of $FADH_2$ (which in turn produces one molecule of $QH_2$) and one molecule of acetyl-CoA. The acetyl-CoA, the end product of each round of $\beta$-oxidation, then enters the TCA cycle, where it is further oxidised to $CO_2$ with the concomitant generation of three molecules of NADH, one molecule of $FADH_2$ ($QH_2$) and one molecule of ATP. The NADH and $FADH_2$ ($QH_2$) generated during fat oxidation and acetyl-CoA oxidation in the TCA cycle then enter the mitochondrial respiratory pathway for the production of ATP.

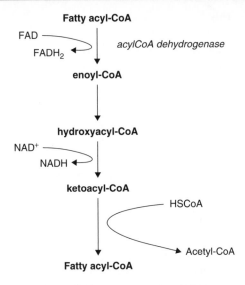

**Figure 2.7** The β-oxidation of fats.

**Clinical problems related to fatty acid metabolism.** Deficiencies in carnitine lead to an inability to transport fatty acids into the mitochondria for oxidation. This can occur in newborns and particularly in pre-term infants. Treatment is by oral carnitine administration. Carnitine palmitoyltransferase I (CPT I) deficiency primarily affects the liver and leads to reduced fatty acid oxidation and ketogenesis. CPT II deficiency results in recurrent muscle pain, fatigue and myoglobinuria following strenuous exercise.

**Deficiencies in acyl-CoA dehydrogenases.** Acetyl-CoA is generated from fatty acids through repeated β-oxidation cycles. Sets of four enzymes (an acyl dehydrogenase, a hydratase, a hydroxyacyl dehydrogenase and a lyase) specific for different chain lengths (very long chain, long chain, medium chain and short chain) are required to catabolise a long-chain fatty acid completely. Inheritance for all fatty acid oxidation defects is autosomal recessive. Medium-chain acyl dehydrogenase deficiency (MCADD) is the most common form of acyl-CoA dehydrogenase deficiency. In the first years of life this deficiency will become apparent following a prolonged fasting period. Symptoms include vomiting, lethargy and frequently coma. Excessive urinary excretion of medium-chain dicarboxylic acids, as well as their glycine and carnitine esters, is diagnostic of this condition.

**Disorders of glycerol metabolism.** Glycerol is converted to glycerol-3-phosphate by the hepatic enzyme glycerol kinase; deficiency results in episodic vomiting, lethargy and hypotonia. Glycerol kinase deficiency is X-linked.

Fatty acid synthesis is discussed in Chapter 5.

## 2.12 Amino acid catabolism

Whereas carbohydrates and fat fuels contain only the elements carbon, hydrogen and oxygen, amino acids also contain nitrogen. Therefore, the first step in amino acid catabolism is the

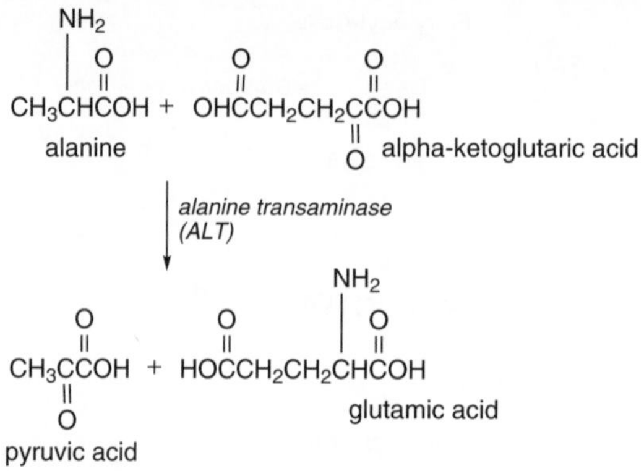

NH$_2$
$|$  O        O        O
$|$  $||$        $||$        $||$
CH$_3$CHCOH + OHCCH$_2$CH$_2$CCOH
alanine                $||$  alpha-ketoglutaric acid
                        O

alanine transaminase
(ALT)

                        NH$_2$
O            O            $|$  O
$||$            $||$            $|$  $||$
CH$_3$CCOH + HOCCH$_2$CH$_2$CHCOH
$||$                        glutamic acid
O
pyruvic acid

**Figure 2.8**  The interconversion of amino acids by transamination.

removal (deamination) of the nitrogen (the amino group). Transamination reactions (Figure 2.8) enable both the interconversion of amino acids and a catabolic route for their degradation.

Eventually, transfer of an amino group (from any amino acid) to $\alpha$-ketoglutarate (the most abundant $\alpha$-keto acid) produces glutamate.

Glutamate dehydrogenase (in the liver) converts glutamate to ammonia and $\alpha$-ketoglutarate; ammonia enters the urea cycle, which ensures its detoxification.

$$\text{glutamate} + \text{NAD}^+ + \text{H}_2\text{O} \underset{glutamate\ dehydrogenase}{\rightleftharpoons} \text{NH}_4 + \text{alpha-ketoglutarate} + \text{H}^+$$

Deamination of an amino acid invariably yields an $\alpha$-keto acid; glucogenic amino acids can be deaminated to pyruvate, oxaloacetate, $\alpha$-ketoglutarate or other 4- or 5-carbon intermediates of the TCA cycle, which are precursors for gluconeogenesis (hence the term glucogenic) (Figure 2.9). Glucogenic amino acids are a major carbon source for gluconeogenesis when glucose levels are low. They can also be catabolised for energy or converted to glycogen or fatty acids for energy storage.

A smaller number of amino acids are degraded to acetyl-CoA or acetoacetyl-CoA. Neither acetyl-CoA nor acetoacetyl-CoA can yield a net production of oxaloacetate, the precursor for the gluconeogenesis pathway (because for every 2-carbon acetyl residue entering the TCA cycle, two carbon atoms leave as $CO_2$). These are referred to as the ketogenic amino acids; they can be catabolised for energy in the TCA cycle, or converted to ketone bodies or fatty acids, but they cannot be converted to glucose.

# 2.13  Blood glucose homeostasis

There are at least four independent sensors that monitor blood glucose concentration in the body and contribute to the overall regulation of food intake.

1. pancreatic $\beta$-cells

2. nucleus of the solitary tract

3. hypothalamus

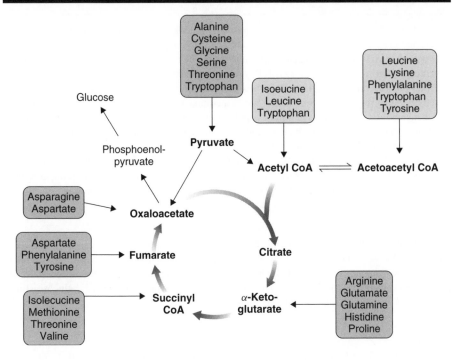

**Figure 2.9**  Amino acid and carbohydrate integration.

4. carotid bodies (a small cluster of chemoreceptors and supporting cells located near the fork of the carotid artery)

5. portal vein glucose sensor (uncertain role in humans).

In addition, sweet detectors, similar to those on the tongue, have recently been documented in the epithelial lining of the duodenum.

Primary sensors are located in the pancreatic islets, and also in the carotid bodies, medulla and the hypothalamus. There are inputs from the eyes, nose, taste buds and gut which signal food is on the way, while fear and stress help prepare the body to face difficult times ahead.

The first glucose sensors to be discovered were the pancreatic $\beta$-cells in the islets of Langerhans; these manufacture the hormone insulin and release it into the blood when glucose concentration rises. Islet tissue also contains $\alpha$-cells, which manufacture the antagonistic hormone glucagon. Insulin secretion is a complex process, and the islet cells receive additional signals from the gut and the autonomic nervous system, which modulate the insulin release to match the food that has been eaten.

The hypothalamus and the solitary tract nucleus largely control the autonomic nervous system, and manage a spectrum of sensations ranging from a sense of well-being to one of terror. Overall the brain has a very active glucose-uptake mechanism with a low $K_m$ (high affinity). This is required for normal brain function at all physiological levels of blood glucose. However, in contrast to the rest of the central nervous system (CNS), the cells of the arcuate nucleus contain GK. Given its high $K_m$ (low affinity) for glucose, GK activity changes in tack with normal variations in blood sugar levels, and so provides a 'true' monitor. As in pancreatic $\beta$-cells, the GLUT2-GK system measures blood sugar levels (see Sections 2.14 and 2.15).

The glucose transporter GLUT2, glucokinase and glucokinase regulatory protein have been detected in several regions of the brain, including the ventromedial and arcuate nuclei of the hypothalamus. Increased glycaemia after meals may be recognised by specific hypothalamic neurones due to the high $K_m$ of GLUT2 and glucokinase. This enzyme is considered to be a true glucose sensor because it catalyses the rate-limiting step of glucose catabolism, its activity being regulated by interaction with glucokinase regulatory protein, which functions as a metabolic sensor. The glucokinase regulatory protein is best documented in hepatocytes, where it has been shown to bind to and move glucokinase, controlling both the activity and intracellular location of this key enzyme.

In rats fed a high-fat diet, the level of GLUT2 and glucokinase mRNA in pancreas has been shown to be significantly decreased, suggesting that high-fat feeding impairs the signal transduction mechanism in pancreatic $\beta$-cells to reduce insulin secretion.

While the mechanism for coupling of GLUT2-GK to appetite control is not yet clear, increased glucose concentrations sensed and 'reported' by GLUT2-GK appear to stimulate neuropeptide Y/agouti-related peptide (NPY/AgRP)-producing neurons in the fasted state and pro-opiomelanocortinergic (POMC)-secreting neurons in the fed state. That is, glucose stimulates hunger between meals and inhibits hunger after meals.

Neuropeptide Y (NPY) receptors are a class of G-protein-coupled receptors that are activated by the closely related peptide hormones NPY, peptide YY and pancreatic polypeptide. These receptors are involved in control of a diverse set of behavioural processes, including appetite, circadian rhythm and anxiety. AgRP is a receptor antagonist of CNS melanocortin receptors and appears to have an important role in the control of food intake. Hypothalamic POMC neurons are important mediators in the regulation of feeding behaviour, insulin levels and ultimately body weight.

Additional signals, adrenocorticotrophic hormone (ACTH) and pituitary growth hormone (pGH), act to increase blood glucose by inhibiting uptake by extrahepatic tissues. Cortisol, the major glucocorticoid released from the adrenal cortex, is secreted in response to the increase in circulating ACTH, and also acts to increase blood glucose levels by inhibiting glucose uptake. The adrenal medullary hormone, adrenaline (epinephrine), stimulates production of glucose by activating glycogenolysis in response to stressful stimuli.

The predominant tissue responding to signals that indicate reduced or elevated blood glucose levels is the liver. Elevated or reduced levels of blood glucose trigger hormonal responses to initiate pathways designed to restore glucose homeostasis (Table 2.3).

Glucagon binding to receptors on the surface of liver cells triggers an increase in cAMP production, leading to an increased rate of glycogenolysis, by activating glycogen phosphorylase via the PKA-mediated cascade (as does adrenaline/epinephrine). The resultant increased level of G-6-P in hepatocytes is hydrolysed to free glucose by glucose-6-phosphatase, which then diffuses to the blood. The glucose enters extrahepatic cells, where it is re-phosphorylated by hexokinase to G-6-P. Since muscle and brain cells lack glucose-6-phosphatase, the G-6-P is retained and oxidised by these tissues for energy.

**Table 2.3**   Blood glucose control depends on concerted action between several circulating hormones

|  | Insulin | Glucagon/ adrenalin | Growth hormone stimulate | Cortisol | TNF-$\alpha^a$ |
|---|---|---|---|---|---|
| Sugars | Glycolysis | Gluconeogenesis | Gluconeogenesis | Gluconeogenesis | Glycolysis |
| Glycogen | Synthesis | Breakdown | Synthesis | Redistribution | Breakdown |
| Proteins | Synthesis | Breakdown | Synthesis | Breakdown | Breakdown |
| Fats | Synthesis | Breakdown | Breakdown | Redistribution | Breakdown |

$^a$TNF-$\alpha$ (tumour necrosis factor alpha) is the best known member of a group of pro-inflammatory cytokines, including interleukin-1 (IL-1) and interleukin-6 (IL-6), collectively responsible for the fever and inflammation associated with infections and serious disease.

In opposition to glucagon, insulin stimulates extrahepatic uptake of glucose from the blood and inhibits glycogenolysis in extrahepatic cells, conversely stimulating glycogen synthesis. As glucose enters hepatocytes, it binds to glycogen phosphorylase $a$, stimulating the dephosphorylation of phosphorylase $a$ and thereby inactivating it. That glucose is not immediately converted to glucose-6-phosphate in the liver is due to the presence in hepatocytes of the isoenzyme glucokinase, rather than hexokinase.

## 2.14   Glucokinase and hexokinase

Glucokinase (GK) has a much lower affinity for glucose than hexokinase; therefore it is not fully active at the physiological ranges of blood glucose. Additionally, GK is not inhibited by its product G-6-P, whereas hexokinase is.

---

Different internal glucose concentrations require different genes to be expressed in different cells. Most tissues use hexokinase to phosphorylate glucose to G-6-P, so that it can be metabolised within the cells. Hexokinase has a low $K_m$ for glucose, allowing it to work efficiently at the low internal glucose concentrations in the majority of tissues. In contrast, cells which routinely export glucose to the bloodstream, such as liver, gut wall and kidney tubules, express a different gene, for the isoenzyme GK. This enzyme has a much higher $K_m$ for glucose, so that most of the intracellular glucose escapes phosphorylation and is available for export, or to inhibit phosphorylase $a$ and glycogenolysis (see also Section 9.7 and Figure 9.5).

The $K_m$ of an enzyme is a measure of the affinity of the enzyme for its substrate; the higher the $K_m$, the lower the affinity.

---

## 2.15   Glucose transporters

A major response of non-hepatic tissues to insulin is the recruitment to the cell surface of glucose transporter complexes. Glucose transporters compose a family of at least 14 members. The best characterised are GLUT1, GLUT2, GLUT3, GLUT4 and GLUT5. Glucose transporters are facilitative transporters; they carry hexose sugars across the membrane down their concentration gradients.

- GLUT1 is ubiquitously distributed in various tissues.

- GLUT2 is found primarily in intestine, pancreatic $\beta$-cells, kidney and liver. GLUT2 transports both glucose and fructose. When the concentration of blood glucose increases in response to food intake, pancreatic GLUT2 molecules mediate an increase in glucose uptake, which leads to increased insulin secretion. For this reason, GLUT2 is considered to be a 'glucose sensor'.

- GLUT3 is found primarily in neurons but also in the intestine. GLUT3 binds glucose with high affinity (has the lowest $K_m$ of the GLUT transporters), which allows neurons to have enhanced access to glucose especially under conditions of low blood glucose.

- Insulin-sensitive tissues, such as skeletal muscle and adipose tissue, contain GLUT4, whose mobilisation to the cell surface is stimulated by insulin action.

- GLUT5, along with the closely related transporter GLUT7, is involved in fructose transport. GLUT5 is expressed in intestine, kidney, testes, skeletal muscle, adipose tissue and brain. Although GLUT2, -5, -7, 8, -9, -11 and -12 can all transport fructose, GLUT5 is the only transporter that exclusively transports fructose.

## 2.16 Diabetes mellitus

Diabetes mellitus refers to the group of diseases that lead to high blood glucose levels, due to defects in either insulin secretion or insulin action in the body. Diabetes develops due to a diminished production of insulin (type 1) or a resistance to its effects (type 2), including gestational diabetes. This can lead to hyperglycaemia, which is largely responsible for the acute signs of diabetes, namely:

- excessive urine production (polyuria)

- thirst and increased fluid intake (polydipsia)

- blurred vision

- unexplained weight loss (in type 1)

- lethargy

- changes in energy metabolism.

> Gestational diabetes mellitus resembles type 2 diabetes, but is transient, occurring in about 2–5% of pregnancies. While it is fully treatable, about 20–50% of affected women develop type 2 diabetes later in life.

Diabetes mellitus is characterised by recurrent or persistent hyperglycaemia, and is diagnosed by demonstrating any one of the following:

- fasting plasma glucose level at or above 126 mg/dl (7.0 mmol/l)

- plasma glucose at or above 200 mg/dl (11.1 mmol/l), 2 hours after a 75 g oral glucose load in a glucose tolerance test

- symptoms of hyperglycaemia and casual plasma glucose at or above 200 mg/dl (11.1 mmol/l).

---

## The mechanism of insulin release

The secretion of insulin from pancreatic $\beta$-cells is a complex process involving the integration and interaction of multiple external and internal stimuli. The primary stimulus for insulin secretion is the $\beta$-cell response to changes in ambient glucose. Normally, glucose induces a biphasic pattern of insulin release. First-phase insulin release occurs within the first few minutes after exposure to an elevated glucose level; this is followed by a more enduring second phase of insulin release. Of particular importance is that first-phase insulin secretion is lost in patients with type 2 diabetes. The generally accepted sequence of events involved in glucose-induced insulin secretion is as follows:

1. Glucose is transported into $\beta$-cells through facilitated diffusion on GLUT2 glucose transporters.

2. Intracellular glucose is metabolised to ATP.

3. Elevation in the ATP/ADP ratio induces closure of cell-surface ATP-sensitive $K^+$ channels, leading to cell-membrane depolarisation.

4. Cell-surface voltage-dependent $Ca^{2+}$ channels are opened, facilitating extracellular $Ca^{2+}$ influx into the $\beta$-cell.

5. A rise in free cytosolic $Ca^{2+}$ triggers the exocytosis of insulin.

---

## 2.17 Type 1 diabetes

Type 1 diabetes has been predicted to double by 2020, in children under five years, with a predicted 24 400 new cases in Europe.

The cause of type 1 diabetes is not fully understood. An autoimmune attack (to the $\beta$-cells of the pancreas) may be triggered by reaction to an infection, for example by one of the viruses of the Coxsackie virus family or German measles, although the evidence is inconclusive. Individuals may display genetic vulnerability; an observed inherited tendency to develop type 1 diabetes has been traced to particular human leukocyte antigen (HLA) genotypes (the major histocompatibility complex (MHC) in humans is known as the HLA system). Environmental factors can also strongly influence expression of type 1 diabetes.

Type 1 diabetes is a polygenic disease (different genes contribute to its expression); it can be dominant, recessive or intermediate. The gene IDDM1, located in the MHC class II region on chromosome 6, is believed to be responsible for the histocompatibility disorder characteristic of type 1 diabetes. Insulin-producing pancreas cells ($\beta$-cells) display improper antigens to T-cells, which lead to the production of antibodies that attack those $\beta$-cells. Other associated genes are located on chromosomes 11 and 18. Pancreatic $\beta$-cells in the islets of Langerhans are destroyed or damaged sufficiently to effectively abolish endogenous insulin production. This aetiology distinguishes type 1 origin from type 2; that is, whether the patient is insulin resistant (type 2) or insulin deficient without insulin resistance (type 1).

---

Prior research has shown that both $\beta$-cells and T-cells act to initiate type 1 diabetes. T-cells attack and destroy the insulin-producing $\beta$-cells. $\beta$-cells don't directly attack insulin-producing cells, but may trigger the T-cells to attack. Rituximab, a chimeric monoclonal antibody against CD20, found primarily on the surface of $\beta$-cells, has shown potential in delaying the development of type 1 diabetes.

---

Type 1 diabetes, formerly known as 'childhood', 'juvenile' or 'insulin-dependent' diabetes, is not exclusively a childhood problem. Adults who contract type 1 diabetes may be misdiagnosed with type 2 diabetes. A subtype of type 1 (identifiable by the presence of antibodies against $\beta$-cells) typically develops slowly and is often confused with type 2. In addition, a small proportion of type 1 cases have the hereditary condition maturity onset diabetes of the young (MODY), which can also be confused with type 2.

Type 1 diabetes is treated with insulin replacement therapy, usually by insulin injection or insulin pump, along with attention to dietary management and careful monitoring of blood glucose levels. Today most insulin is produced using genetic recombination techniques; insulin analogues are a form of modified insulin with different onset-of-action times or duration-of-action times.

The most definitive laboratory test to distinguish type 1 from type 2 diabetes is the C-peptide assay, which is a measure of endogenous insulin production. With type 2 diabetes, proinsulin can be split into insulin and C-peptide; lack of C-peptide indicates type 1 diabetes. The presence of anti-islet antibodies (to glutamic acid decarboxylase, insulinoma associated peptide-2 or insulin) or absence of insulin resistance (determined by a glucose tolerance test) is also suggestive of type 1.

## 2.18 Type 2 diabetes

Type 2 diabetes (non-insulin-dependent diabetes mellitus (NIDDM) or adult-onset diabetes) is a metabolic disorder characterised of two processes: a slowly developing resistance to insulin signalling and a compensatory increase in $\beta$-cell release of the hormone. With time $\beta$-cells no longer produce enough insulin to maintain control of metabolism and type 2 diabetes results. With prevalence rates doubling in the USA between 1990 and 2005, type 2 diabetes has been described as an epidemic. Traditionally considered a disease of adults, it is increasingly diagnosed in children in parallel to rising obesity rates.

While the underlying cause of insulin resistance is unknown, there is a striking correlation between obesity, increased plasma lipids and resistance. Insulin resistance is generally 'post-receptor', meaning it is a problem with the cells that respond to insulin rather than a problem with production of insulin. The molecular mechanism underlying defective insulin-stimulated glucose transport may be attributed to increases in intracellular lipid metabolites (fatty acyl-CoAs and diacylglycerol), which in turn activate a serine/threonine kinase cascade, leading to defects in insulin signalling through Ser/Thr phosphorylation of insulin receptor substrate (IRS-1). Central obesity (fat concentrated around the waist in relation to abdominal organs, but not subcutaneous fat) is known to predispose individuals to insulin resistance. Abdominal fat is especially active hormonally, secreting a group of hormones called adipokines, which may possibly impair glucose tolerance. Obesity is found in approximately 55% of patients diagnosed with type 2 diabetes.

There is also a strong inheritable genetic connection in type 2 diabetes. Having relatives (especially first degree) with this disorder substantially increases the risk of developing type 2 diabetes. Additionally there is a mutation to the islet amyloid polypeptide gene that results in an earlier-onset, more severe form of diabetes.

Environmental exposures may contribute to recent increases in the rate of type 2 diabetes. For example, there is a positive correlation between concentration in the urine of bisphenol A, a constituent of polycarbonate plastic, and the incidence of type 2 diabetes.

---

The term 'diabetes' is usually taken to refer to diabetes mellitus, which is associated with excessive sweet urine (known as 'glycosuria'). Rarer diabetic conditions include diabetes insipidus, where the urine is not sweet; this can be caused by either kidney or pituitary gland damage.

---

Type 2 diabetes may go unnoticed for years; visible symptoms are typically mild, non-existent or sporadic, and usually there are no ketoacidotic episodes. However, severe long-term complications can result from unnoticed type 2 diabetes, including:

- renal failure due to diabetic nephropathy

- vascular disease (including coronary artery disease)

- vision damage due to diabetic retinopathy

- loss of sensation or pain due to diabetic neuropathy

- liver damage from non-alcoholic steatohepatitis

- heart failure from diabetic cardiomyopathy.

Type 2 diabetes may be first treated by increasing physical activity, decreasing carbohydrate intake and weight loss; insulin sensitivity can be restored with only moderate weight loss.

The use of oral antidiabetic drugs may be necessary. Insulin production is initially only moderately impaired in type 2 diabetes, so oral medication can often be used to improve insulin production (e.g. sulphonylureas), to regulate inappropriate release of glucose by the liver and attenuate insulin resistance to some extent (e.g. metformin), and to substantially attenuate insulin resistance (e.g. thiazolidinediones). A final resort is insulin therapy to maintain normal or near-normal glucose levels.

**Table 2.4** A comparison and explanation of the common symptoms of types 1 and 2 diabetes

| Symptom | Type 1 diabetes | Type 2 diabetes |
|---|---|---|
| Tiredness | Inefficient utilisation of fuels | Inefficient utilisation of fuels |
| Thirst/polyuria | High glucose (osmotic diuresis) | – |
| Very low insulin | Damage to insulin-producing $\beta$-cells | – |
| Raised insulin | – | Suggests insulin resistance – linked with obesity |
| Weight loss | Protein catabolism to provide amino acids for gluconeogenesis, and utilisation of fats for energy | – |
| Raised HbA1c | High – blood glucose constantly high | Moderate – blood glucose often higher than normal |
| Ketonuria | Increased metabolism of fats, raised acetyl CoA and increased ketogenesis | – |

---

Metformin (trade names Glucophage, Riomet, Fortamet and others), is an oral anti-diabetic drug for the treatment of type 2 diabetes. It is a biguanide drug. Biguanides do not affect the output of insulin, unlike the sulphonylureas and meglitinides, and can therefore also be effective in type 1 patients in concert with insulin therapy. Their exact mode of action is not fully elucidated. They can lower fasting levels of insulin in plasma, through their tendency to reduce gluconeogenesis in the liver. They also have effects in reducing insulin resistance in tissues; metformin has been shown to stimulate AMP-activated protein kinase (AMPK), an enzyme responsible for glucose and fat metabolism as well as insulin signalling. It also helps reduce LDL cholesterol and triglyceride levels, and may aid weight loss. As of 2008, metformin is one of only two oral anti-diabetics in the World Health Organization Model List of Essential Medicines.

See Table 2.4 for a comparison of the common symptoms of types 1 and 2 diabetes.

---

# 2.19 Insulin/Glucagon effects on metabolism

Insulin has a range of effects on a number of different metabolic pathways, either stimulatory or inhibitory; in many cases glucagon has an opposing effect. It is the insulin to glucagon ratio that is of most importance. Table 2.5 summarises the major effects of insulin; the actions of insulin and glucagon are indicated in Figure 2.10.

With reference to Figure 2.10, the actions of insulin are to:

- promote the synthesis of glycogen from glucose (stimulation of glycogen synthase) and inhibit glycogenolysis (inhibition of glycogen phosphorylase)

- increase the breakdown of glucose through glycolysis by stimulating GK/hexokinase, phosphofructokinase (PFK) and pyruvate kinase

• stimulate fat synthesis by promoting the conversion of acetyl-CoA to malonyl-CoA and by inhibiting lipolysis at the level of hormone-sensitive lipase.

**Table 2.5**  Insulin stimulation or inhibition of metabolic pathways

| Promoted by insulin | Inhibited by insulin |
| --- | --- |
| Glucose uptake by tissues | Gluconeogenesis |
| Glycolysis | Glycogenolysis |
| Glycogen synthesis | Lipolysis |
| Protein synthesis | Ketogenesis |
| Fatty acid synthesis | Proteolysis |

On the other hand, glucagon actively promotes gluconeogenesis by stimulating PEP carboxykinase, and also promotes glycogenolysis and lipolysis.

Ketogenesis, the formation of 'ketone bodies' (i.e. acetoacetate, acetone, $\beta$-hydroxybutyrate), is commonly associated with type 1 diabetes; ketosis can be fatal. Lack of insulin results in an increased lipolysis (from glucagon) with increasing acetyl-CoA concentrations; glucagon actively promotes gluconeogenesis (PEP carboxykinase), which along with low insulin allows for increased metabolism of amino acids (cause of weight loss in type 1 diabetes), meaning a deficiency of the TCA intermediate oxaloacetate. With a lower oxaloacetate concentration, acetyl-CoA is prevented from entering the TCA cycle, adding to an increase in its cellular concentration. The result is increased ketogenesis.

## 2.20   Hyperglycaemia and associated pathology

Serious long-term complications of diabetes and raised glucose levels include cardiovascular disease, chronic renal failure, retinal damage (leading to blindness), neural damage and microvascular damage (with associated poor wound healing). In the developed world, diabetes is the most significant cause of adult blindness in the non-elderly and the leading cause of non-traumatic amputation in adults, while diabetic nephropathy is the main illness requiring renal dialysis in the USA.

## 2.21   Glycation

Many of the pathological effects of diabetes arise from the process of glycation. Glycation is the non-enzymatic and haphazard condensation of the aldehyde and ketone groups in sugars with amino groups in proteins, leading to their functional impairment (the enzyme-controlled addition of sugars to protein or lipid molecules is termed glycosylation). These may undergo further chemical reactions to produce 'advanced glycation end products', or AGEs. Glycation damages collagen in blood vessel walls, leading to inflammation and atherosclerosis. This process is now considered to be a major contributor to diabetic pathology and has resulted in greater clinical emphasis on good glycaemic control. Clinical measurement of glycated haemoglobin (HbA1c) and serum albumin is used to assess the adequacy of blood sugar regulation in diabetic patients (Table 2.6). Normal (non-diabetic) values of glycated haemoglobin are 4.0–6.5%; that is, approximately 6 red cells out of every 100 will have glucose attached.

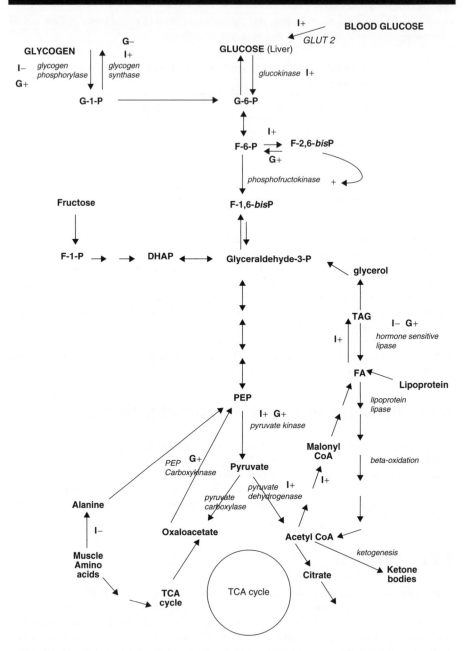

**Figure 2.10** The opposing effects of insulin and glucagon on energy metabolism. Insulin stimulation (I+) or inhibition (I−); glucagon stimulation (G+) or inhibition (G−).

**Table 2.6**  Clinical HbA1c levels

| HbA1c (%) | Normal/abnormal | Average blood glucose (mM) |
|-----------|-----------------|----------------------------|
| 4–6.5 | Normal (without diabetes) | 3–8 |
| 6.5–7.5 | Target range (with diabetes) | 8–10 |
| 8–9.5 | High | 11–14 |
| >9.5 | Very high | >15 |

US guidelines are to have an HbA1c of less than 7.5%. Long-term damage caused by protein glycation includes ulcers, kidney failure, blindness, strokes and ischaemic heart disease.

## 2.22   The polyol pathway

Excess glucose can enter the polyol pathway, where it is reduced to sorbitol (by aldose reductase and the reductant NADPH). Sorbitol dehydrogenase will oxidise sorbitol to fructose, which also produces NADH from $NAD^+$. Hexokinase will return fructose to the glycolysis pathway by phosphorylating it to fructose-6-phosphate. However, in uncontrolled diabetics with high blood glucose, the production of sorbitol is favoured.

Activation of the polyol pathway results in a decrease of NADPH and $NAD^+$; these are necessary cofactors in redox reactions throughout the body. The decreased concentration of these cofactors leads to decreased synthesis of reduced glutathione, nitric oxide, myoinositol and taurine. Myoinositol is particularly required for the normal function of nerves. Sorbitol may also glycate the amino nitrogen on proteins such as collagen, forming AGEs.

In diabetic neuropathy, nerves may be affected by damage to small blood vessels surrounding the sheath, but also by an accumulation of sorbitol and fructose in Schwann cells, leading to de-myelination. Schwann cells are a type of glial cell that are necessary for the maintenance of peripheral nerve fibres (both myelinated and unmyelinated). In myelinated axons, Schwann cells produce the myelin sheath.

# CHAPTER 3
# Regulating body weight

Our hunger is intermittently relieved by eating, and our need to eat is suppressed by inhibitory impulses, so-called 'satiety factors'. Satiation is the feeling of 'fullness' and well-being that controls the interval to the next meal. The 'front end' of the body (eyes, nose, taste buds and forebrain) tends to increase food consumption, but the 'back end' (liver and lower GI tract), which must cope with the ingested food, mainly signals satiation to terminate eating behaviour.

The 'metabolic syndrome', characterised by abdominal obesity, insulin resistance, dislipidaemia, low-grade inflammation, hypertension and cardiovascular disease, is a common and serious medical problem throughout the developed world that merits particular attention. Obesity is a growing problem in children.

## 3.1  Obesity

Obesity is often defined in terms of the body mass index (BMI):

$$BMI = \text{weight (kg) / height (m)}^2$$

A BMI of 25–30 is considered overweight; above 30 is obese. This is useful as a guide, but BMI does not adequately distinguish fat from lean muscle mass. Separate norms should be used for men, women, children and different ethnic groups.

The pathogenic mechanisms in diabetes seem to involve the non-enzymic glycation of connective tissue proteins, leading to microangiopathy followed by kidney, retinal and neurological problems. Diabetics also have an adverse blood lipid profile that is associated with atherosclerosis and large vessel disease. Numerous other conditions are allegedly associated with obesity.

## 3.2  Weight regulation

Weight regulation is normally a very precise process. Ignoring short-term fluctuations (which are mostly gain and loss of water), body weight stays remarkably constant. Weight regulation

*Essential Biochemistry for Medicine*   Dr Mitchell Fry
© 2010 John Wiley & Sons, Ltd

is achieved through precise and effective negative feedback loops, making it very difficult to lose weight by dieting.

- Hunger varies inversely with body weight; if you lose weight the desire to eat more food becomes more powerful.

- Metabolism varies directly with body weight; as you lose weight your basal metabolic rate (BMR) declines to conserve the limited fat stores that remain.

For these reasons the body 'defends' a 'set weight'.

---

BMR is the amount of energy expended while at rest in a neutrally temperate environment, in the post-absorptive state (expressed in kcal/day). BMR decreases with age and with the loss of lean body mass. Increasing muscle mass increases BMR. Illness, previously consumed food and beverages, environmental temperature and stress levels can affect overall energy expenditure as well as the BMR. BMR is accurately determined by gas analysis (direct or indirect calorimetry); an estimation can be found using the equation:

$$P = 9.99\,m + 6.25\,h + 4.92\,a \text{ kcal/day}$$

where $P$ is the total heat production at complete rest, $m$ is weight (kg), $h$ is height (cm) and $a$ is age (years). Such formulae are based on body weight, which does not take into account the difference in metabolic activity between lean body mass and body fat. To calculate daily calorie needs, the BMR value is multiplied by 'activity factors', between 1.2 and 1.9, depending on the person's activity level.

---

## 3.3    Controlling food intake

The control of food intake occurs at a number of levels, including:

1. pre-gastric factors

2. gastrointestinal and post-absorptive factors

3. the enteric nervous system

4. the central nervous system

5. long-term controls.

## 3.4    Pre-gastric factors

Pre-gastric factors can include the appearance of food, its taste and odour, learned preferences, aversions and psychological factors. Mental states such as fear, depression and social interactions can all affect food intake. Such factors are of particular importance to clinicians because they can be manipulated to manage anorectic (anorexigenic) patients.

## 3.5   Gastrointestinal and post-absorptive factors

A distensible stomach allows opportunistic feeding when food is available, but there are limits to the amount of food that can be stored and the rate at which it can be processed. The degree of gastrointestinal fill is the most important signal from the digestive tract; a full stomach and intestine induce satiety, probably via the vagus nerve relay to the hypothalamus. Additionally, the enteric hormone cholecystokinin (CCK) is well documented to induce satiety in experimental settings, while the hormone ghrelin seems to be a potent stimulator of appetite. As nutrients are absorbed, their concentrations in blood rise, with changes in the concentration of several hormones, CCK, as well as insulin and glucagon.

---

**Cholecystokinin (CCK):** CCK is the best known member of a group of hormones (incretins) secreted by the duodenum in response to the partially digested output from the stomach (chyme); fatty meals are particularly effective. CCK delays gastric emptying, stimulates pancreatic enzyme production, causes the gall bladder to contract, promotes insulin release by the pancreas and produces a sensation of fullness or satiation. CCK signals to the brain via the vagus nerve. Other incretins in the same group are gastric inhibitory peptide (GIP), glucagon-like peptide 1 and glucagon-like peptide 2. All have been identified as drug development targets.

**Insulin:** In addition to its role in blood glucose homeostasis, insulin reduces food intake and plays a major part in appetite regulation. Gene knockout experiments have confirmed that lack of either brain insulin receptors or insulin receptor substrate 2 (IRS2) results in hyperphagia, obesity and female infertility. It is believed that insulin promotes phosphorylation of leptin receptors and JAK2 (janus kinase 2), which enhances the phosphorylation of STAT3 (signal transducer and activator of transcription 3) in the presence of leptin (see Table 3.1). Insulin levels are often raised in type 2 diabetes, which is associated with insulin resistance and obesity.

**Ghrelin:** The best documented 'hunger hormone', ghrelin is a 28-residue peptide secreted by endocrine cells within the gastric sub-mucosa. It acts on the hypothalamus to stimulate growth hormone release by the pituitary. Ghrelin is also produced locally by neurons within the hypothalamus, and in other parts of the intestine. Ghrelin receptors are found in most tissues, including many other parts of the brain, and ghrelin signalling is widely distributed throughout the vertebrate phylum. Ghrelin antagonises leptin, increases metabolic efficiency and stimulates eating behaviour, resulting in weight gain.

---

## 3.6   Enteric nervous system

Mostly the gut functions semi-autonomously; peristalsis, secretion and absorption are coordinated by the enteric nervous system. Peristalsis is mainly controlled by the myenteric plexus (Auerbach's plexus) between the circular and longitudinal muscles in the gut wall, and secretion by the sub-mucous plexus (Meissner's plexus), which is closer to the lumen of the gut. The gastrointestinal tract has several distinct modes of operation; in one of these, coordinated patterns of electrical and muscular activity, known as migrating motility complexes, originate from the stomach and propagate through the small intestine, gently massaging the contents along the gut. The stomach releases chyme through the pyloric sphincter in small squirts that are adjusted to match the processing capacity of the duodenum. Consumption of fresh food stimulates a gastrocolic reflex that moves previous meals through the hindgut.

In humans, and in all other vertebrates, information about smell, taste and gastric fullness is conveyed by the cranial nerves. Smell receptors transmit information via the olfactory nerve; taste buds are mainly innervated by the facial nerve and the glossopharyngeal; while the liver, stomach and duodenum are served by the vagus nerve. This system does not merely detect that the gut is full, but can also monitor what kinds of food have been eaten, their digestibility and their likely calorific yield. The flow of information is a two-way process, and the brain is able to modulate the volume and composition of the digestive juices, the rate of passage through the pylorus, and the total residence time within the gut. Fundamental feeding assessments are made by the hindbrain, in the nucleus of the solitary tract, or possibly in the parabrachial nucleus nearby. Corresponding emotional and physical sensations are communicated to the higher centres and influence our conscious behaviour.

## 3.7  The central nervous system

While the brain uses the autonomic nervous system to supervise the operation of the local gut hormones and the enteric nervous system, it is the hypothalamus that has access to all significant sources of information and controls the major effectors. The hypothalamus is a small volume of nervous tissue surrounding the third ventricle, with important neural connections to the hindbrain, and also to the pituitary gland. Some of this tissue lies outside the blood–brain barrier, and is able to respond directly to circulating hormones, and to sense the glucose concentration in the blood. It is uniquely placed to control both energy inputs and outputs. Fat cells produce leptin (see below), which reflects total fat stores. Pancreatic $\beta$-cells produce insulin, which reflects the recent food supply, and the gut signals its current contents via the vagus nerve and the hindbrain. There are sensory pathways from the nucleus of the solitary tract to the hypothalamus. All these signals are integrated by the hypothalamus to regulate physical activity, thermogenesis and feeding behaviour. Within the arcuate nucleus of the hypothalamus the various input signals drive an appetite-stimulating (orexigenic) system, based on neuropeptide Y/agouti-related peptide (NPY/AgRP)-secreting neurones, which is balanced by an appetite-suppressing (anorexigenic) system, based on pro-opiomelanocortin (POMC)-secreting neurones.

---

The appetite centre in the arcuate nucleus appears to be composed of at least two classes of neurons: primary neurons that sense metabolite levels and regulating hormones, and secondary neurons that synchronise information from primary neurons and coordinate bodily functions through vagal signalling. There are those which stimulate appetite through secretion of NPY and the AgRP, and those which depress appetite through secretion of POMC (see also Section 3.9).

---

Output signals from the hypothalamus involve internal tracts to the forebrain, probably involving orexin signalling and affecting conscious behaviour; there are also internal tracts to the nucleus of the solitary tract that affect feeding behaviour. There is control of the anterior pituitary gland, and a direct control over the autonomic nervous system.

> Orexins A and B are an important pair of neurotransmitters (otherwise known as hypocretins 1 and 2), derived from a common precursor. They have a major role in arousal and food-seeking behaviour. Damage to the orexin signalling system leads to narcolepsy (a disorder that causes excessive sleepiness during the day and frequent and uncontrollable episodes of falling asleep).

## 3.8   Long-term control

Defining the feedback signals, the so-called 'satiety factors', is an ongoing (pharmaceutical) quest. Some of the best-known and most likely candidates are shown in Table 3.1.

## 3.9   CNS factors

Other factors within the CNS that influence food intake and energy expenditure continue to be identified, including:

- **POMC**, a protein that is cleaved to yield a variety of important signalling molecules. It is the precursor of the pituitary hormones MSH and ACTH. At least four distinct G-protein-linked receptors recognise the core heptapeptide sequence of the melanocortins. MC1R controls skin pigmentation, MC2R is the ACTH receptor, while MC3R and MC4R control appetite and energy expenditure. Gene knockout experiments show that both MC3R and MC4R transduce anorexic (appetite-reducing) signals.

- **NPY**, a 36-residue, highly conserved peptide that is widely distributed throughout the vertebrate nervous system. NPY delivers a powerful orexigenic (appetite-promoting) signal within the hypothalamus, which activates a neural pathway leading to the nucleus of the solitary tract. NPY is over-produced in leptin-deficient mice and may mediate the overfeeding observed in leptin-deficient animals.

- **AgRP**, a powerful antagonist of the MC3R and MC4R melanocortin receptors in the hypothalamus. Obese patients have elevated plasma levels of AgRP, and over-expression of AgRP in animal models leads to obesity.

- **Serotonin**. Serotonin central receptors appear to play a major role in glucose homeostasis. Experiments in obese mice have demonstrated that small doses of a classical serotonin agonist, metachlorophenylpiperazine (mCPP), markedly lower plasma insulin levels and increase insulin sensitivity, without affecting food intake, body weight or fat mass. The downstream target of the involved serotonin receptor appears to be melanocortin-4 receptors, in the arcuate nucleus of the hypothalamus.

- **Metabolic pituitary hormones**. The hypothalamus controls energy expenditure via the anterior pituitary and the sympathetic nervous system. An important output signal to the pituitary is thyrotropin-releasing hormone (TRH), a small peptide secreted into the hypophyseal portal system which causes the anterior pituitary to release thyroid-stimulating hormone (TSH). Thyroid hormone acts on peripheral tissues, increasing alertness, heat production and BMR. Thyroid hormone production and a 'normal' level of alertness and physical activity require adequate food intake. Thyroxin output falls during prolonged starvation, when metabolism and physical activity are severely restricted in order to conserve fuel stocks. Two other pituitary hormones are associated with fasting and gluconeogenesis: corticosteroid (which participates in a feedback loop involving pro-inflammatory cytokines) and growth hormone (which defends the body's protein and glycogen reserves and promotes the breakdown of fat).

- **Autonomic outputs**. Hypoglycaemia and hypothermia both lead to sustained sympathetic responses. Subjects feel hungry and eat if possible, but they refine their other actions to suit the circumstances. Hypoglycaemia requires hepatic glycogenolysis and gluconeogenesis, while hypothermia requires increased heat production and a redistribution of blood flow. Sympathetic activity is controlled by the hypothalamus, which instructs the adrenal medulla to secrete adrenalin. This is a rather blunt control, and so localised sympathetic responses (such as blood flow regulation) are mediated by individual nerves. Parasympathetic activity can also respond to the hypothalamus, which controls the nucleus of the solitary tract.

- **Fatty acids**. Normal brain tissue does not take up or metabolise fatty acids (hence its dependence on glucose, or the switch to ketone body utilisation). However, the arcuate nucleus converts fatty acids to long-chain fatty acyl-CoA intermediates. The exact mechanism of appetite control is unclear, but it is known that the long-chain fatty acyl-CoA intermediates formed in the arcuate nucleus dampen appetite and reduce food intake.

- **Prokineticin 2**. Recently discovered in the suprachiasmatic nucleus, prokineticin 2 is a signalling molecule that appears to help control hunger. Both lean and obese mice treated with prokineticin for five days lost almost 5% of their body weight.

## 3.10    Lifestyle changes

Weight loss is clearly beneficial for patients suffering from the metabolic syndrome. Highly motivated pathologically obese individuals have achieved excellent results, but it is very difficult for most people to achieve a weight loss greater than 10% for an extended period, although even this modest reduction is considered to be worthwhile. To achieve sustained weight loss it is normally recommended to adopt a low-fat, physically bulky diet, rich in fresh fruit, green vegetables and unrefined carbohydrates; the diet should have a low glycaemic index, reducing the need for insulin. This diet should be supplemented by an exercise programme to increase energy demand, leading to a gradual weight loss over several months. This altered lifestyle needs to be maintained indefinitely. 'Crash' diets should be avoided; they chiefly affect body water content rather than fat. High-fat 'Atkins' diets are also effective in producing weight loss, particularly because they induce rapid satiation. They are not widely recommended by nutritionists, although the evidence against them is largely speculative and anecdotal. High-fat and 'Mediterranean' diets seem more effective than conventional low-carbohydrate diets for sustained weight loss. The feared dislipidaemia and cardiovascular problems with such diets have so far failed to materialise.

**Table 3.1** Candidate satiety factors

| Factor | Description | Comments |
|---|---|---|
| Leptin | A cytokine, produced predominantly by fat cells (adipocytes); plasma levels of leptin rise and fall in parallel to body fat content (> fat > leptin). | A major site of leptin receptors is in the hypothalamus (the arcuate nucleus). The janus kinase (JAK)–signal transducer and activator of transcription (STAT) pathway play a critical role in the signalling of a wide array of cytokines and growth factors, leading to various cellular functions, including proliferation, growth, haematopoiesis and immune response. It also has an effect on ATP-sensitive potassium channels in glucose-responsive neurones, which affect the neuronal firing rate. Leptin has major effects on reproductive behaviour (sexual maturation is delayed by lack of food). Starving women, female athletes and anorexics with low fat stores experience secondary amenorrhea. Leptin signalling defects lead to gross obesity, but these are very rare in humans. |
| Peptide tyrosine tyrosine (PYY) | A gut hormone present in endocrine cells in the lower intestine that can be released by the presence of luminal free fatty acids. | Shown to inhibit gut motility and gastrointestinal and pancreatic secretions. It inhibits 'appetite-stimulating' NPY/AgRP-producing neurons (see below), thus signalling food intake and damping hunger. Many of these gut peptides are incretin hormones, which also stimulate insulin release. |
| Resistin | A peptide hormone produced by adipocytes (and probably by other tissues). | Polymorphism of the resistin gene is associated with obesity. Resistin has an anti-insulin action, and is itself suppressed by insulin and the pro-inflammatory cytokines. Output is increased by thyroid hormone T4 but the physiological function is not yet understood. |
| Adiponectin | A mixture of anti-inflammatory peptide hormones secreted by adipocytes, which also regulate energy homeostasis and the metabolism of glucose and lipids. | By increasing glucose catabolism, adiponectin achieves a reduction of glucose levels in vivo. Adiponectin increases insulin sensitivity in target tissues, but also stimulates fatty acid oxidation and blocks the differentiation of new adipocytes in bone marrow. |

(*continued overleaf*)

**Table 3.1** (*continued*)

| Factor | Description | Comments |
|---|---|---|
| Pro-inflammatory cytokines | TNF-$\alpha$, IL-6 and IL-1 act on the hypothalamus to reduce appetite and raise body temperature in response to infection and other illnesses. First identified as products of the immune system (macrophages), it is now known that many other tissues (including adipocytes) can secrete these compounds. | There is a major negative feedback loop involving the hypothalamus, corticotrophin-releasing hormone, corticotropin (=ACTH) secreted by the pituitary and corticosteroids from the adrenal cortex, which dampen pro-inflammatory cytokine production. This loop normally acts to stabilise immune system activity, but it also has spillover effects on appetite and weight regulation. |
| Amylin | Pancreatic $\beta$-cells co-release a second polypeptide hormone called amylin at the same time as they release insulin. | Amylin produces a feeling of satiation, and may assist in the regulation of food intake. A modified amylin, Pramlintide, is being investigated as a hypoglycaemic agent in early type 2 diabetes. It potently reduces glucagon secretion and therefore postprandial hyperglycaemia. |
| Apolipoprotein A-IV | A glycoprotein synthesised by enterocytes in response to long-chain dietary fat. | Apolipoprotein A-IV may regulate PYY (see above). It is thought to regulate food intake, possibly by stimulating CCK production. It may be effective in its own right because it is also present in the brain. |
| Endocannabinoids, anandamide (orexigenic) and oleoylethanolamide (anorexigenic) | May be important for gastrointestinal function and the regulation of food intake. | Rimonabant is an inverse agonist for CB1 cannabinoid receptors, which is approved in Europe for weight-loss therapy, but has significant side effects on the central nervous system. |

## 3.11 The basics of dieting

Dieting has much to do with understanding what the body does with the excess food ingested, for example:

- **High-carbohydrate (low-fat) diet.** In a diet consisting of 70% carbohydrates and 30% protein with no fat, some protein will be used for body building and repair, and some will be converted into glucose. All the carbohydrates will be converted to glucose initially. This will result in a rapid and sustained elevation in blood glucose levels, stimulating insulin production. Insulin stimulates cells to uptake glucose, as well as increasing appetite, causing most people to eat again not long after eating a high-carbohydrate meal. Insulin stimulates the body to store fat. Thus, a high-carbohydrate diet will provide excess of what is necessary for immediate energy usage. Some will be converted to glycogen and stored in the liver, but most is converted into fat for storage in the body tissues.

- **High-fat (low-carbohydrate) diet.** In a diet consisting of 30% protein, 70% fat with no carbohydrates, proteins will be used as before, but in the absence of carbohydrates the body must 'burn' the fat it consumes. This causes the body to 'convert' to a fat-burning engine instead of being primarily a glucose-burning engine. Fats, unlike carbohydrates, have a high satiety factor; fats make you feel full, and the satiety lasts for hours. Therefore, you tend to consume fewer calories on a high-fat diet than on a high-carbohydrate diet. Also, with a lower carbohydrate intake, the levels of insulin are low. Therefore, the fat you eat tends not to be stored. Thus a high-fat diet, in the absence of carbohydrates, typically results in weight loss.

## 3.12   Medical and surgical treatment

There is no entirely satisfactory drug treatment for obesity, and it is normally recommended that moderately overweight patients (BMI <30) are treated with diet and exercise alone. Some existing 'anti-obesity' drugs are given in Table 3.2.

None of these drugs are particularly effective and a 10% weight loss is considered to be good; re-accumulation of lost weight is the tendency once therapy is stopped.

Liposuction to remove excess fat is not currently recommended, but gastric surgery is proving effective and popular. 'Stomach stapling' operations include gastric bypass and a variety of banding procedures. Surgery normally produces much larger weight losses than drug therapy, but does carry significant risk.

**Table 3.2**   Anti-obesity drugs

| Drug | Target | Side effects | Comments |
|---|---|---|---|
| Orlistat | Blocks pancreatic lipase | Anal leakage | Fatty stools |
| Sibutramine | Blocks serotonin uptake | Dry mouth, headaches | FDA (USA) approved |
| Glucagon-like peptide 1 | Delays gastric emptying; more insulin produced | Natural product | Used in type 2 diabetes |
| Rimonabant | Cannabinoid CB1 receptor antagonist | Depression, suicide | Approved in Europe, not USA |
| Topiramate | Multiple CNS effects; antiepileptic | Psychomotor problems | Poor patient compliance |
| Butabindide | Blocks cholecystokinin breakdown | Not fully tested | Developmental |

# CHAPTER 4

# Digestion and absorption

## 4.1 The gastrointestinal tract

The gastrointestinal tract comprises the stomach, duodenum, jejunum, ileum, colon, rectum and anal canal. The gastrointestinal tract and oesophagus form the alimentary canal. The stomach is both a reservoir and a digestive organ. The mucosa (epithelium, lamina propria and muscularis mucosae) forms longitudinal folds (gastric folds or rugae), which disappear when the stomach is fully distended. On the mucosal surface are small, funnel-shaped depressions (gastric pits). Almost the entire mucosa is occupied by simple, tubular gastric glands which open into the bottom of the gastric pits (Figure 4.1). The surface epithelium (simple, tall columnar) does not change throughout the stomach. It contains mucus-producing cells which form a secretory sheath (glandular epithelium). The mucus is alkaline and adheres to the epithelium, forming a protective layer. The surface epithelium is renewed approximately every third day. The source of the new cells is the isthmus; that is, the upper part of the neck of the gastric glands, where cells divide and then migrate towards the surface epithelium and differentiate into mature epithelial cells. In contrast to the surface epithelium, the cellular composition and function of the gastric glands are specialised in the different parts of the stomach. In the principal (or corpus-fundic) glands, there are four cell types: chief cells, parietal cells, mucous neck cells and endocrine cells.

Parietal cells (or oxyntic cells) occur most frequently in the necks of the glands, where they reach the lumen. Parietal cells secrete the hydrochloric acid of the gastric juice.

Parietal cells also secrete intrinsic factor, which is necessary for the absorption of vitamin B12. Vitamin B12 is a cofactor of enzymes which synthesise tetrahydrofolic acid, which in turn is needed for the synthesis of DNA components. An impairment of DNA synthesis will affect rapidly dividing cell populations, among them the haematopoietic cells of the bone marrow, which may result in pernicious anaemia. This condition may result from a destruction of the gastric mucosa by, for example, autoimmune gastritis or the resection of large parts of the lower ileum, which is the main site of vitamin B12 absorption, or of the stomach.

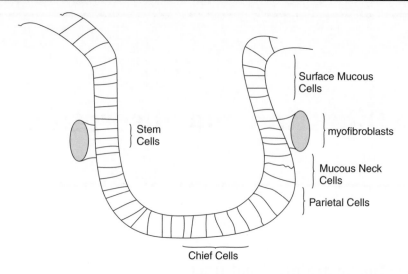

**Figure 4.1**  Components of a gastric gland. The pit and the luminal surface are lined by surface mucous cells. The isthmus contains stem and progenitor cells and is enclosed by a sheath of myofibroblastic cells. Mucous neck cells are found in the neck, while chief and endocrine cells are present in the base of the gland. Parietal cells are actually scattered throughout the gland.

## 4.2   Gastric acid production

Parietal cells possess receptors for three stimulators of acid secretion: neural (acetylcholine, muscarinic-type receptor), paracrine (gastrin) and endocrine control (histamine, $H_2$ type receptor) (Figure 4.2).

Histamine from nearby enterochromaffin-like (ECL) cells is probably the primary modulator, but the magnitude of the stimulus depends upon an interaction of signals of each type; low levels of only histamine or gastrin weakly stimulate acid secretion, but together they strongly reinforce it. Pharmacologic antagonists of each of these molecules can block acid secretion.

Gastrin is a linear peptide hormone produced by G cells of the duodenum and in the pyloric antrum of the stomach. It is secreted into the bloodstream (its effect is paracrine). Gastrin is released in response to certain stimuli, including stomach distension, vagal stimulation, the presence of partially digested proteins (amino acids) and hypercalcaemia. Its release is inhibited by the presence of acid in the stomach (negative feedback), somatostatin, along with secretin, GIP (gastroinhibitory peptide), glucagon and calcitonin.

Parietal cells contain a proton pump, the $H^+K^+$-ATPase, located in the plasma membranes of the canaliculi (canaliculi folding increases the surface area for secretion of acid). This ATPase is magnesium-dependent and is not inhibited by ouabain (which distinguishes it from the $Na^+K^+$-ATPase).

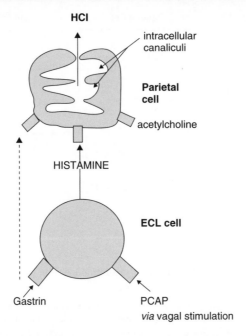

**Figure 4.2** Gastric acid production. Two cell types in the mucosa of the corpus of stomach are principally responsible for secretion of acid. Histamine secreted from nearby enterochromaffin-like (ECL) cells stimulates parietal cells to secrete acid. A variety of substances can stimulate the ECL cell to secrete histamine, including PCAP, pituitary adenyl cyclase-activating peptide (released from enteric nervous system interneurones in the gastric mucosa), and gastrin, both stimulating ECL cells via adenyl cyclase to raise intracellular levels of cAMP that lead to histamine secretion.

Bicarbonate ion (production catalysed by carbonic anhydrase) exits the cell on the basolateral surface, in exchange for chloride. The outflow of bicarbonate into blood results in a slight alkalinity of the blood, known as the 'alkaline tide'.

Chloride and potassium ions are transported into the lumen of the cannaliculi by conductance channels.

Hydrogen ions are pumped out of the cell, into the gut lumen, in exchange for potassium, through the action of the proton pump; potassium is thus effectively recycled (Figure 4.3).

# Cell surface polarity

The apical membrane of a polarised cell is that part of the plasma membrane that forms its luminal surface, particularly so in the case of epithelial and endothelial cells. The basolateral membrane of a polarised cell refers to that part of the plasma membrane that forms its basal and lateral surfaces. Proteins are free to move from the basal to lateral surfaces, but not to the apical surface; tight junctions, which join epithelial cells near their apical surfaces, prevent migration of proteins to the apical surface. The apical surface is therefore distinct from the basal/lateral surfaces.

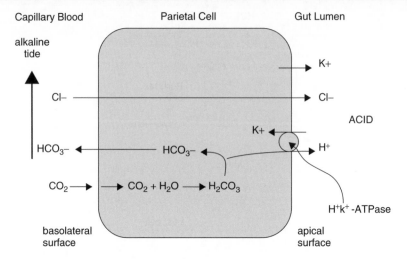

**Figure 4.3**   Scheme of gastric acid production in parietal cells.

Accumulation of osmotically-active chloride (which is required to maintain electroneutrality with hydrogen ions) in the canaliculi generates an osmotic gradient that results in outward diffusion of water; the resulting gastric juice is about 155 mM HCl and 15 mM KCl, with a small amount of NaCl. The highly acidic environment causes denaturation of proteins, making them susceptible to proteolysis by pepsin (which is itself acid-stable).

> Gastrin and vagus nerve stimulation trigger the release of pepsinogen from chief cells in the gastric glands. Pepsinogen (inactive) is a zymogen which under acidic conditions autocatalytically cleaves itself to form pepsin (active), an enzyme with a pH optimum of 1.5–2. It cleaves peptide bonds on the N-terminal side of aromatic amino acids; peptides are further digested by proteases in the duodenum.

## 4.3   Proton pump inhibitors

The gastric $H^+K^+$-ATPase proton pump is a target for rational drug design. Safe and effective inhibition of gastric acid secretion is a goal of clinicians in treatments of gastro-oesophageal reflux disease and peptic ulcer. Two classes of drug, the histamine $H_2$-receptor antagonists and the proton pump inhibitors (PPIs), achieve this goal with a high level of success.

PPIs, including omeprazole (Figure 4.4), esomeprazole, lansoprazole, pantoprazole and rabeprazole, cause a dose-dependent inhibition of gastric acid secretion by inhibiting the proton pump in actively secreting gastric parietal cells. They have a greater efficacy and longer effect than $H_2$-receptor antagonists. NICE (National Institute for Clinical Excellence) guidelines for dyspepsia recommend that all PPIs are well tolerated and have similar efficacy, so the least expensive PPI should be used:

- for short-term treatments of gastric and duodenal ulcers

- in combination with antibacterials for the eradication of *Helicobacter pylori*.

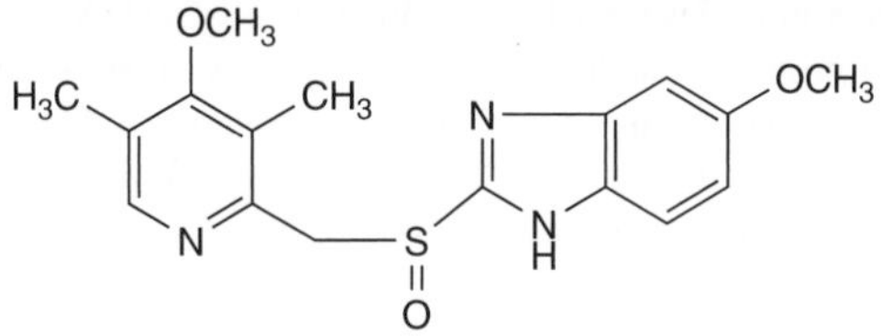

**Figure 4.4** Structure of omeprazole.

An initial short course of a PPI is the treatment of choice in gastro-oesophageal reflux disease with severe symptoms. PPIs are superior to $H_2$-antagonists in empirical treatment of typical gastro-oesophageal reflux disease symptoms. Patients with endoscopically confirmed erosive or ulcerative oesophagitis, or oesophagitis complicated by stricture, usually need to be maintained on a PPI. PPIs are also used in the prevention and treatment of non-steroidal anti-inflammatory drug (NSAID)-associated ulcers. Omeprazole and lansoprazole are effective in the treatment of the Zollinger–Ellison syndrome.

> Zollinger–Ellison syndrome is characterised by increased levels of the hormone gastrin, causing the stomach to produce excess hydrochloric acid. Often the cause is a tumour (gastrinoma) of the duodenum or pancreas. Peptic ulcers occur in almost 95% of patients.

Additional mediators of gastric acid secretion include calcium, gastrin-releasing hormone and enkephalin, while opiate receptors have also been identified on parietal cells.

For further information, see Focus on: Controlling Gastric Acid Production.

> A $Ca^{2+}$-sensing receptor (CaSR) located at the basolateral membrane of the parietal cell is involved in the regulation of the $H^+K^+$-ATPase.
>
> Gastrin-releasing hormone stimulates gastrin release through G-protein-coupled receptors. Together with cholecystokinin, it is the major source of negative-feedback signals that suppress feeding.
>
> Another group of G-protein receptors (opioid receptors) are affected by enkephalins.

## 4.4 Helicobacter pylori

This Gram-negative, microaerophilic bacterium infects areas of the stomach and duodenum; it can result in peptic ulcers, gastritis, duodenitis and cancers. *H. pylori*'s helical shape and flagella favour its motility in the mucus layer. Adhesins are produced by the bacterium, which binds to membrane-associated lipids and carbohydrates to maintain its attachment to epithelial cells. Large amounts of the enzyme urease are produced, both inside and outside of the bacterium. Urease metabolises urea (which is normally secreted into the stomach) to carbon dioxide and ammonia (which neutralises gastric acid), and is instrumental in the survival of the bacterium in the acidic environment.

$$(NH_2)_2CO + H_2O \rightarrow CO_2 + 2NH_3$$

The ammonia that is produced is toxic to the epithelial cells, and along with the other products of *H. Pylori*, including protease, catalase and certain phospholipases, causes damage to these cells. Some strains of the bacterium 'inject' the inflammatory inducing agent peptidoglycan from their own cell wall into epithelial stomach cells. The risk of developing stomach cancer is thought to be increased with long-term infection with *H. pylori*.

---

*H. pylori* infection can be diagnosed by:

- a 'breath test' (for $CO_2$) in which the patient drinks $^{14}$C- or $^{13}$C-labelled urea, which the bacterium metabolises, producing labelled $CO_2$ that can be detected in the breath

- a 'stool antigen test'

- a blood test to detect antibodies to *H. pylori*

- a biopsy of the lining of the stomach (the most reliable).

---

*H. pylori* infections are normally treated by a 'triple therapy' course over a week, namely two antibiotics plus an acid inhibitor, for example amoxicillin, clarithromycin and a PPI such as omeprazole.

## 4.5   The small intestine

This is the main site of absorption of virtually all nutrients into blood. It consists of the duodenum, a short section that receives secretions from the pancreas and liver via the pancreatic and common bile ducts, the jejunum and the ileum. Enterocytes mature into absorptive cells of the epithelium. Two other major cell types are present: enteroendocrine cells, which secrete hormones such as cholecystokinin and gastrin into blood, and goblet cells, which secrete lubricating mucus. Villi (Figure 4.5) are projections into the lumen covered predominantly with mature, absorptive enterocytes, along with occasional mucus-secreting goblet cells. These cells live only for a few days, die and are shed into the lumen. Crypts (of Lieberkuhn) are moat-like invaginations of the epithelium around the villi, and are lined largely with younger epithelial cells, which are involved primarily in secretion. Toward the base of the crypts are stem cells, which continually divide and provide the source of all the epithelial cells in the crypts and on the villi.

## 4.6   The gastrointestinal barrier

The gastrointestinal barrier is considered to comprise two components: the intrinsic barrier, composed of epithelial cells lining the digestive tube and the tight junctions that tie them together; and the extrinsic barrier, consisting of secretions and other influences that are not physically part of the epithelium but which affect the epithelial cells and maintain their barrier function (Table 4.1).

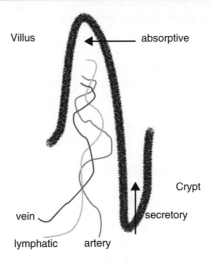

**Figure 4.5**   Villi absorption and secretion.

**Table 4.1**   The gastrointestinal barrier

| Intrinsic | Extrinsic |
| --- | --- |
| Tight junctions between epithelial cells seal the paracellular spaces and establish the basic gastrointestinal barrier. | Defined by the layer of mucus. |
| Gastric parietal cells and chief cells have low permeability to protons, preventing back-diffusion of acid. | Reduces shear stress on the epithelium. |
| Cells have rapid turnover rates, usually a few days. | Abundant carbohydrates in mucin bind bacteria, preventing colonisation and causing aggregation, which accelerates their clearance. |
| Stem cells, in the middle of gastric pits and crypts, provide continual replenishment. | The effects of toxins are minimised by their reduced diffusion in mucus. |
| | Gastric and duodenal epithelial cells secrete bicarbonate to their apical faces to maintain a neutral pH along the epithelial plasma membrane. |

## 4.7   Paneth cells

Paneth cells contribute to the maintenance of the gastrointestinal barrier, by secreting a number of antimicrobial molecules (alpha-defensins or cryptidins), as well as lysozyme and phospholipase $A_2$. Their location, adjacent to crypt stem cells, suggests they have a role in defending epithelial cell renewal.

# 4.8   The enteric endocrine system

Digestive function is affected by hormones produced in many endocrine glands, but the most profound control is exerted by hormones produced within the gastrointestinal tract itself. The gastrointestinal tract is the largest endocrine organ in the body and the endocrine cells within it are referred to collectively as the enteric endocrine system. Three of the best-studied enteric hormones are gastrin, secreted from the stomach, which plays an important role in control of gastric acid secretion, cholecystokinin, which stimulates secretion of pancreatic enzymes and bile, and secretin, which stimulates secretion of bicarbonate-rich fluids from the pancreas and liver. Normal proliferation of gastric and intestinal epithelial cells, as well as proliferation in response to such injury as ulceration, is known to be affected by a large number of endocrine and paracrine factors.

Prostaglandins, particularly prostaglandin E2 and prostacyclin, have 'cytoprotective' effects on the gastrointestinal epithelium. A common clinical correlate is that use of aspirin and other non-steroidal anti-inflammatory drugs (NSAIDs), which inhibit prostaglandin synthesis, is associated with gastric erosions and ulcers. Prostaglandins are synthesised within the mucosa from arachidonic acid through the action of cyclooxygenases. Their cytoprotective effect appears to result from stimulation of mucosal mucus and bicarbonate secretion, increasing mucosal blood flow and, particularly in the stomach, limiting back-diffusion of acid into the epithelium. Considerable effort is being made to develop NSAIDs that fail to inhibit mucosal prostaglandin synthesis.

Two peptides that have received attention for their potential role in barrier maintenance are epidermal growth factor (EGF) and transforming growth factor-alpha (TGF-alpha). EGF is secreted in saliva and from duodenal glands, while TGF-alpha is produced by gastric epithelial cells. Both peptides bind to a common receptor and stimulate epithelial cell proliferation. In the stomach they also enhance mucus secretion and inhibit acid production.

Cytokines, such as fibroblast growth factor and hepatocyte growth factor, have been shown to enhance healing of gastrointestinal ulcers in experimental models.

Trefoil proteins are a family of small peptides that are secreted by goblet cells in the gastric and intestinal mucosa, and coat the apical face of the epithelial cells. Their distinctive molecular structure appears to render them resistant to proteolytic destruction. They appear to play an important role in mucosal integrity, repair of lesions and limiting epithelial cell proliferation, as well as in protecting the epithelium from a broad range of toxic chemicals and drugs. Trefoil proteins also appear to be central players in the restitution phase of epithelial damage repair, where epithelial cells flatten and migrate from the wound edge to cover denuded areas. Mice with targeted deletions in trefoil genes showed exaggerated responses to mild chemical injury and delayed mucosal healing.

Nitric oxide (NO) appears to play a crucial role in mucosal integrity and barrier function, but paradoxically also contributes to mucosal injury in a number of digestive diseases. Nitric oxide is synthesised from arginine through the action of one of three isoenzyme forms of nitric oxide synthetase (NOS). Research in this area has focused on understanding the effects of applying NO donors, such as glyceryltrinitrate, or NOS inhibitors. In several models, NO donors significantly reduced the severity of mucosal injury induced by toxic chemicals (e.g. ethanol), or that associated with ischaemia and reperfusion. Similarly, healing of gastric ulcers in rats has been accelerated by application of NO donors. Co-administration of NO donors and NSAIDs results in anti-inflammatory properties comparable to NSAIDs alone, but with less damage to the gastrointestinal mucosa. NOS inhibitors are under investigation for treatment of situations in which NO overproduction appears to contribute to mucosal injury.

The synthesis of NO from dietary nitrite, under the acid conditions of the stomach and the presence of reductants such as ascorbic acid, has been proposed to have an antimicrobial function.

An important part of barrier function is to prevent transit of bacteria from the lumen through the epithelium. Paneth cells are epithelial granulocytes located in small intestinal crypts of many mammals. They synthesise and secrete several antimicrobial peptides, chief among them isoforms of alpha-defensins, also known as cryptdins ('crypt defensin'). These peptides have antimicrobial activity against a number of potential pathogens, including several genera of bacteria, some yeasts and Giardia trophozoites. Their mechanism of action is likely similar to neutrophilic alpha-defensins, which permeabilise target cell membranes.

Barrier function is also supported by the gastrointestinal immune system; much of the epithelium is bathed in immunoglobulin A (IgA), which is secreted from sub-epithelial plasma cells and transcytosed across the epithelium into the lumen. IgA provides an antigenic barrier by binding bacteria and other antigens, although this barrier function is specific for particular antigens and requires previous exposure for development of the response.

For further information, see Focus on: Controlling Gastric Acid Production.

## 4.9 The pancreas

The pancreas is both exocrine (secreting pancreatic juice containing digestive enzymes) and endocrine (producing several important hormones, including insulin, glucagon and somatostatin). Pancreatic secretions are secreted into the lumen of the acinus and accumulate in intralobular ducts that drain to the main pancreatic duct, then directly into the duodenum. Control of the exocrine function of the pancreas is via the hormones gastrin, cholecystokinin and secretin. Major proteases are secreted as their inactive (zymogen) forms (e.g. trypsinogen and chymotrypsin); secreted to a lesser degree are pancreatic lipase and pancreatic amylase.

Pancreatic secretions from ductal cells contain bicarbonate ions that neutralise the acidic chyme from the stomach and are important in protecting the pancreas from recurrent acute and chronic pancreatitis by quickly sweeping zymogens out of it. The key molecule in the pancreatic duct is the cystic fibrosis transmembrane conductance regulator (CFTR). Activation of anion ($Cl^-/HCO_3^-$) exchange at the apical plasma membrane drives anion secretion and a concomitant flow of $Na^+$ and $H_2O$ into the lumen (see Figure 4.12 and Focus on: Cystic Fibrosis (CF)). Loss of anion transport, due to decreased CFTR expression or protein function, therefore leads to reduced bicarbonate and fluid secretion. The pancreas is usually one of the first organs to fail in cystic fibrosis (CF) because this molecular channel has such a central role in pancreatic physiology.

## 4.10 Absorption in the small intestine: general principles

The single most important process that takes place in the small gut, which makes nutrient absorption possible, is the establishment of an electrochemical and concentration gradient of sodium across the epithelial cell boundary of the lumen.

To remain viable, all cells of the body are required to maintain a low intracellular concentration of sodium. Low intracellular sodium concentration is maintained by a large number of

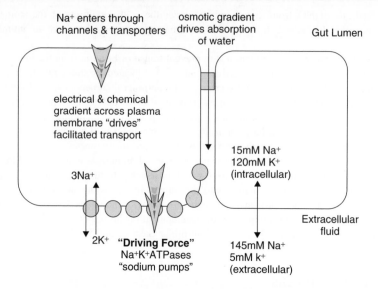

**Figure 4.6** Central role of sodium–potassium pumps.

$Na^+K^+$ ATPases, so-called sodium–potassium pumps (Figure 4.6), embedded in the basolateral membrane. The sodium–potassium pump is a highly conserved integral membrane protein, expressed in virtually all animal cells. The transport of sodium creates both an electrical and a chemical gradient across the plasma membrane. In turn this provides:

- a cell's resting membrane potential, the gradient of which is the basis for excitability in nerve and muscle cells

- export of sodium from the cell, providing the driving force for several facilitated transporters, which import glucose, amino acids and other nutrients into the cell

- translocation of sodium from one side of an epithelium to the other, creating an osmotic gradient that drives absorption of water.

Cells contain huge numbers of these pumps. Abnormalities in the number or function of $Na^+K^+$-ATPases are thought to be involved in several pathologic states, in particular heart disease and hypertension.

---

The $Na^+K^+$-ATPase is composed of two subunits. An alpha subunit binds ATP and both $Na^+$ and $K^+$ ions, and contains a phosphorylation site. A beta glycoprotein subunit appears critical in facilitating plasma membrane localisation and activation of the alpha subunit. There are 8–10 transmembrane domains; alpha and beta subunits exist in several isoforms. The alpha subunit of the $Na^+K^+$-ATPase is the receptor for cardiac glycosides such as digitalis and ouabain. Different isoforms of the alpha subunit have different affinities for such glycosides. Binding of these widely-used drugs to sodium pumps specifically inhibits their activity. Cardiac glycosides are widely used to increase the strength of contraction of the heart. Inhibition of sodium pump activity in cardiac myocytes results in an increase in intracellular sodium concentration; in turn this leads to an increase in intracellular calcium concentration by sodium–calcium exchange, which appears to be the proximal mechanism for enhancing cardiac contractility.

Chronic or sustained changes in pump activity within cells are usually caused by increases in transcription rate or mRNA stability. The major hormonal controls over pump activity can be summarised as follows:

- Thyroid hormones appear to stimulate subunit gene transcription.

- Aldosterone stimulates rapid and sustained increases in pump numbers in a number of tissues.

- Catecholamines have varied effects: dopamine inhibits pump activity in kidney, while adrenaline (epinethrine) stimulates pump activity in skeletal muscle.

- Insulin is a major regulator of potassium homeostasis and has multiple effects on sodium pump activity. Within minutes of elevated insulin secretion, pumps containing alpha-1 and 2 isoforms have increased affinity for sodium and increased turnover rate. Sustained elevations in insulin cause up-regulation of alpha-2 synthesis. In skeletal muscle, insulin may also recruit pumps stored in the cytoplasm or activate latent pumps already present in the membrane.

## 4.11 Crossing the gastrointestinal barrier

There are two possible routes across the epithelium of the gut (Figure 4.7), the transcellular route (crossing the plasma membrane) and the paracellular route (crossing the tight junctions between cells). Some molecules, water for instance, are transported by both routes, but the tight junctions are impermeable to large organic molecules from the diet (e.g. amino acids and glucose). Such molecules are transported exclusively by the transcellular route, by absorptive enterocytes equipped with specific transporter molecules that facilitate their entry into and out of the cells. Within the intestine, there is a proximal-to-distal gradient in osmotic permeability.

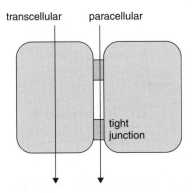

**Figure 4.7** Crossing the gut epithelium.

Passing down the intestinal tube, the effective pore size through the epithelium decreases, so the 'upper' (proximal) gut is more permeable to water than the 'lower' (distal) gut; the distal intestine can actually absorb water better than the proximal gut. The observed differences in permeability to water across the epithelium are due almost entirely to differences in conductivity across the paracellular path; tight junctions vary considerably in 'tightness' along the length of the gut.

## 4.12    Absorption and secretion of water and electrolytes

Large quantities of water are both secreted into and reabsorbed from the lumen of the small intestine. Water flows across the mucosa in response to osmotic gradients. In the case of secretion, two distinct processes establish an osmotic gradient that pulls water into the lumen of the intestine:

- **Increases in lumen osmotic pressure resulting from influx and digestion of foodstuffs**. The chyme that passes into the intestine from the stomach typically is not hyperosmotic, but as its macromolecular components are digested, the osmolarity of that solution increases (e.g. glucose is osmotically more active than sucrose).

- **Crypt cells actively secrete electrolytes, leading to water secretion (Figure 4.8)**. The apical membranes of crypt epithelial cells contain an ion channel of immense medical significance, a cyclic adenosine monophosphat (cAMP)-dependent chloride channel, known as the cystic

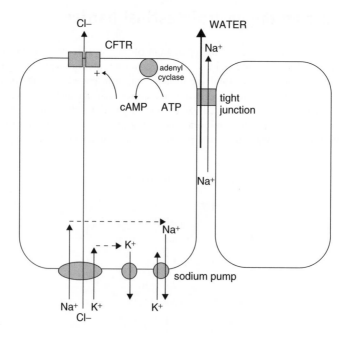

**Figure 4.8**   Crypt cells actively secrete electrolytes.

fibrosis transmembrane conductance regulator (CFTR). Defects in this channel are the basis of the disease cystic fibrosis (see Focus on: Cystic Fibrosis (CF)).

The CFTR chloride channel is responsible for secretion of water through the following steps:

1. Chloride ions enter the crypt epithelial cell by co-transport with sodium and potassium; sodium is pumped back out via sodium pumps and potassium is exported via a number of channels on the basolateral surface.

2. Elevated intracellular concentrations of cAMP in crypt cells activate the CFTR, resulting in secretion of chloride ions into the lumen.

3. Accumulation of negatively charged chloride anions in the crypt creates an electric potential that attracts sodium ions, pulled into the lumen apparently across tight junctions; the net result is secretion of NaCl.

4. Secretion of NaCl into the crypt creates an osmotic gradient across the tight junction and water is drawn into the lumen by the paracellular route.

# 4.13   Pathophysiology of diarrhoea

Diarrhoea is an increase in the volume of stool or frequency of defecation. It is one of the most common clinical signs of gastrointestinal disease, but can also reflect primary disorders outside of the digestive system. There are numerous causes of diarrhoea, but in almost all cases this disorder is a manifestation of one of the following four basic mechanisms:

- **Osmotic diarrhoea**. Occurs if osmotically active solutes are retained in the intestinal lumen; water will not be absorbed and diarrhoea will result. Ingestion of poorly absorbed substrates, such as mannitol, sorbitol, Epson salts ($MgSO_4$) and some antacids ($MgOH_2$), can occur in a number of malabsorption syndromes. For example, a failure to digest lactose (lactose intolerance) means that lactose remains in the intestinal lumen and osmotically 'holds' water.

- **Secretory diarrhoea**. Occurs when secretion of water into the lumen exceeds absorption. The causative agent of cholera, *Vibrio cholera*, produces cholera toxin (CTX), which strongly activates adenyl cyclase, causing a prolonged increase in intracellular concentration of cAMP within crypt enterocytes. This change results in prolonged opening of the chloride channels, leading to uncontrolled secretion of water. Additionally, CTX affects the enteric nervous system, resulting in an independent stimulus of secretion. The *E. coli* heat-labile ST-toxin induces a similar effect; in this case it is a guanyl cyclase that is activated, raising cGMP levels and activating the chloride channel. Secretory diarrhoea can also result from laxatives, hormones secreted by certain types of tumour (e.g. vasoactive intestinal peptide), drugs (e.g. asthma medications, antidepressants, cardiac drugs), certain metals, organic toxins and plant products (e.g. arsenic, insecticides, mushroom toxins and caffeine).

- **Inflammatory and infectious diarrhoea**. Examples of pathogens frequently associated with infectious diarrhoea include bacteria, *Salmonella*, *E. coli*, *Campylobacter*, viruses, rotaviruses, coronaviruses, parvoviruses (canine and feline), norovirus and protozoa, coccidia species, *Cryptosporium* and *Giardia*. The immune response to inflammatory conditions in the bowel contributes substantively to development of diarrhoea. Activation of white blood cells leads them to secrete inflammatory mediators and cytokines, which can stimulate secretion, effectively producing both secretory and inflammatory diarrhoea.

- **Diarrhoea associated with deranged motility.** In order for nutrients and water to be efficiently absorbed, the intestinal contents must be adequately exposed to the mucosal epithelium and retained long enough to allow absorption. Disorders in motility that accelerate transit time can decrease absorption, resulting in diarrhoea. Alterations in intestinal motility (usually increased propulsion) are observed in many types of diarrhoea. What is not usually clear, and is very difficult to demonstrate, is whether primary alterations in motility are the cause of diarrhoea or simply an effect.

## 4.14   Rehydration therapy

Oral rehydration therapy is a simple, cheap and effective treatment for diarrhoea-related dehydration, such as that caused by cholera or rotavirus. It consists of a solution of salt and glucose and is administered orally. It has saved millions of children from diarrhoea, still a leading cause of death, particularly in the developing world. The standard manufactured WHO/UNICEF solution contains:

- sodium chloride 2.6 g/l

- anhydrous glucose 13.5 g/l

- potassium chloride 1.5 g/l

- trisodium citrate, dihydrate 2.9 g/l

   The provision of salt and glucose allows for the symport transport of glucose and $Na^+$ (see Figure 4.9) and therefore restoration of a higher osmotic potential in extracellular fluid that 'draws' water in from the gut lumen.

## 4.15   Absorption of sugars and amino acids

Final digestion of dietary carbohydrates and proteins occurs on the lumen face of small intestinal enterocytes by so-called 'brush border enzymes', including maltase, sucrose, lactase and peptidases.

   Monosaccharides, such as glucose and related hexoses, enter the cell along with sodium, on symporters (SGLT transporters), and glucose passes out of the cell (down its concentration gradient) by facilitated diffusion, mediated by a different glucose-carrier protein in the basal and lateral membrane domains (glucose transporters (GLUT)) (Figure 4.9); like the liver and kidney, enterocytes are among the few tissues that can both import and export glucose. The cellular $Na^+$ gradient, which 'drives' the $Na^+$ glucose symport, is maintained by the $Na^+K^+$-ATPase in the basal and lateral plasma membrane domains, which keeps the internal concentration of $Na^+$ low. The transport of glucose must be accompanied by $Na^+$ (symport) and is the basis of rehydration therapy (see Section 4.14).

---

*Symporters* transport substances in the same direction; antiporters transport substances in opposite directions.

---

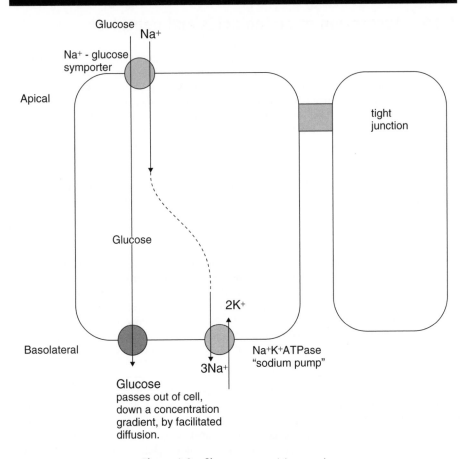

**Figure 4.9** Glucose symport transport.

Adjacent cells are connected by impermeable junctions (tight junctions). The junctions have a dual function: preventing solutes from crossing the epithelium between cells and allowing a concentration gradient of glucose to be maintained across the cell sheet; and acting as diffusion barriers within the plasma membrane, which help confine the various carrier proteins to their respective membrane domains.

SGLT-1 and SGLT-2 are $Na^+$-glucose symporters; they concentrate glucose (and related hexoses) inside the cell using the energy provided by co-transport of $Na^+$ ions down their electrochemical gradient.

Fructose is not co-transported with sodium; it enters the cell on a GLUT5 transporter. Glucose, galactose and fructose are transported out of the enterocyte through the GLUT2 transporter in the basolateral membrane.

GLUT isoforms play a specific role in glucose metabolism determined by their pattern of tissue expression, substrate specificity, transport kinetics and regulated expression in different physiological conditions.

# 4.16   Absorption of amino acids and peptides

Dietary proteins, with very few exceptions, are not absorbed; rather they must be digested into amino acids, or di- and tripeptides. Protein digestion begins in the stomach, where proenzyme pepsinogen is autocatalytically converted to pepsin A. Most proteolysis takes place in the duodenum via enzymes secreted by the pancreas, including trypsinogen, chymotrypsinogen and pro-carboxypeptidase A. These serine and zinc proteases are produced in the form of their respective proenzymes; they are both endopeptidase and exopeptidase, and their combined action leads to the production of amino acids, dipeptides and tripeptides.

> Endopeptidases are proteolytic peptidases that break peptide bonds of nonterminal amino acids (i.e. within the molecule), in contrast to exopeptidases, which break peptide bonds from their terminal end-pieces.

Enterokinase, a brush-border enzyme, activates trypsinogen into trypsin, which in turn converts a number of precursor pancreatic proteases into their active forms.

The mechanism by which amino acids are absorbed is conceptually identical to that of monosaccharides. The lumen plasma membrane of the absorptive cell bears a number of different $Na^+$ amino acid symporters. Again it is the $Na^+$ gradient across the cell that drives this process. $Na^+$-independent transporters on the basolateral membrane export amino acids to the extracellular space. The resulting osmotic gradient contributes to water absorption.

> Amino acid transporters belong to the SLC (solute carrier) group of membrane protein transporters, and include over 300 members. The SLC group includes examples of transport proteins that are facilitative transporters, which allow solutes to flow downhill with their electrochemical gradients, for example those located in the basolateral membrane, and secondary active transporters, which allow solutes to flow uphill against their gradient by coupling to transport of a second solute that flows downhill with its gradient such that the overall free energy change is still favourable, for example those located in the apical membrane.

There is virtually no absorption of peptides longer than four amino acids, but there is absorption of di- and tripeptides in the small intestine. Such peptides are co-transported with $H^+$ ions via an SLC protein (SLC15) comprising four members, namely SLC15A1 (PEPT1), SLC15A2 (PEPT2), SLC15A3 (PHT2, PTR3) and SLC15A4 (PHT1, PTR4). Such transporters are of proven pharmaceutical utility for enhancing oral absorption.

Neonates have the ability to absorb intact proteins. This ability, which is rapidly lost, is of immense importance because it allows the newborn animal to acquire passive immunity by absorbing immunoglobulins in colostral milk.

> Hartnup disorder is an autosomal recessive impairment of neutral amino acid transport affecting the kidney tubules and small intestine. It is believed that the defect is in a specific system responsible for neutral amino acid transport across the brush-border membrane of renal and intestinal epithelium, but the defect has not yet been characterised.

Kwashiorkor is a type of malnutrition associated with insufficient protein intake, usually affecting children aged 1–4 years, although it can also occur in older children and adults. It is likely caused by a combination of factors (protein deficiency, energy and micronutrient deficiency). The absence of lysine in low-grade cereal proteins (used as a dietary mainstay in many underdeveloped countries) can lead to kwashiorkor.

## 4.17  Absorption of lipids

The bulk of dietary lipid is neutral fat or triacylglycerol, but it also includes phospholipids, sterols (such as cholesterol) and many minor lipids, including fat-soluble vitamins. Bile acids and pancreatic lipase, mixed with chyme, act in the lumen to emulsify and digest triacylglycerols into their monoacylglycerols and free fatty acids. Bile, stored in the gall bladder and released into the duodenum, contains the bile salts, sodium glycocholate and sodium taurocholate. Such amphipathic molecules have a 'detergent effect' on lipids, intercalating and breaking them down to smaller aggregates, and eventually to micelles (4–8 nm diameter), thereby enabling the action of pancreatic lipase.

The drug orlistat (Xenical), which is promoted for the treatment of obesity, acts by inhibiting pancreatic lipase, thereby reducing the digestion and absorption of fat in the small intestine.

Fatty acids and monoacylglycerols enter the enterocyte by diffusion and are transported into the endoplasmic reticulum, where they are used to re-synthesise triacylglycerol. Beginning in the endoplasmic reticulum and continuing in the Golgi, triacylglycerol is packaged with cholesterol, lipoproteins and other lipids into particles called chylomicrons. Chylomicrons are extruded from the Golgi into exocytotic vesicles, which are transported to the basolateral aspect of the enterocyte. The vesicles fuse with the plasma membrane and undergo exocytosis, placing the chylomicrons into the extracellular space. Chylomicrons enter the blood via the lymph system. Blood-borne chylomicrons are rapidly disassembled and their constituents are utilised throughout the body.

Cholesterol is absorbed by intestinal epithelial cells via a Niemann-Pick C1-Like 1 (NPC1L1) protein. This is the target of a number of anti-hyperlipidaemic drugs used to lower cholesterol levels. Examples include miglustat, allopregnanolone, oxysterols and cyclodextrins; all are able to slow the progress of the disease, but none as yet provides an effective long-term treatment.

## 4.18  Absorption of minerals and metals

Most mineral absorption occurs in the small intestine. Calcium and iron absorption are the most studied; deficiencies in these are significant health problems throughout the world. In many cases intestinal absorption is a key regulatory step in mineral homeostasis.

Calcium is absorbed by two distinct mechanisms. *Active* transcellular absorption occurs only in the duodenum when calcium intake has been low. Key factors are calcium-binding proteins (calbindin), whose synthesis is dependent upon vitamin D (1, 25-dihydroxycholecalciferol). *Passive* paracellular absorption occurs in the jejunum and ileum, and to a much lesser extent the colon, when dietary calcium levels have been moderate or high. Calcium mucosal transport has been shown to have both a saturating and a non-saturating component. In the ileum there appears to be a significant secretory component. Calcium exits the cell through the basolateral membrane, by exchange with $Na^+$ or a $Ca^{2+}Mg^{2+}$ATPase. The ATPase is stimulated by calmodulin and inhibited by vanadate. The ATPase transporter has a high affinity for $Ca^{2+}$ ($K_m$ of $0.02-0.25\,\mu M$), in line with the low intracellular concentration of calcium ($<1\,\mu M$). This active transport system operates to pump $Ca^{2+}$ against a concentration gradient (the extracellular concentration of calcium is around 1 mM).

Iron homeostasis is regulated at the level of intestinal absorption by villus enterocytes in the proximal duodenum (Figure 4.10). Efficient absorption requires a slightly acidic environment; antacids and other conditions that interfere with gastric acid secretion can interfere with iron absorption. Ferric iron ($Fe^{3+}$) in the duodenal lumen is reduced to its ferrous form through the action of a brush-border ferrireductase; $Fe^{2+}$ is co-transported with a proton into the enterocyte via the divalent metal transporter DMT-1. This transporter is not specific for iron and also transports many other divalent metal ions. Inside the enterocyte, iron follows one of two routes. In an 'iron-abundant state', iron within the enterocyte is trapped by incorporation into ferritin; when the enterocyte is shed this iron is lost. In an 'iron-limiting state', iron is exported from the cell via a transporter (ferroportin) located in the basolateral membrane. It then binds to the iron-carrier protein transferrin for transport throughout the body. Iron in the form of haem, from ingestion of haemoglobin or myoglobin, is also readily absorbed by a separate 'folate' carrier protein.

Phosphorus is predominantly absorbed as inorganic phosphate in the upper small intestine. Phosphate is transported into the epithelial cells by co-transport with sodium; expression of transport is enhanced by vitamin D.

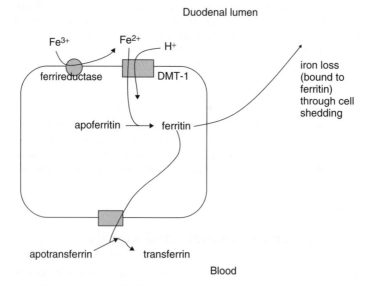

Duodenal lumen

**Figure 4.10** Iron absorption.

Copper absorption appears to occur through both a rapid, low-capacity system, and a slower, high-capacity system, which may be similar to the two processes seen with calcium absorption. Inactivating mutations in the gene encoding an intracellular copper ATPase have been shown to be responsible for the failure of intestinal copper absorption in Menkes disease.

---

Ceruloplasmin is an enzyme synthesised in the liver which contains six atoms of copper in its structure. Ceruloplasmin carries 90% of plasma copper; the other 10% is carried by albumin. Ceruloplasmin exhibits a copper-dependent oxidase activity, which is associated with possible oxidation of $Fe^{2+}$ (ferrous iron) into $Fe^{3+}$ (ferric iron), therefore assisting in its transport in the plasma in association with transferrin, which can only carry iron in the ferric state.

---

Zinc homeostasis is largely regulated by its uptake and loss through the small intestine. Although a number of zinc transporters and binding proteins have been identified in villus epithelial cells, a full understanding of zinc absorption is not yet at hand.

## 4.19  Malabsorption syndromes

Malabsorption is defined as an inadequate assimilation of dietary substances due to defects in digestion, absorption or transport. Malabsorption can affect macronutrients (proteins, carbohydrates, fats), micronutrients (vitamins, minerals) or both, causing excessive faecal excretion and producing nutritional deficiencies and GI symptoms. Digestion and absorption occur in three phases, namely (i) the intra-lumen hydrolysis of fats, proteins and carbohydrates by enzymes, and emulsification by bile salts, (ii) digestion by brush-border enzymes and uptake of end-products and (iii) lymphatic transport of nutrients. Malabsorption can occur when any of these phases is impaired.

Malabsorption has many causes. Some malabsorptive disorders, for example coeliac sprue, impair the absorption of most nutrients, vitamins and trace minerals (global malabsorption); others, for example pernicious anaemia, are more selective. Pancreatic insufficiency causes malabsorption if >90% of function is lost. Increased lumen acidity (e.g. Zollinger–Ellison syndrome) inhibits lipase and fat digestion. Cirrhosis and cholestasis reduce hepatic bile synthesis or delivery of bile salts to the duodenum, causing malabsorption. Some causes are summarised in Table 4.2.

The effects of unabsorbed substances in the gut lead to diarrhoea, steatorrhoea, abdominal bloating and gas. Other symptoms result from nutritional deficiencies. Patients often lose weight despite adequate food intake. Chronic diarrhoea is the most common symptom and is what usually prompts evaluation of the patient. Steatorrhoea (fatty stool), the hallmark of malabsorption, occurs when >7 g/day of fat is excreted, associated with foul-smelling, pale, bulky and greasy stools.

Malabsorption is suspected in a patient with chronic diarrhoea, weight loss and anaemia. Often the aetiology is obvious. For example, malabsorption due to chronic pancreatitis is usually linked to bouts of acute pancreatitis; diarrhoea in coeliac sprue is exacerbated by gluten products, abdominal distension, excessive flatus and watery diarrhoea occurring 30–90 minutes after carbohydrate ingestion suggest deficiency of a disaccharidase enzyme, usually lactase. Macrocytic anaemia should prompt measurement of serum folate and vitamin $B_{12}$ levels. Folate deficiency is common in mucosal disorders involving the proximal small bowel (e.g. coeliac

**Table 4.2**  Causes of malabsorption

| Cause | Example |
|---|---|
| Inadequate gastric mixing, rapid emptying or both | Billroth II gastrectomy<br>Gastrocolic fistula<br>Gastroenterostomy |
| Insufficient digestive agents | Biliary obstruction<br>Chronic liver failure<br>Chronic pancreatitis<br>Cystic fibrosis<br>Lactase deficiency<br>Pancreatic cancer<br>Pancreatic resection<br>Sucrase-isomaltase deficiency |
| Improper milieu | Abnormal motility secondary to diabetes, scleroderma, hyperthyroidism<br>Bacterial overgrowth – blind loops (deconjugation of bile salts), diverticula<br>Zollinger–Ellison syndrome (low duodenal pH) |
| Acutely abnormal epithelium | Acute intestinal infections<br>Alcohol<br>Neomycin |
| Chronically abnormal epithelium | Amyloidosis<br>Coeliac disease<br>Crohn's disease<br>Ischaemia<br>Radiation enteritis<br>Tropical sprue<br>Whipple's disease |
| Short bowel | Intestinal resection (e.g. for Crohn's disease, volvulus, intussusception or infarction)<br>Jejunoileal bypass for obesity |
| Impaired transport | Abetalipoproteinaemia<br>Addison's disease<br>Blocked lacteals – lymphoma, tuberculosis<br>Lymphangiectasia |

sprue, tropical sprue, Whipple's disease). Low vitamin $B_{12}$ levels can occur in pernicious anaemia, chronic pancreatitis, bacterial overgrowth and terminal ileal disease. A combination of low vitamin $B_{12}$ and high folate levels is suggestive of bacterial overgrowth, because intestinal bacteria use vitamin $B_{12}$ and synthesise folate. Microcytic anaemia suggests iron deficiency, which may occur with coeliac sprue. Albumin is a general indicator of nutritional state. Low albumin can result from poor intake, decreased synthesis in cirrhosis or protein wasting. Low serum carotene (a precursor of vitamin A) suggests malabsorption if intake is adequate.

# Focus on: cystic fibrosis (CF)

CF is caused by a defect in the CFTR protein. This protein is a cAMP- and phosphorylation-regulated chloride ion channel, important in maintaining osmotic gradients and the movement of water across epithelial membranes.

The CFTR gene is located on the long arm of chromosome 7; it belongs to a family of genes called ABC (ATP-binding cassette transporters). Genes in the ABC family provide instructions for making transporter proteins that carry many types of molecule, such as fats, sugars and amino acids, as well as drugs, across cell membranes. More than 1000 mutations in the human CFTR gene have been identified. Most mutations change single amino acids in the CFTR protein or delete a small amount of DNA from the CFTR gene. The most common mutation, called delta F508, is a deletion of one amino acid at position 508 in the CFTR protein. The resulting abnormal channel breaks down shortly after it is made and is not incorporated into the cell membrane. The disorder is autosomal recessive.

CF affects mainly the exocrine (mucus) glands of the lungs, liver, pancreas and intestines, causing progressive disability due to multisystem failure. Thick mucus production, as well as a less competent immune system, results in frequent lung infections. Diminished secretion of pancreatic enzymes is the main cause of poor growth, fatty diarrhoea (steatorrhoea) and deficiency in fat-soluble vitamins. Meconium ileus (see below) is a typical finding in newborn babies with CF. CF may be diagnosed by newborn screening (immunoreactive trypsinogen (IRT), an enzyme produced by the pancreas), sweat testing (salty sweat) or genetic testing. CF is one of the most common life-shortening, childhood-onset inherited diseases. In the United States, 1 in 3900 children are born with CF. It is most common among Europeans and Ashkenazi Jews; 1 in 22 people of European descent are carriers of one gene for CF, making it the most common genetic disease in these populations. Ireland has the highest rate of CF carriers in the world (1 in 19).

The thick mucus seen in the lungs has its counterpart in thickened secretions from the pancreas that can block the movement of the digestive enzymes into the duodenum and result in irreversible damage to the pancreas, often with painful inflammation (pancreatitis). The lack of digestive enzymes results in poor absorption of nutrients, leading to malnutrition and poor growth. Individuals with CF also have difficulties absorbing the fat-soluble vitamins A, D, E and K. Thickened secretions may cause liver problems in patients with CF; viscous bile secreted by the liver to aid in digestion may block the bile ducts, leading to liver damage and cirrhosis.

---

Meconium is the earliest stool of an infant, composed of materials ingested during the time the infant spends in the uterus. It should be completely passed by the end of the first few days of postpartum life, with the stools progressing toward yellow (digested milk). Sometimes the meconium becomes thickened and congested in the ileum, a condition known as meconium ileus. Meconium ileus is often the first symptom of CF. In CF patients, the meconium can form a bituminous black-green mechanical obstruction in a segment of the ileum. About 20% of cases of CF present with meconium ileus.

A defective CFTR channel will disturb electrochemical gradients in both intestinal crypt cells and pancreatic epithelial cells, with a common end result, namely the inability to secrete water into the lumen. In crypt cells, chloride is normally transported to the lumen, drawing sodium ions across the tight junctions and creating an osmotic potential that in turn draws water into the lumen (Figure 4.11). A defective CFTR will prevent the gradient forming and hence halt water movement to the lumen.

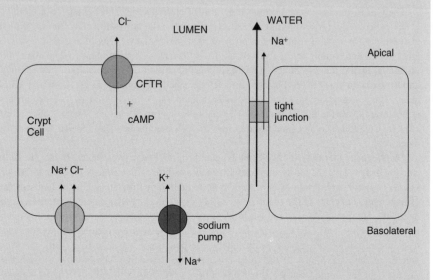

**Figure 4.11**   Electrolyte movement in secretory crypt cells of the small intestine.

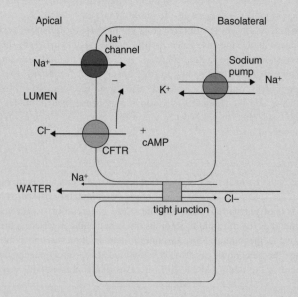

**Figure 4.12**   Electrolyte movement in pancreas or lung epithelial cells.

In the pancreas and lung, a similar situation exists. In 'secretory' mode, the CFTR passes chloride to the lumen (while a sodium channel is inactivated); this draws sodium ions across the tight junctions, in turn pulling water into the lumen (Figure 4.12). Again, a defective CFTR will halt water movement into the lumen, leading to viscous secretions.

In the epithelial cells lining the sweat ducts, the situation is reversed (Figure 4.13). Salt is normally removed from the ducts; both Na$^+$ and Cl$^-$ enter the cells down a concentration gradient (Cl$^-$ through the CFTR channel). Water cannot follow this osmotic gradient because of the very 'tight' cell junctions between cells of the sweat duct. With a defective CFTR channel, neither Na$^+$ nor Cl$^-$ is able to enter the cell (Na$^+$ will not move against an electrochemical gradient), and so the duct solution remains 'salty'.

It is postulated that there is an evolutionary advantage to carrying the CF gene since it may afford some protection against cholera. As water secretion is increased when CTX 'switches on' the adenyl cyclase, in turn activating the CFTR channel, it is reasoned that a defective CFTR channel (in CF) would protect individuals against death by dehydration.

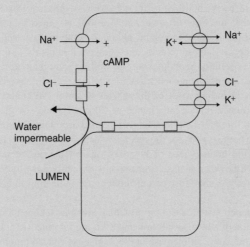

**Figure 4.13** Electrolyte movement in sweat cells.

Cholera toxin (CTX) is an oligomeric complex made up of six protein subunits: a single copy of the A subunit and five copies of the B subunit. The A subunit contains a globular enzyme that ADP-ribosylates (activates) G-proteins. This structure is similar in shape, mechanism and sequence to the heat-labile enterotoxin secreted by some strains of the *E. coli* bacterium. The five B subunits form a five-membered ring structure that binds to ganglioside receptors on the surface of the host cell. After binding, the entire CTX complex is internalised by the cell. An A1 chain fragment is released from the complex, by the reduction of a disulphide bridge, which binds with a human partner protein called ADP-ribosylation factor 6, driving a conformational change and exposing an active site on the A1 fragment. The A1 fragment 'complex' controls the GDP to GTP conversion of G-protein; the G-protein is effectively 'locked on' in the G-GTP state, continuously activating adenyl cyclase to produce cAMP.

## 4.20 Steatorrhoea

Steatorrhoea is the formation of non-solid faeces. Floating stools, due to excess fat, are oily in appearance and foul smelling. There is increased fat excretion, which can be measured by determining the faecal fat level. Possible biological causes can be lack of bile acids (due to liver damage or hypolipidaemic drugs), defects or a reduction in pancreatic enzymes (lipase), and defective mucosal cells. The absence of bile acids will cause the faeces to turn grey or pale.

## 4.21 Lactose intolerance

Lactose intolerance is the inability to digest lactose (a dissacharide of glucose and galactose), due to a relative deficiency of the epithelial brush border enzyme lactase. Mammals have evolved a developmental pattern of small-intestinal gene expression that promotes high-level production of lactase early in life (to digest lactose in milk), followed by a turnoff of lactase expression around the time of weaning. 'Turnoff' is generally complete by 5–10 years and individuals become lactose intolerant, although some remain lactose tolerant into adulthood. Lactase persistence (lactose tolerance) is seen predominantly in individuals with northern European ancestry, especially Scandinavian, as well as some nomadic peoples of the Middle East and Africa. Lactase non-persistence (lactose intolerance) is observed in a majority of the world's populations, including most of those with Asian or African ancestry. Lactose persistence and non-persistence reflect inheritance of different alleles of the lactase gene. Lactose persistence, and therefore lactose tolerance, is inherited as a dominant trait. Lactose intolerance is the result of being homozygous for the recessive lactase allele, which is poorly expressed after early childhood. Being homozygous or heterozygous for the persistence allele allows lactase expression after the time when lactase expression is normally down-regulated. In some circumstances, heterozygotes can manifest partial intolerance, indicating that this is an incompletely dominant gene.

The lactose tolerance test is a simple test that involves taking blood samples at intervals following consumption of a lactose solution and assaying for glucose; individuals that are unable to digest lactose do not show an increase in blood glucose concentrations. Symptoms of lactose intolerance can include nausea, cramping, bloating, diarrhoea and flatulence following a lactose challenge. These signs reflect the osmotic effects of unassimilated lactose in the intestinal lumen, plus the fermentation products generated in the large intestine. The condition is easily managed by avoiding lactose.

## 4.22 Glucose–galactose malabsorption

The SLC5A1 gene (on chromosome 22) codes for the sodium–glucose co-transporter protein, SGLT1. More than 40 mutations have been characterised, some resulting in a protein that is too short or not folded properly. The defective transporter prevents uptake of glucose and galactose; the increased osmotic potential of the lumen causes dehydration and severe osmotic diarrhoea. Glucose–galactose malabsorption is a rare disorder, only a few hundred cases have been identified worldwide. This condition is inherited in an autosomal recessive pattern.

## 4.23 Coeliac disease

Coeliac disease is an autoimmune disorder of the small bowel that occurs in genetically pre-disposed people of all ages from middle infancy. Symptoms may include chronic diarrhoea, failure to thrive (in children) and fatigue. It is estimated to affect about 1% of Indo-European populations, but is thought to be significantly under-diagnosed. Coeliac disease is caused by a reaction to gliadin, a gluten protein found in wheat (and similar proteins of the tribe Triticeae, which include other cultivars such as barley and rye). Upon exposure to gliadin, the enzyme tissue transglutaminase modifies the protein, causing the immune system to cross-react with the bowel tissue and inducing an inflammatory reaction. Over time this leads to flattening of the lining of the small intestine and loss of villi, leading to general symptoms of malnutrition. The only effective treatment is a lifelong gluten-free diet. Symptoms are very varied but often include diarrhoea (osmotic diarrhoea).

---

Tissue transglutaminase crosslinks proteins between an $\varepsilon$-amino group of a lysine residue and the $\gamma$-carboxamide group of a glutamine residue, creating an inter- or intramolecular bond that is highly resistant to proteolysis. This generates the autoantigen in coeliac disease, but it is also known to play a role in apoptosis and cellular differentiation.

Gliadins are prolamins, a group of plant storage proteins with a high proline content, found in the seeds of cereal grains: wheat (gliadin), barley (hordein), rye (secalin), corn (zein) and, as a minor protein, avenin in oats.

---

## 4.24 Crohn's disease

Crohn's disease is generally classified as an autoimmune disease. Diarrhoea (osmotic) is a common symptom. Individuals with Crohn's disease are at risk of malnutrition for many reasons, including decreased food intake and malabsorption. The exact cause of Crohn's disease is unknown. However, genetic and environmental factors have been invoked in the pathogenesis of the disease. Mutations in the CARD15 gene are associated with Crohn's disease, but over eight other genes have been shown to play a role.

## 4.25 The large intestine

The surface of the mucosa is relatively smooth as there are no intestinal villi. Crypts of Lieberkühn are present. Goblet cells account for more of the epithelial cells than in the small intestine. The mammalian large intestine is important for the maintenance of water and electrolyte balance. Its primary function is the reabsorption of water, sodium, chloride and volatile fatty acids; it secretes potassium and bicarbonate.

Sodium ion is the main cation entering the colon by active transport. This active process is the primary driving force for the movement of fluid and other electrolytes through the paracellular pathway. The negative mucosal potential difference should favour chloride absorption and $K^+$ secretion. Although there is general agreement that active $Na^+$ transport is present, considerable controversy exists about the overall mechanism of $K^+$, chloride and bicarbonate transport in the large intestine.

# Focus on: controlling gastric acid production

Gastric acid production is regulated by both the autonomic nervous system and several hormones. The parasympathetic nervous system, via the vagus nerve and the hormone gastrin, stimulates the parietal cell to produce gastric acid, acting both directly on parietal cells and indirectly through the stimulation of the secretion of the hormone histamine from ECL cells. Vasoactive intestinal peptides, cholecystokinin and secretin all inhibit acid production.

The production of gastric acid in the stomach is tightly regulated by positive regulators and negative-feedback mechanisms. Four types of cell are involved in this process: parietal cells, G cells, D cells and ECL cells. Besides this, the endings of the vagus nerve and the intramural nervous plexus in the digestive tract influence the secretion significantly.

Nerve endings in the stomach secrete two stimulatory neurotransmitters: acetylcholine and gastrin-releasing peptide. Their action is both direct on parietal cells and mediated through the secretion of gastrin from G cells and histamine from ECL cells. Gastrin acts on parietal cells directly and indirectly too, by stimulating the release of histamine.

The release of histamine is the most important positive regulation mechanism of the secretion of gastric acid in the stomach; its release is stimulated by gastrin and acetylcholine and inhibited by somatostatin.

There are three main classes of drug used against acid-related gastrointestinal conditions:

- $H_2$ antagonists

- PPIs

- prostaglandin analogues.

$H_2$ antagonists, such as cimetidine and ranitidine, block the action of histamine at the histamine receptor on parietal cells. They have a long duration of effect, 6–10 hours, and can be used as a prophylactic before meals to reduce the chance of heartburn. Their use has waned however since the advent of the more effective PPIs.

PPIs, such as omeprazole, lansoprazole and rabeprazole, inhibit the $H^+K^+$-ATPase (proton pump) in parietal cells. Their action is pronounced and long-lasting, reducing gastric acid secretion by up to 99%, with only minimal side effects. The PPIs are given in an inactive form. The inactive form is neutrally charged and readily crosses cell membranes into intracellular compartments (like the parietal cell canaliculi) that have acidic environments. In an acid environment, the inactive drug is protonated and rearranges into its active form.

Misoprostol is a synthetic prostaglandin $E_1$ analogue. It inhibits gastric acid secretion and stimulates the production of mucus and bicarbonate. Misoprostol is as effective as $H_2$ antagonists in healing duodenal ulcers but is less effective in the treatment of gastric ulcers and gastro-oesophageal reflux disease. It is particularly effective in preventing NSAID-associated ulcers. Its use is mainly restricted because of its potential side effects; it is also used to induce abortion or labour.

Prostaglandin $E_2$ is known to be synthesised in gastric mucosa. Prostaglandin $E_2$ regulates various gastric cytoprotective actions, including acid secretion, mucin synthesis, mucin secretion, $HCO_3^-$ secretion, blood flow and gastric motility. Of these actions, mucin synthesis and secretion are of particular importance in view of mucosal surface protection. NSAIDs (e.g. aspirin, ibuprofen), which inhibit $PGE_2$ synthesis, can induce gastric lesions, and supplementation with $PGE_2$ prevents NSAIDs-induced lesions.

Prostaglandins are synthesised by two isoenzymes of cyclo-oxygenase (COX), COX-1 and COX-2 (Figure 4.14). COX-1 is always present in intact gastric mucosa (it is constitutive), but COX-2 is not. COX-2 synthesises prostaglandins in the gastric mucosa in response to injury, ulceration and inflammatory lesions (it is inducible). COX-2 plays an important role in maintaining gastric mucosal integrity.

It is proposed that the damaging effects of NSAIDs on the gastric mucosa (through inhibition of COX-1) could be limited by using compounds more specific for COX-2. Selectivity for COX-2 is the main feature of celecoxib and rofecoxib. However, the risk of peptic ulceration with these drugs still remains, as well as a number of other side effects.

Rofecoxib was removed from the market in 2004 because of concerns. Different distributions and types of $PGE_2$ receptors may partly explain the undesirable side effects of COX-2 inhibitors.

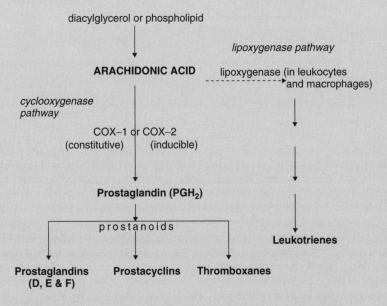

**Figure 4.14** Synthesis of prostaglandins, prostacyclins and thromboxanes.

COX-1 and COX-2 are of similar molecular mass (approximately 70 and 72 kDa respectively), with 65% amino acid sequence homology and near-identical catalytic sites. The most significant difference between the isoenzymes, which allows for selective inhibition, is the substitution of isoleucine at position 523 in COX-1 with valine in COX-2. The relatively smaller $Val_{523}$ residue in COX-2 allows access to a hydrophobic side-pocket in the enzyme (which $Ile_{523}$ sterically hinders). Drugs, such as the coxibs, bind to this alternative site and are considered to be selective inhibitors of COX-2.

See Figure 4.15 for a proposed general scheme for anti-gastric drugs and NSAIDs.

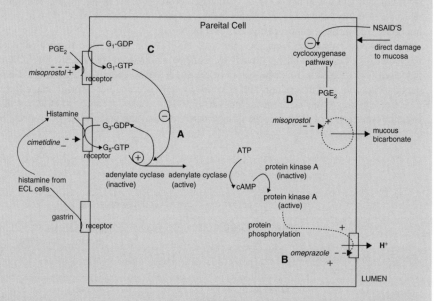

**Figure 4.15** Proposed general scheme of action for anti-gastric drugs and NSAIDs. A) Histamine binding to its receptor triggers the GTP (active) form of a $G_s$ (stimulatory) protein; this activates adenyl cyclase, raising cAMP levels and initiating an intracellular signalling cascade that results in the activation of the proton pump. Binding of histamine, and therefore initiation of this response, is prevented by the $H_2$ antagonist, cimetidine. B) Omeprazole directly inhibits the $H^+K^+$-ATPase proton pump. C) Misoprostol, a PGE agonist, is assumed to bind to a $PGE_2$ receptor, triggering the activation of a $G_i$ (inhibitory) protein, which in turn acts to prevent the activation of adenyl cyclase, thereby modulating the effects of histamine. This would explain the observation that misoprostol directly inhibits gastric acid production. Misoprostol has also been shown to stimulate mucous and bicarbonate production, presumably by working alongside $PGE_2$ in D), as well as protecting against NSAID damage by replacing the $PGE_2$ lost as a result of COX inhibition.

# CHAPTER 5

# Synthesis, mobilisation and transport of lipids and lipoproteins

## 5.1 Fatty acid synthesis

Fatty acids are predominantly formed in the liver and adipose tissue, as well as the mammary glands during lactation. Fatty acid synthesis occurs in the cytosol (fatty acid oxidation occurs in the mitochondria; compartmentalisation of the two pathways allows for distinct regulation of each). Oxidation or synthesis of fats utilises an activated two-carbon intermediate, acetyl-CoA, but the acetyl-CoA in fat synthesis exists temporarily bound to the enzyme complex as malonyl-CoA. Acetyl-CoA is mostly produced from pyruvate (pyruvate dehydrogenase) in the mitochondria; it is condensed with oxaloacetate to form citrate, which is then transported into the cytosol and broken down to yield acetyl-CoA and oxaloacetate (ATP citrate lyase).

Synthesis of malonyl-CoA is the first committed step of fatty acid synthesis (Figure 5.1); the enzyme that catalyses this reaction, acetyl-CoA carboxylase (ACC), is the major site of regulation of fatty acid synthesis (this reaction also requires a biotin prosthetic group).

ACC exists in equilibrium between monomeric and polymeric forms; the active form is polymeric. This conformational change is enhanced by citrate and inhibited by long-chain fatty acids (Figure 5.2).

*Essential Biochemistry for Medicine*   Dr Mitchell Fry
© 2010 John Wiley & Sons, Ltd

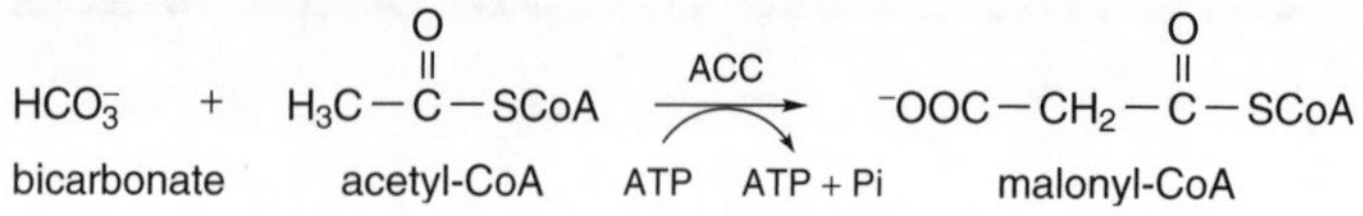

$$HCO_3^- \; + \; H_3C-\overset{\overset{\textstyle O}{\|}}{C}-SCoA \quad \xrightarrow[\text{ATP} \;\; \text{ATP}+\text{Pi}]{\text{ACC}} \quad {}^-OOC-CH_2-\overset{\overset{\textstyle O}{\|}}{C}-SCoA$$

bicarbonate       acetyl-CoA                        malonyl-CoA

**Figure 5.1**   Acetyl CoA carboxylase reaction.

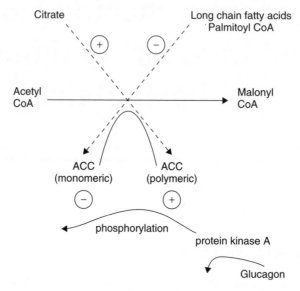

**Figure 5.2**   Control of acetyl-CoA carboxylase (ACC) by both the levels of citrate (+) and of fatty acids (−), as well as by phosphorylation (−).

---

There are two major isoenzymes of ACC in mammalian tissues, ACC1 and ACC2. ACC1 is strictly cytosolic and is enriched in liver, adipose tissue and lactating mammary tissue. ACC2 is expressed in heart and skeletal muscle, and liver. Both isoenzymes are allosterically activated by citrate and inhibited by palmitoyl-CoA and other short- and long-chain fatty acyl-CoAs. Citrate triggers the polymerisation of ACC1, which leads to significant increases in its activity.

---

ACC is also controlled through hormone-mediated phosphorylation. Phosphorylation of ACC1 at three serine residues (S79, S1200 and S1215) by AMP-activated protein kinase (AMPK) leads to inhibition of the enzyme. Glucagon-stimulated increases in cyclic adenosine monophosphate (cAMP), with increase in protein kinase A (PKA) activity, also lead to phosphorylation of ACC.

---

Protein kinase A (PKA) refers to a family of enzymes whose activity is dependent on the level of cAMP in the cell. PKA is also known as cAMP-dependent protein kinase.

---

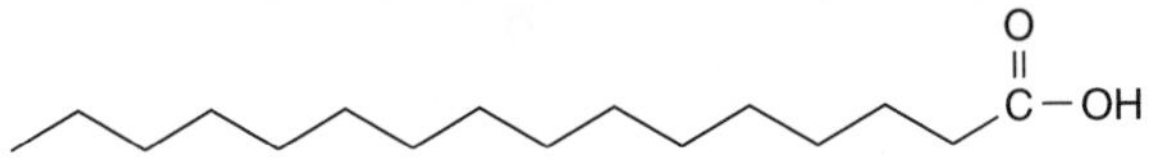

**Figure 5.3**   Palmitic acid.

The activating effects of insulin on ACC are complex and not completely resolved but are related at least in part to changes in cAMP levels.

Continued condensation of malonyl-CoA with acetyl-CoA units is catalysed by fatty acid synthase, eventually leading to the 16-carbon palmitic acid (Figure 5.3); this is then released and may undergo separate elongation and/or unsaturation reactions, to yield other fatty acid molecules. The active form of fatty acid synthase is a dimer of identical subunits.

## 5.2   Long-term control of fatty acid synthase

Long-term control of both ACC and fatty acid synthase is through their synthesis and turnover. Insulin is known to stimulate synthesis of both enzymes, whereas starvation leads to decreased synthesis.

Fatty acid synthase is transcriptionally regulated by upstream stimulatory factor and sterol regulatory element binding protein 1c (SREBP-1c), in response to feeding/insulin.

---

SREBPs are transcription factors that bind to the sterol regulatory element DNA sequence TCACNCCAC. Unactivated SREBPs are attached to the nuclear envelope and endoplasmic reticulum membranes. In cells with low levels of sterols, SREBPs are cleaved to a water-soluble N-terminal domain that is translocated to the nucleus. These activated SREBPs then bind to specific sterol regulatory element DNA sequences, thus up-regulating the synthesis of enzymes involved in sterol biosynthesis. Sterols in turn inhibit the cleavage of SREBPs and therefore synthesis of additional sterols is reduced through a negative-feedback loop.

Carbohydrate-responsive element-binding protein (ChREBP) is a central regulator of lipid synthesis in the liver. It was first identified as a glucose-responsive transcription factor (it is required for the glucose-induced expression of the hepatic isoenzyme of glycolysis, pyruvate kinase). ChREBP acts together with SREBPs to induce 'lipogenic genes' such as those for ACC and fatty acid synthase.

Expression of the ChREBP hepatic gene is induced in response to increasing glucose. ChREBP has also been shown to be a direct target of liver X receptors (LXRs); LXRs are members of the steroid/thyroid hormone cytosolic ligand-binding receptors that migrate to the nucleus upon ligand binding and regulate gene expression by binding to specific target sequences.

---

## 5.3   Triacylglycerols

Triacylglycerols (Figure 5.4) consist of a glycerol backbone to which three fatty acids are esterified ($R_1$, $R_2$, $R_3$). Fatty acids are stored as triacylglycerol in all cells, but primarily in adipocytes of adipose tissue. The fatty acids in triacylglycerols are predominantly saturated. The major building block for the synthesis of triacylglycerols, in tissues other than adipose,

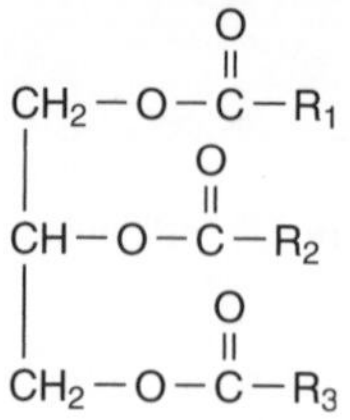

$$
\begin{array}{l}
\phantom{CH_2-O-}\overset{\displaystyle O}{\overset{\displaystyle \|}{\phantom{C}}} \\
CH_2-O-C-R_1 \\
\phantom{CH_2-O-}\overset{\displaystyle O}{\overset{\displaystyle \|}{\phantom{C}}} \\
CH-O-C-R_2 \\
\phantom{CH_2-O-}\overset{\displaystyle O}{\overset{\displaystyle \|}{\phantom{C}}} \\
CH_2-O-C-R_3
\end{array}
$$

**Figure 5.4** Triacylglycerol general structure.

is glycerol, but adipocytes lack glycerol kinase, and instead use dihydroxyacetone phosphate (from glycolysis) as the precursor. This means that adipocytes must have glucose to oxidise in order to store fatty acids in the form of triacylglycerols.

## 5.4 Mobilisation of lipid stores

In response to energy demands, the fatty acids of stored triacylglycerols must be mobilised for use by peripheral tissues. This release is controlled by a complex series of interrelated cascades that result in the activation of hormone-sensitive lipase. In adipocytes this stimulus can come from glucagon, adrenaline (epinephrine) or $\beta$-corticotropin. These hormones bind cell-surface receptors that are coupled to the activation of adenyl cyclase; the resultant increase in intracellular cAMP leads to activation of cAMP-dependent protein kinase, which in turn phosphorylates and activates hormone-sensitive lipase (Figure 5.5).

Hormone-sensitive lipase hydrolyses fatty acids from carbon atoms 1 or 3 of triacylglycerols. The resulting diacylglycerols are substrates for either hormone-sensitive lipase or the non-inducible enzyme diacylglycerol lipase. Finally, monoacylglycerols are substrates for monoa-cylglycerol lipase.

Free fatty acids diffuse from adipose cells, combine with albumin in the blood and are thereby transported to other tissues.

The mobilisation of adipose lipid stores is inhibited by numerous stimuli, the most significant being insulin (through the inhibition of adenyl cyclase activity). In a well-fed individual, insulin release prevents the inappropriate mobilisation of stored lipid; instead any excess fat and carbohydrate are incorporated into the triacylglycerol pool within adipose tissue.

## 5.5 Transport of lipids

The transport of lipids and cholesterol is accomplished by packaging them into lipoprotein complexes; this is undertaken by both hepatocytes and enterocytes. Dietary triacylglycerols and cholesterol are packaged by enterocytes into chylomicrons, whereas *de novo* synthesised triacylglycerols in the liver are packaged in very low-density lipoproteins (VLDLs).

## 5.6 Intestinal uptake of lipids

Solubilisation (or emulsification) of dietary lipids is accomplished by means of bile salts, which are synthesised from cholesterol in the liver and then stored in the gallbladder; they are emptied into the gut following the ingestion of fat. Emulsification of dietary fats renders them accessible

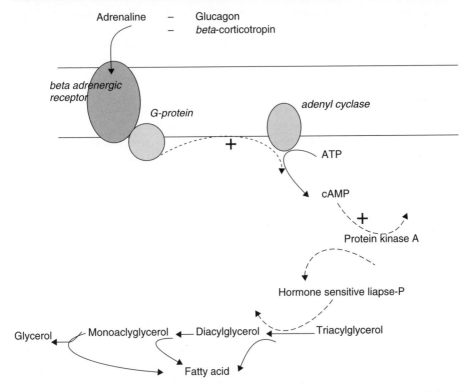

**Figure 5.5**  Hormone-induced fatty acid mobilisation in adipocytes. Through hormone-induced rises in intracellular cAMP levels, phosphorylation (activation) of hormone-sensitive lipase initiates the mobilisation of fatty acids from triacylglycerol.

to pancreatic lipases (primarily lipase and phospholipase A$_2$); these enzymes generate free fatty acids, mono- and diacylglycerols. Phospholipids are degraded at the 2 position by pancreatic phospholipase A$_2$ releasing a free fatty acid and the lysophospholipid. These products may diffuse into the intestinal epithelial cells, where the re-synthesis of triacylglycerols occurs. Both dietary triacylglycerols and cholesterol are packaged into chylomicrons; these enter the blood (left subclavian vein) via the lymph system.

## 5.7  Chylomicrons

Chylomicrons consists of triacylglycerols, cholesterol and cholesteryl esters, surrounded by phospholipids and proteins, identified as apolipoproteins.

## 5.8  Apolipoproteins

Apolipoproteins help maintain the structural integrity and solubility of the lipoprotein complexes, and aid in lipoprotein receptor recognition and regulation of certain enzymes in lipoprotein metabolism.

Apolipoproteins that predominate prior to chylomicrons entering the circulation include apo-B-48, apo-A-I, -A-II and -IV. Apo-B-48 combines only with chylomicrons (see Table 5.1). In the bloodstream, chylomicrons further acquire apo-C-II and apo-E from plasma HDLs (high-density lipoproteins).

**Table 5.1**    Major types of apoprotein

| Apoprotein | Lipoprotein association | Function |
|---|---|---|
| Apo-A-I | Chylomicrons, HDL | Major protein of HDL, activates lecithin:cholesterol acyltransferase, LCAT |
| Apo-A-II | Chylomicrons, HDL | Primarily in HDL, enhances hepatic lipase activity |
| Apo-A-IV | Chylomicrons and HDL | Present in triacylglycerol-rich lipoproteins |
| Apo-B-48 | Chylomicrons | Exclusively found in chylomicrons, derived from apo-B-100 gene by RNA editing in intestinal epithelium; lacks the LDL receptor-binding domain of apo-B-100 |
| Apo-B-100 | VLDL, IDL and LDL | Major protein of LDL, binds to LDL receptor; one of the longest known proteins in humans |
| Apo-C-I | Chylomicrons, VLDL, IDL and HDL | May also activate LCAT |
| Apo-C-II | Chylomicrons, VLDL, IDL and HDL | Activates lipoprotein lipase |
| Apo-C-III | Chylomicrons, VLDL, IDL and HDL | Inhibits lipoprotein lipase |
| Apo-D | HDL | Closely associated with LCAT |
| Cholesterol ester transfer protein (CETP) | HDL | Exclusively associated with HDL, cholesteryl ester transfer |
| Apo-E (at least three alleles (E-2, E-3, E-4), each of which has multiple isoforms) | Chylomicron remnants, VLDL, IDL and HDL | Binds to LDL receptor; apo-E$\varepsilon$-4 allele amplification associated with late-onset Alzheimer's disease |
| Apo-H (also known as $\beta$-2-glycoprotein I) | Chylomicrons | Triacylglycerol metabolism |
| Apo(a) – at least 19 different alleles | LDL | Disulfide bonded to apo-B-100; forms a complex with LDL identified as lipoprotein(a), Lp(a); strongly resembles plasminogen; may deliver cholesterol to sites of vascular injury; high-risk association with premature coronary artery disease and stroke |

In the capillaries of adipose tissue and muscle, fatty acids of chylomicrons are removed from the triacylglycerols by the action of lipoprotein lipase (LPL); LPL is present on the surface of the endothelial cells of the capillaries. Chylomicron apo-C-II activates LPL in the presence of phospholipid.

---

LPL catalyses the hydrolysis of triacylglycerols to liberate free fatty acids and glycerol. LPL is specifically found in endothelial cells lining the capillaries. LPL has different isoenzyme forms in different tissues; that form found in adipocytes is activated by insulin (which helps to explain why adipose cells gain fat in a well-fed state).

Adipocytes are found mostly in the abdominal cavity and subcutaneous tissue. They are metabolically very active; their stored triacylglycerol is constantly hydrolysed and re-synthesised.

---

Free fatty acids may be absorbed directly by tissues, or bound to albumin for transport; human serum albumin possesses multiple fatty acid binding sites of various affinities. Glycerol is returned via the blood to the liver (and kidneys), where it is converted to the glycolytic intermediate dihydroxyacetone phosphate (glycerol is an important source of glucose in gluconeogenesis).

Long-chain fatty acids were previously believed to enter cells by diffusion, but evidence now implicates a family of protein-mediated transporters, the fatty acid transport protein (FATP) family. Internally, fatty acid-binding proteins (FABPs) constitute a family of low-molecular-weight proteins that function as intracellular transporters of fatty acids.

---

The FABPs are a family of carrier proteins for fatty acids and other lipophilic substances, such as eicosanoids and retinoids. These proteins are thought to facilitate the transfer of fatty acids between extra- and intracellular membranes. Adipocyte fatty acid-binding protein (aP2; FABP4) is expressed in adipocytes and macrophages, and integrates inflammatory and metabolic responses. Studies in aP2-deficient mice have shown that this lipid chaperone has a significant role in several aspects of the metabolic syndrome, including type 2 diabetes and atherosclerosis. FABP has also been introduced as a plasma marker of acute myocardial infarction.

---

During the removal of fatty acids, a substantial portion of the phospholipids and apolipoproteins (A and C) of chylomicrons is transferred to HDLs. In particular, the loss of apo-C-II prevents LPL from further degrading the chylomicron.

Chylomicron remnants, containing primarily cholesterol, apo-E and apo-B-48, are then taken up by the liver through interaction with the chylomicron remnant receptor (this recognition requires apo-E).

Chylomicrons therefore function to:

- deliver dietary triacylglycerols to adipose tissue and muscle

- deliver dietary cholesterol to the liver.

## 5.9    Export of fat from the liver

Triacylglycerols and cholesterol are exported from the liver as nascent VLDL complexes, destined primarily to muscle and adipose tissues. The VLDL complex contains apolipoprotein B-100 and acquires C-I, C-II, C-III and E from circulating HDL complexes. Fatty acids are released from VLDLs in the same way as chylomicrons, through the action of LPL. This action, coupled to a loss of certain apoproteins (the apo-Cs), converts VLDLs to intermediate-density lipoproteins (IDLs), also termed VLDL remnants. The apo-Cs are transferred to HDLs. The predominant remaining proteins are apo-B-100 and apo-E. Further loss of triacylglycerols converts IDLs to LDLs.

*IDL complexes* are formed as triacylglycerols and are further removed from VLDLs. The fate of IDLs is either conversion to LDLs or direct uptake by the liver. Conversion of IDLs to LDLs occurs as more triacylglycerols are removed. The liver takes up IDLs after they have interacted with the LDL receptor to form a complex, which is endocytosed by the cell. For LDL receptors in the liver to recognise IDLs requires the presence of both apo-B-100 and apo-E (the LDL receptor is also called the apo-B-100/apo-E receptor). The importance of apo-E in cholesterol uptake by LDL receptors has been demonstrated in transgenic mice lacking functional apo-E genes. These mice develop severe atherosclerotic lesions at 10 weeks of age.

*LDL complexes* are the primary plasma carriers of cholesterol for delivery to all tissues. LDLs are taken up by cells via LDL receptor-mediated endocytosis. The uptake of LDLs occurs predominantly in liver (75%), adrenals and adipose tissue. As with IDLs, the interaction of LDLs with LDL receptors requires the presence of apo-B-100. The endocytosed membrane vesicles (endosomes) fuse with lysosomes, in which the apoproteins are degraded and the cholesterol esters are hydrolysed to yield free cholesterol.

Insulin and tri-iodothyronine (T3) increase the binding of LDLs to liver cells, whereas glucocorticoids (e.g. dexamethasone) have the opposite effect. The precise mechanism for these effects is unclear, but it may be mediated through the regulation of apo-B degradation. The effects of insulin and T3 on hepatic LDL binding may explain the hypercholesterolaemia and increased risk of atherosclerosis that have been shown to be associated with uncontrolled diabetes and hypothyroidism.

## 5.10    Role of HDL in lipid metabolism

HDLs are synthesised *de novo* in the liver and small intestine, as primarily protein-rich particles. Newly formed HDLs are essentially devoid of cholesterol and cholesteryl esters. The primary apoproteins of HDLs are apo-A-I, apo-C-I, apo-C-II and apo-E. One major function of HDLs is to act as circulating stores of apo-C-I, apo-C-II and apo-E.

HDLs gradually accumulate cholesteryl esters, converting nascent HDLs to $HDL_2$ and $HDL_3$. Any free cholesterol present in chylomicron remnants and VLDL remnants (IDLs) can be esterified through the action of the HDL-associated enzyme, lecithin : cholesterol acyltransferase (LCAT). LCAT is synthesised in the liver and so named because it transfers a fatty acid from the C-2 position of lecithin to the C-3-OH of cholesterol, generating a cholesteryl ester and lysolecithin. The activity of LCAT requires interaction with apo-A-I, which is found on the surface of HDLs.

Cholesterol-rich HDLs return to the liver, where they are endocytosed. Hepatic uptake of HDLs, so-called reverse cholesterol transport, may be mediated through an HDL-specific apo-A-I receptor. Macrophages also take up HDLs through an apo-A-I receptor interaction; the HDLs acquire cholesterol and additional apo-E from the macrophages. When secreted from the macrophages, the added apo-E in these HDLs leads to an increase in their uptake and catabolism by the liver.

HDLs also acquire cholesterol by extracting it from cell-surface membranes. This process has the effect of lowering the level of intracellular cholesterol, since the cholesterol stored within cells as cholesteryl esters will be mobilised to replace the cholesterol removed from the plasma membrane.

The cholesterol esters of HDLs can also be transferred to VLDLs and LDLs through the action of the HDL-associated enzyme, cholesterol ester transfer protein (CETP). This has the added effect of allowing the excess cellular cholesterol to be returned to the liver through the LDL-receptor pathway as well as the HDL-receptor pathway.

Figure 5.6 shows a general scheme of lipid metabolism.

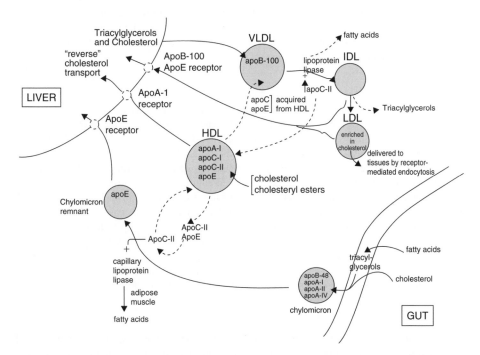

**Figure 5.6** General scheme of lipoprotein metabolism. Triacylglycerols and cholesterol are exported from the liver in VLDLs, containing apolipoprotein B-100; they further acquire apo-C-I, II, III and apo-E from circulating HDL. Apo-C-II activates lipoprotein lipase to remove fatty acids from VLDLs. As triacylglycerols are removed, VLDLs transform to IDLs and finally LDLs. LDLs are the main vehicle for transfer of cholesterol to the tissues; uptake of LDL occurs primarily in the liver through LDL-receptor-mediated endocytosis, which requires the presence of apo-B-100. HDLs are synthesised essentially devoid of cholesterol or triacylglycerol and provide a circulating source of apo-C-I, II and apo-E. HDLs gradually accumulate cholesteryl esters, eventually returning these to the liver, mediated by an apo-A-I receptor; this is referred to as 'reverse cholesterol transport'.

## 5.11    Apoprotein classes

There are six major classes of apoproteins; several sub-classes and hundreds of genetic poly-morphisms have been described (Table 5.1).

## 5.12    LDL receptors

The cellular uptake of cholesterol from LDLs occurs following the interaction of LDLs with the LDL receptor. The sole apoprotein present in LDLs is apo-B-100, which is required for interaction with the LDL receptor.

> The LDL receptor is a polypeptide of 839 amino acids that spans the plasma membrane. An extracellular domain is responsible for apo-B-100/apo-E binding. The intracellular domain is responsible for the clustering of LDL receptors into regions of the plasma membrane termed coated pits.

Once bound to the receptor, the complexes are rapidly internalised (endocytosed). ATP-dependent proton pumps lower the pH in the endosomes, resulting in dissociation of the LDL from the receptor (the receptor is recycled to the plasma membrane). Lysosomal enzymes degrade the apoproteins and release free fatty acids and cholesterol.

The level of intracellular cholesterol is regulated through cholesterol-induced suppression of LDL-receptor synthesis and cholesterol-induced inhibition of cholesterol synthesis. The increased level of intracellular cholesterol that results from LDL uptake has the additional effect of activating acyl-CoA cholesteryl acyl transferase (ACAT) (see below), thereby allow-ing the storage of excess cholesterol within cells. However, the effect of cholesterol-induced suppression of LDL-receptor synthesis is a decrease in the rate at which LDLs and IDLs are removed from the serum. This can lead to excess circulating levels of cholesterol and cholesteryl esters when the dietary intake of fat and cholesterol is excessive. Excess cholesterol tends to be deposited in the skin and tendons and within the arteries, which can lead to atherosclerosis.

> ACAT transfers amino-acyl groups from one molecule to another. ACAT is an important enzyme in bile acid synthesis, and catalyses the intracellular esterification of cholesterol and formation of cholesteryl esters. ACAT-mediated esterification of cholesterol limits its solubility in the cell membrane and thus promotes accumulation of cholesterol ester in the fat droplets within the cytoplasm; this process is important in preventing the toxic accumulation of free cholesterol that would otherwise damage cell-membrane structure and function. Most of the cholesterol absorbed during intestinal transport undergoes ACAT-mediated esterification before incorporation into chylomicrons. In the liver, ACAT-mediated esterification of cholesterol is involved in the production and release of apo-B-containing lipoproteins.
>
> Uptake of cholesterol by enterocytes is mediated by the transporter NPC1L1 (Niemann–Pick C1-like 1), located on the lumen surface. ACAT orchestrates the

esterification of cholesterol, which in turn stimulates synthesis of apo-B48 (see Table 5.1), with assembly into chylomicrons being mediated by the microsomal triacylglycerol transfer protein (MTP).

## 5.13   Disorders of lipoprotein metabolism

Specific lipoprotein disorders (hyper- or hypolipoproteinaemia) are rare but there is increasing knowledge and awareness of the importance of apolipoproteins and their relevance to a variety of clinical disorders.

Individuals suffering from diabetes mellitus, hypothyroidism or kidney disease often exhibit abnormal lipoprotein metabolism as a result of secondary effects of their disorders. For example, because LPL synthesis is regulated by insulin, LPL deficiencies leading to type I hyperlipoproteinaemia may occur as a secondary outcome of diabetes mellitus.

Insulin and thyroid hormones positively affect hepatic LDL-receptor interactions; therefore, the hypercholesterolaemia and increased risk of atherosclerosis associated with uncontrolled diabetes or hypothyroidism is likely due to decreased hepatic LDL uptake and metabolism.

Of the many disorders of lipoprotein metabolism (Tables 5.2 and 5.3), familial hypercholesterolaemia type II may be the most prevalent in the general population. It is an autosomal dominant disorder that results from mutations affecting the structure and function of the cell-surface receptor that binds plasma LDLs and removes them from the circulation. The defects in LDL-receptor interaction result in lifelong elevation of LDL cholesterol in the blood. The resultant hypercholesterolaemia leads to premature coronary artery disease and atherosclerotic plaque formation. Familial hypercholesterolaemia was the first inherited disorder recognised as being a cause of myocardial infarction (heart attack).

## 5.14   Alzheimer's disease

An apolipoprotein E variant is the only unequivocal genetic risk factor for late-onset Alzheimer's disease in a variety of ethnic groups. Caucasians and Japanese with the apo-E-$\varepsilon$4 isoform have between 10 and 30 times the risk of developing Alzheimer's by 75 years of age. While the exact mechanism is unknown, evidence suggests an interaction with amyloid. Alzheimer's disease is characterised by plaques consisting of the peptide beta-amyloid. Apolipoprotein E enhances proteolytic breakdown of this peptide. However, the isoform apo-E-$\varepsilon$4 is much less effective, which might result in an increased vulnerability to Alzheimer's in individuals with that gene variation.

## 5.15   Pharmacologic intervention

Drug treatment to lower plasma lipoproteins and/or cholesterol is primarily aimed at reducing the risk of atherosclerosis and subsequent coronary artery disease that exists in patients with elevated circulating lipids. Drug therapy is considered as an option only if non-pharmacologic interventions (altered diet and exercise) have failed to lower plasma lipids.

**Table 5.2**　Hyperlipoproteinaemia disorders

| Disorder | Defect | Observation |
|---|---|---|
| Type I (familial LPL deficiency, familial hyperchylomicronaemia) | A deficiency of LPL, production of abnormal LPL, and apo-C-II deficiency | Chylomicron clearance is slow, LDL and HDL levels are reduced. Does not appear to be any increased risk of coronary artery disease |
| Type II (familial hypercholesterolaemia) | Four classes of LDL receptor defect | Reduced LDL clearance leads to hypercholesterolaemia; can result in atherosclerosis and coronary artery disease |
| Type III (familial dysbetalipoproteinaemia, remnant removal disease, broad beta disease, apolipoprotein E deficiency) | Hepatic remnant clearance impaired due to apo-E abnormality; patients only express the apo-E2 isoform that interacts poorly with the apo-E receptor | Can cause xanthomas, hypercholesterolaemia and atherosclerosis in peripheral and coronary arteries due to elevated levels of chylomicrons and VLDLs |
| Type IV (familial hypertriacylglycerolaemia) | Elevated production of VLDL associated with glucose intolerance and hyperinsulinaemia | Frequently associated with type II non-insulin-dependent diabetes mellitus, obesity, alcoholism and administration of progestational hormones; elevated cholesterol as a result of increased VLDLs |
| Type V (familial) | Elevated chylomicrons and VLDLs due to unknown cause | Hypertriacylglycerolaemia and hypercholesterolaemia with decreased LDLs and HDLs |
| Familial hyperalphalipoproteinaemia | Increased level of HDLs | A rare condition that is beneficial for health and longevity |
| Type II familial hyperbetalipoproteinaemia | Increased LDL production and delayed clearance of triacylglycerols and fatty acids | Strongly associated with increased risk of coronary artery disease |
| Familial ligand-defective apo-B | Reduced affinity of LDL for LDL receptor | Increase in LDL levels; no effect on HDL, VLDL or plasma triglyceride levels; significant cause of hypercholesterolemia and premature coronary artery disease |
| Familial LCAT deficiency | Absence of LCAT leads to inability of HDLs to take up cholesterol (reverse cholesterol transport) | Decreased levels of plasma cholesteryl esters and lysolecithin; abnormal LDLs (Lp-X) and VLDLs; symptoms also found associated with cholestasis |
| Wolman disease (cholesteryl ester storage disease) | Defect in lysosomal cholesteryl ester hydrolase; affects metabolism of LDLs | Reduced LDL clearance leads to hypercholesterolaemia, resulting in atherosclerosis and coronary artery disease |
| Heparin-releasable hepatic triglyceride lipase deficiency | Deficiency of the lipase leads to accumulation of triacyiglycerol-rich HDLs and VLDL remnants (IDLs) | Causes xanthomas and coronary artery disease |

**Table 5.3**  Hypolipoproteinaemia disorders

| Disorder | Defect | Observation |
|---|---|---|
| Abetalipoproteinaemia (acanthocytosis, Bassen–Kornzweig syndrome) | No chylomicrons, VLDLs or LDLs, due to defect in apo-B expression | Rare defect; intestine and liver fat accumulate; malabsorption of fat; retinitis pigmentosa; ataxic neuropathic disease |
| Familial hypobetalipoproteinaemia | At least 20 different apo-B gene mutations identified; LDL concentrations 10–20% of normal; VLDL slightly lower; HDL normal | Mild or no pathological changes |
| Familial alpha-lipoprotein deficiency (Tangier disease, fish-eye disease, apo-A-I and C-III deficiencies) | All have reduced HDL concentrations; no effect on chylomicron or VLDL production | Tendency to hypertriacylglycerolaemia; some elevation in VLDLs; fish-eye disease characterised by severe corneal opacity |

- **Statins**. Statins, such as atorvastatin (Lipotor), simvastatin (Zocor) and lovastatin (Mevacor), are fungal-derived HMG-CoA reductase inhibitors. Treatment results in an increased cellular uptake of LDLs, since the intracellular synthesis of cholesterol is inhibited and cells are therefore dependent on extracellular sources of cholesterol. However, since mevalonate (the product of the HMG-CoA reductase reaction) is also required for the synthesis of other important isoprenoid compounds besides cholesterol, long-term treatments carry some risk of toxicity.

- **Nicotinic acid**. This reduces the plasma levels of both VLDLs and LDLs by inhibiting hepatic VLDL secretion, as well as suppressing the flux of free-fatty-acid release from adipose tissue by inhibiting lipolysis. Because of its ability to cause large reductions in circulating levels of cholesterol, nicotinic acid is used to treat Type II, III, IV and V hyperlipoproteinaemias.

- **Fibrates**. Fibrates, such as gemfibrozil (Lopid) and fenofibrate (TriCor), lead to an increased $\beta$-oxidation of fatty acids, thereby decreasing secretion of triacylglycerol- and cholesterol-rich VLDLs, and increasing clearance of chylomicron remnants, increasing levels of HDLs and increasing LPL activity, which in turn promotes rapid VLDL turnover.

- **Cholestyramine or colestipol (resins)**. These are compounds that bind bile acids; the drop in hepatic reabsorption of bile acids releases a feedback inhibition, resulting in a greater amount of cholesterol being converted to bile acids to maintain a steady level in the circulation. Additionally, synthesis of LDL receptors increases to allow for the increased cholesterol uptake for bile acid synthesis; the overall effect is a reduction in plasma cholesterol.

- **Proprotein convertase subtilisin/kexin type 9 (PCSK9)**. PCSK9 is involved in the degradation of LDL receptors and appears to be a key regulator of LDL cholesterol. Elevated PCSK9 activity produces familial hypercholesterolaemia, while lowered activity results in a decrease of LDL cholesterol and lowered cardiovascular risk. PCSK9 may be a prime therapeutic target for the prevention and treatment of cardiovascular diseases. Mice studies show that annexin A2 binds strongly to PCSK9 and inhibits its function (annexin A2, a member of the annexin family, is a calcium-dependent phospholipid-binding protein that is generally considered to be involved with regulation of cell growth, signal transduction pathways and osteoclast formation and bone resorption).

# Focus on: atherosclerosis

'Arteriosclerosis' is a general term describing any hardening or loss of elasticity of medium or large arteries, and refers to the formation of an atheromatous plaque. It is a syndrome that affects arterial blood vessels. It is believed to be initiated when LDLs become oxidised. Oxidation of LDL is stimulated chiefly by lipoprotein-associated phospholipase $A_2$ and oxygen-reactive species in the blood-vessel endothelium. Lipoprotein-associated phospholipase $A_2$ is an emerging cardiovascular risk marker. Following LDL oxidation, lipoprotein-associated phospholipase $A_2$ generates oxidised non-esterified fatty acids and lysophosphatidylcholine, both of which have demonstrated proinflammatory and proapoptotic activities.

Oxidised LDL in the artery wall initiates a series of reactions designed to repair the damage it causes. Initial damage triggers an inflammatory response. Monocytes enter the artery wall from the bloodstream, with platelets adhering to the area of insult. This may be promoted by induction of factors such as vascular cell adhesion molecule 1 (VCAM-1), a cell-surface sialoglycoprotein that is expressed by cytokine-activated endothelium. This membrane protein mediates leukocyte-endothelial cell adhesion and signal transduction, and may play a role in the development of atherosclerosis and rheumatoid arthritis.

Monocytes differentiate into macrophages, which ingest oxidised LDL, slowly turning into large 'foam cells' (so called because of their numerous internal cytoplasmic vesicles and high lipid content). Under the microscope, the lesion appears as a fatty streak. Foam cells eventually die and further propagate the inflammatory process. There is also smooth-muscle proliferation and migration from tunica media to intima, responding to cytokines secreted by damaged endothelial cells. This leads to the formation of a fibrous capsule covering the fatty streak.

Scavenger receptors recognise modified LDL, as well as other macromolecules with a negative charge. They are thought to participate in the removal of many foreign substances and waste materials in the body. In atherosclerotic lesions, macrophages that express scavenger receptors aggressively uptake the oxidised LDL deposited in the blood-vessel wall, forming foam cells that secrete various inflammatory cytokines and accelerate the development of atherosclerosis.

Atherogenesis is the developmental process of atheromatous plaques. It is characterised by a remodelling of arteries involving the concomitant accumulation of fatty substances called plaques. The bulk of these lesions are made of excess fat, collagen and elastin. As the plaques grow, artery wall thickening occurs without any narrowing of the artery lumen; stenosis, the narrowing of the artery opening, is a late event, which may or may not occur, and is likely the result of repeated plaque rupture and healing responses. The atherosclerosis syndrome is slowly progressive and cumulative. Most commonly a plaque will rupture, forming a thrombus, which can rapidly slow or stop blood flow, leading to death of the tissues fed by the artery: an infarction. One of the most commonly recognised scenarios is the coronary thrombosis of a coronary artery, causing myocardial infarction (a heart attack); the same process in an artery to the brain can lead to a stroke.

There is some evidence that atherosclerosis may be caused by an infection of the vascular smooth-muscle cells. Chickens, for example, develop atherosclerosis when infected with the Marek's disease herpesvirus. Herpesvirus infection of arterial smooth-muscle cells has been shown to cause cholesteryl ester accumulation, which is associated with atherosclerosis.

Various anatomical, physiological and behavioural risk factors for atherosclerosis are known. Many of these are recognised within the 'metabolic syndrome', a combination of disorders that increases the risk of developing cardiovascular disease and diabetes. Prevalence increases with age, affecting up to 25% of the population in the USA. Risk factors include:

- diabetes or impaired glucose tolerance

- dyslipoproteinaemia

- high serum LDL and/or very low VLDL

- low serum HDL

- an LDL : HDL ratio greater than 3 : 1

- smoking (increases risk by up to 200%)

- high blood pressure (increases risk by up to 60%)

- elevated serum C-reactive protein

- advancing age

- male sex

- having close relatives who have had some complication of atherosclerosis (e.g. coronary heart disease or stroke)

- genetic abnormalities, for example familial hypercholesterolaemia.

HDL has an important role in atherosclerosis:

- A limited study, using the genetically rare apo-A-1 Milano HDL, produced significant reduction in measured coronary plaque volume in patients with unstable angina.

- Increasing HDL particle concentrations in some animal studies has been shown to largely reverse and remove atheromas.

Niacin raises HDL (by 10–30%) and has shown clinical trial benefits. Torcetrapib is the most effective agent currently known for raising HDL (by up to 60%), although undesirable side effects have halted its use.

Statins (see Section 5.15) are a group of medications that have proved popular for the treatment of atherosclerosis. They have few short- or long-term undesirable side-effects. Rosuvastatin is a statin shown to demonstrate regression of atherosclerotic plaque within the coronary arteries. The antioxidant effects of the statins may be partly responsible for their therapeutic success.

# CHAPTER 6

# The liver

## 6.1   General overview

Blood leaves the stomach and intestines, passing through the liver (hepatic portal vein), while oxygenated blood is supplied through the hepatic artery. Two main liver lobes are each made up of thousands of lobules; lobules connect to small ducts that connect to larger ducts, forming the hepatic duct. The hepatic duct transports bile, produced by the hepatocytes, to the gallbladder and duodenum. The liver regulates, synthesises, stores and secretes many important proteins and nutrients, and purifies, transforms and clears toxic or unnecessary compounds from the blood. Hepatocytes are optimised for function through their contact with sinusoids (leading to and from blood vessels) and bile ducts. A special feature of the liver is its ability to regenerate, maintaining function even in the face of moderate damage.

---

*In utero*, energy is provided by glucose, with liver metabolism being directed to glucose degradation; activity of the rate-limiting enzymes of glycolysis, hexokinase and phospho-fructokinase is high. With the onset of post-natal life, and an intake of a fat-rich and carbohydrate-poor diet, infants develop the ability to synthesise glucose *de novo* from non-carbohydrate precursors (gluconeogenesis). Metabolic energy, derived mainly from the $\beta$-oxidation of fatty acids, yields acetyl-CoA (which activates pyruvate carboxylase) and generates a high NADH : NAD$^+$ ratio, shifting the glyceraldehyde 3-phosphate dehy-drogenase reaction in the direction of glucose formation (see Section 2.9).

---

## 6.2   Storage diseases

The liver is designed to store a variety of substances: glucose (as glycogen), fat-soluble vitamins (vitamins A, D, E and K), folate, vitamin $B_{12}$ and minerals such as copper and iron. However,

---

excessive accumulation of certain substances can be harmful. In the inherited condition of Wilson's disease, the secretion of copper into bile is abnormal, resulting in a low blood level of the copper-binding protein ceruloplasmin. Copper accumulates in the liver (leading to cirrhosis) and in the CNS, resulting in neuropsychiatric symptoms.

## 6.3　Glycogen storage diseases

Glycogen storage diseases (GSDs) are the result of inborn errors of metabolism of those enzymes affecting the processing of glycogen, its synthesis or breakdown, within muscle, liver and other cell types. There are nine GSDs (although glycogen synthase deficiency does not result in storage of extra glycogen in the liver, it is often classified with the GSDs as type 0). Glycogen is mainly stored in the liver and muscle cells, but the kidneys and intestines also store some limited amounts of glycogen (Table 6.1).

The inability to degrade glycogen may cause cells to become pathologically engorged, leading to a functional loss of glycogen as an energy source and a blood glucose buffer. GSDs are attributed to specific enzyme deficiencies, although other defects can cause similar symptoms. For example, the enzyme glucose-6-phosphatase is localised on the cisternal (inner) surface of the endoplasmic reticulum, and glucose-6-phosphate must be transported (translocated) across the endoplasmic reticulum to gain access to the enzyme. Mutation of either the phosphatase or the translocase will lead to symptoms characteristic of von Gierke's disease.

## 6.4　General liver metabolism

More than 500 vital functions are associated with the liver; some of the better characterised are listed below:

- production and excretion of bile

- cholesterol metabolism

- drug metabolism and detoxification

- haemoglobin degradation

- protein metabolism; production of albumin

- nitrogen metabolism, ammonia detoxification

- synthesis of coagulation factors

- storage, including glucose (in the form of glycogen), vitamin $B_{12}$, iron and copper

- carbohydrate metabolism

- lipid metabolism

- red blood cell production (first trimester foetus; by the 32nd week of gestation, the bone marrow has almost completely taken over this task)

- Further, the reticuloendothelial system of the liver contains many immunologically active cells, acting as a 'sieve' for antigens carried to it via the portal system.

**Table 6.1** Glycogen storage diseases

| Type | Enzyme deficiency | Common name | Symptoms |
|---|---|---|---|
| I (IA and IB) | Glucose-6-phosphatase | von Gierke's disease | Enlarged liver and kidney; slowed growth; very low blood sugar levels; abnormally high levels of acid, fats and uric acid in blood; growth failure |
| II | Acid maltase | Pompe's disease | Muscle weakness; death by age ~2 years (infantile variant) |
| III | Glycogen debrancher | Cori's disease or Forbe's disease | Enlarged liver; fasting hypoglycaemia; myopathy |
| IV | Glycogen-branching enzyme | Andersen's disease | Failure to thrive; cirrhosis and death at age ~5 years |
| V | Muscle glycogen phosphorylase | McArdle's disease | Exercise-induced cramps; rhabdomyolysis[a]; renal failure by myoglobinuria |
| VI | Liver glycogen phosphorylase | Hers' disease | Enlarged liver; hypoglycaemia |
| VII | Muscle phosphofructokinase | Tarui's disease | Exercise-induced muscle cramps and weakness, sometimes with rhabdomyolysis[a]; haemolytic anaemia |
| VIII | Now classified with VI | – | – |
| IX | Phosphorylase kinase | – | Enlarged liver; hyperlipidaemia; delayed motor development; growth retardation |
| X | Now classified with VI | – | – |
| XI | Glucose transporter, GLUT2 | Fanconi–Bickel syndrome | Similar to Von Gierke's disease, e.g. hypoglycaemia |
| 0 | Glycogen synthase | – | Enlarged liver, accumulation of fat (fatty liver); fasting hypoglycaemia; occasional muscle cramping |

[a] the rapid breakdown of skeletal muscle, caused by physical, chemical or biological factors.

## 6.5 Production and excretion of bile

Bile is a mixture of electrolytes, bile acids, cholesterol, phospholipids and bilirubin. Adults produce between 400 and 800 ml of bile daily. Hepatocytes secrete bile into canaliculi, then into bile ducts, where it is modified by addition of a bicarbonate-rich secretion from ductal epithelial cells. Further modification occurs in the gall bladder, where it is concentrated up to fivefold, through absorption of water and electrolytes. Gallstones, most of which are composed

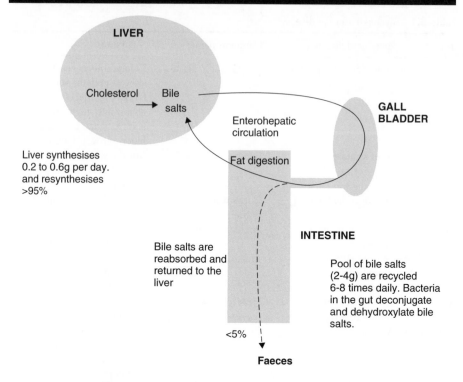

**Figure 6.1**　Enterohepatic recirculation of bile salts.

predominantly of cholesterol, result when cholesterol precipitates from solution. Liver damage or obstruction of a bile duct (e.g. by gallstones) can lead to cholestasis, the blockage of bile flow, resulting in the malabsorption of dietary fats with steatorrhea (foul-smelling diarrhoea caused by non-absorbed fats) and jaundice.

Only relatively small quantities of bile acids are lost from the body; approximately 95% of bile acids delivered to the duodenum are absorbed back into blood within the ileum. Venous blood from the ileum goes straight into the portal vein, and hence through the sinusoids of the liver (enterohepatic circulation). Hepatocytes extract bile acids very efficiently from sinusoidal blood; they are re-secreted into canaliculi. The net result of enterohepatic recirculation is that each bile salt molecule may be reused up to 20 times, and often 2 or 3 times during a single digestive phase (Figure 6.1).

Liver disease, and damage to the canalicular system, can result in escape of bile acids into the systemic circulation. Assay of systemic levels of bile acids is used clinically as a sensitive indicator of hepatic disease.

Bile acids are steroids, characterised by a carbon skeleton with four fused rings, generally arranged in a 6-6-6-5 fashion. Primary bile acids are cholic acid and chenodeoxycholic acid (Figure 6.2). Within the intestines, bacteria convert primary bile acids to secondary bile acids, for example deoxycholate (from cholate) and lithocholate (from chenodeoxycholate). Both primary and secondary bile acids are re-absorbed by the intestines and delivered back to the liver via the portal circulation.

In the duodenum's alkaline environment, bile acids become bile salts (e.g. sodium glyco-cholate). Bile acids are conjugated (joined together) with glycine or taurine, via formation of

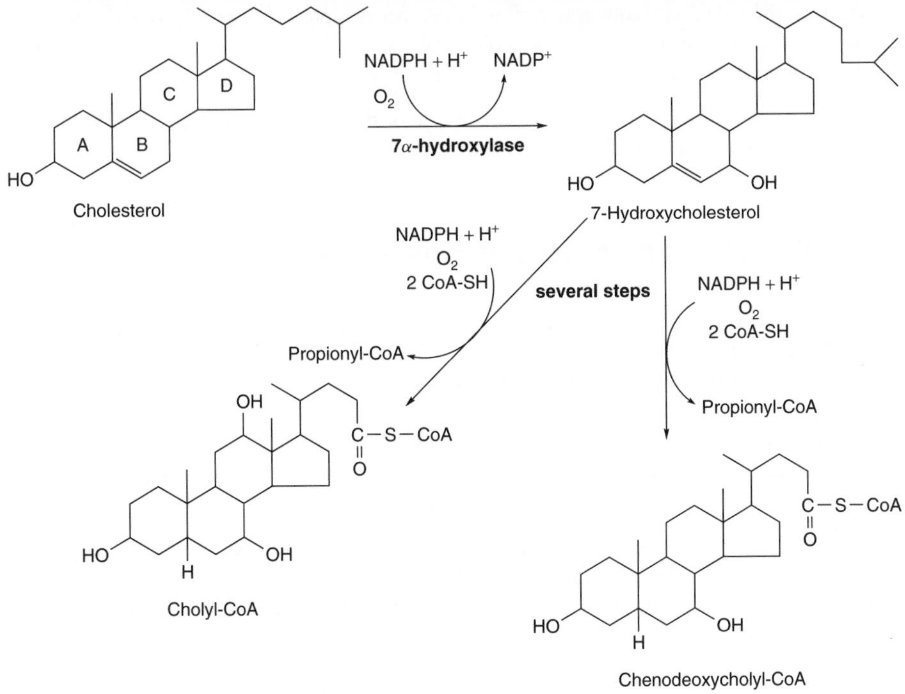

**Figure 6.2** Synthesis of primary bile acids from cholesterol.

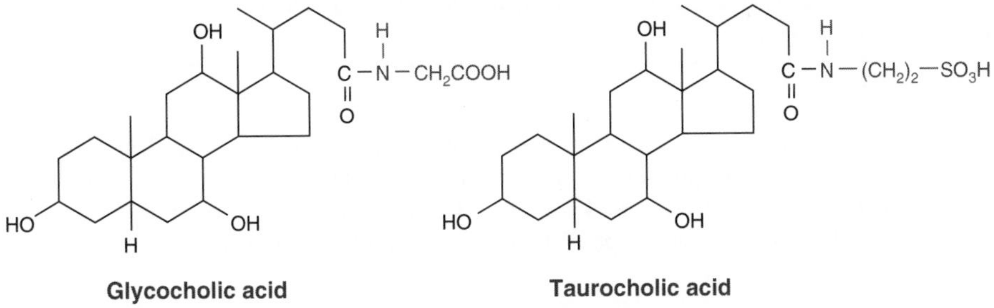

**Figure 6.3** Glycocholic and taurocholic acid structures.

an amide bond, to yield glycocholic acid and taurocholic acid respectively (Figure 6.3); in this form they are stored in the gall bladder.

Bile acids perform four physiologically significant functions:

1. They facilitate the digestion of dietary triacylglycerols by acting as emulsifying agents; emulsification increases the surface area of fat, making it available for digestion by lipases.

2. They facilitate the intestinal absorption of fat-soluble vitamins (vitamin A, retinol; vitamin D, cholecalciferol; vitamin E, tocopherol; and vitamin K).

3. Their synthesis and subsequent excretion in the faeces represents the only significant mechanism for the elimination of excess cholesterol. In humans, roughly 500 mg of cholesterol is converted to bile acids and eliminated in bile every day.

4. Bile acids and phospholipids solubilise cholesterol in the bile, thereby preventing the precipitation of cholesterol in the gall bladder.

# 6.6   Pattern and control of bile secretion

The flow of bile is lowest during fasting, mostly being diverted to the gall bladder for concentrating. When chyme from an ingested meal enters the small intestine, acid and partially digested fats and proteins stimulate secretion of the enteric hormones cholecystokinin and secretin.

- Cholecystokinin (cholecysto = gall bladder and kinin = movement) release is stimulated by the presence of fat in the duodenum; it induces contractions of the gall bladder and common bile duct, resulting in delivery of bile into the gut.

- Secretin is secreted in response to acid in the duodenum. Its effect on the biliary system is similar to that on the pancreas; it simulates biliary duct cells to secrete bicarbonate and water, expanding the volume of bile and increasing the flow rate into the intestine.

---

The processes of gall bladder filling and emptying can be visualised using an imaging technique called scintography. This procedure is utilised as a diagnostic aid in certain types of hepatobiliary disease. Scintography is the process of obtaining a photographic recording of the distribution of an internally administered radiopharmaceutical with the use of a gamma camera.

---

As surfactants (detergents), bile acids are potentially toxic to cells and so their levels are tightly regulated. They function directly as signalling molecules in the liver and the intestines, activating a nuclear hormone receptor known as FXR, the farnesoid X receptor, also known by its gene name NR1H4 (nuclear receptor subfamily 1, group H, member 4).

---

FXR is expressed at high levels in the liver and intestine; chenodeoxycholic acid and other bile acids are natural ligands for FXR. Like other steroid receptors, when activated, FXR translocates to the cell nucleus, forms a dimer and binds to hormone response elements on DNA, eliciting expression or transrepression of gene products. A primary function of FXR activation is the suppression of cholesterol 7 alpha-hydroxylase (CYP7A1), the rate-limiting enzyme in bile acid synthesis from cholesterol. FXR does not directly bind to the CYP7A1 promoter, but rather induces expression of a small heterodimer partner (SHP), which then functions to inhibit transcription of the CYP7A1 gene. In this way a negative-feedback pathway is established in which synthesis of bile acids is inhibited when cellular levels are already high.

# 6.7 Clinical significance of bile secretion

Since bile acids are made from endogenous cholesterol, the enterohepatic circulation of bile acids may be disrupted as a way to lower cholesterol. Bile acid sequestrants bind bile acids in the gut, preventing their re-absorption. In so doing, more endogenous cholesterol is directed to the production of bile acids, thereby lowering cholesterol levels. The sequestered bile acids are excreted in the faeces.

# 6.8 Cholesterol metabolism

Slightly less than half of the cholesterol in the body derives from biosynthesis *de novo*; biosynthesis in the liver accounts for approximately 10% and in the intestines approximately 15% each day. Cholesterol synthesis occurs in the cytoplasm and microsomes (smooth endoplasmic reticulum) (Figure 6.4). The process has five major steps:

1. Acetyl-CoA is converted to 3-hydroxy-3-methylglutaryl-CoA (HMG-CoA).

2. HMG-CoA is converted to mevalonate.

3. Mevalonate is converted to the isoprene based molecule isopentenyl pyrophosphate (IPP), with the concomitant loss of $CO_2$.

4. IPP is converted to squalene.

5. Squalene is converted to cholesterol.

# 6.9 Regulating cholesterol synthesis

HMG-CoA reductase is the rate-limiting step of cholesterol biosynthesis, and is subject to complex regulatory controls. A relatively constant level of cholesterol in the body (150–200 mg/dl) is maintained primarily by controlling the level of *de novo* synthesis. The level of cholesterol synthesis is regulated in part by the dietary intake of cholesterol. Cholesterol from both diet and synthesis is utilised in the formation of membranes and in the synthesis of the steroid hormones and bile acids. The greatest proportion of cholesterol is used in bile acid synthesis.

The cellular supply of cholesterol is maintained at a steady level by three distinct mechanisms:

1. regulation of HMG-CoA reductase (Figure 6.5).

2. regulation of excess intracellular free cholesterol through the activity of acyl-CoA:cholesterol acyltransferase (ACAT).

3. regulation of plasma cholesterol levels via LDL receptor-mediated uptake and HDL-mediated reverse transport.

HMG-CoA reductase activity is the primary means for controlling the level of cholesterol biosynthesis.

The reductase kinase (RK), which phosphorylates (and inhibits) HMG-CoA reductase, is itself controlled through phosphorylation (activated) by reductase kinase kinase (RKK). There

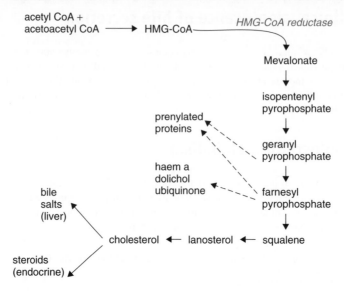

**Figure 6.4** *De novo* synthesis of cholesterol. Pathway of cholesterol biosynthesis. Synthesis begins with the transport of acetyl-CoA from the mitochondrion to the cytosol. The rate-limiting step occurs at the 3-hydroxy-3-methylglutaryl-CoA (HMG-CoA) reductase catalysed step. The phosphorylation reactions are required to solubilise the isoprenoid intermediates in the pathway. Intermediates in the pathway are used for the synthesis of prenylated proteins, dolichol, coenzyme Q and the side chain of haem a.

are two isoforms of RKK, one independent of cAMP and one dependent upon cAMP. The cAMP-dependent RKK is activated in the presence of cAMP. Since the intracellular level of cAMP is regulated by hormonal stimuli, regulation of cholesterol biosynthesis is hormonally controlled. Insulin leads to a decrease in cAMP, which in turn activates cholesterol synthesis. By contrast, glucagon and epinephrine, which increase the level of cAMP, inhibit cholesterol synthesis.

Increases in cAMP further lead to phosphorylation and activation of phosphoprotein phosphatase inhibitor 1 (PPI-1), which inhibits the activity of phosphoprotein phosphatase and so decreases the conversion (dephosphorylation) of inactive to active HMG-CoA reductase.

Long-term control of HMG-CoA reductase activity is exerted primarily through control over the synthesis and degradation of the enzyme. When levels of cholesterol are high, the level of expression of the HMG-CoA gene is reduced. Conversely, reduced levels of cholesterol activate expression of the gene. Insulin also brings about long-term regulation of cholesterol metabolism by increasing the level of HMG-CoA reductase synthesis. The rate of HMG-CoA turnover is also regulated by the supply of cholesterol. When cholesterol is abundant, the rate of HMG-CoA reductase degradation increases.

## 6.10   Regulating sterol synthesis

HMG-CoA reductase is under negative-feedback control by the sterol synthesis end products cholesterol and bile salts. The important indicator of HMG-CoA reductase is the blood cholesterol concentration, which is mostly determined by the presence of LDL particle.

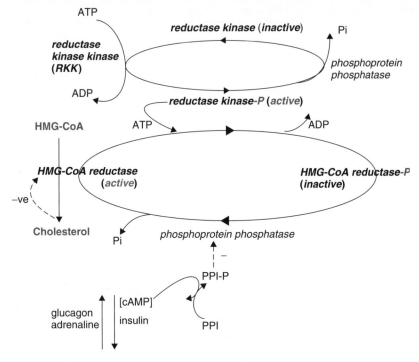

**Figure 6.5** Regulation of HMG-CoA reductase. HMG-CoA reductase is active in the dephospho-rylated state; phosphorylation (inhibition) is catalysed by reductase kinase, an enzyme whose activity is also regulated by phosphorylation by reductase kinase kinase. Hormones such as glucagon and adrenalin (epinephrine) negatively affect cholesterol biosynthesis by increasing the activity of the inhibitor of phosphoprotein phosphatase-1, PPI-1, (by raising cAMP levels) and so reducing the activation of HMG-CoA reductase. Conversely, insulin stimulates the removal of phosphates (and lowers cAMP levels), and thereby activates HMG-CoA reductase activity. Additional regulation of HMG-CoA reductase occurs through an inhibition of synthesis of the enzyme by elevation in intracellular cholesterol levels.

Statins, HMG-CoA reductase inhibitors, form a class of hypolipidaemic drug used to lower cholesterol levels. Inhibition of HMG-CoA reductase stimulates LDL receptors, resulting in an increased clearance of LDL from the bloodstream, with a concomitant decrease in blood cholesterol levels. Drug effect is maximal after four to six weeks of use (see also Section 5.15).

## 6.11 Drug metabolism

Most drugs, and particularly oral drugs, are modified or degraded in the liver. In the liver, drugs may undergo first-pass metabolism, a process in which they are modified, activated or inactivated, before they enter the systemic circulation; alternatively, they may be left unchanged. An oral drug that is absorbed and metabolised in the liver is said to show the 'first-pass effect'.

Medications that are metabolised by the liver must be used with caution in patients with hepatic disease; such patients may need lower doses of the drug. Alcohol is primarily metabolised by the liver, and accumulation of its products can lead to cell injury and death.

**Table 6.2**   Phase I and phase II drug detoxification reactions

| Phase I reactions | Phase II reactions |
|---|---|
| **Oxidations** | Glutathione S-transferases |
| Cytochrome P450 monooxygenase system | Mercapturic acid biosynthesis |
| Flavin-containing monooxygenase system | UDP-glucuronosyltransferases |
| Alcohol dehydrogenase and aldehyde dehydrogenase | N-acetyltransferases |
| Monoamine oxidase | Amino acid N-acyl transferases |
| Co-oxidation by peroxidases | Sulphotransferases |
| **Reduction** | |
| NADPH-cytochrome P450 reductase | |
| Reduced (ferrous) cytochrome P450 | |
| **Hydrolysis** | |
| Esterases and amidases | |
| Epoxide hydrolase | |

The general strategy of xenobiotic metabolism (the metabolism of foreign compounds) is to convert lipophilic compounds into more readily excreted polar products. The rate of this metabolism is an important determinant of the duration and intensity of the pharmacological action of drugs. Drug metabolism can result in either toxication or detoxication. While both do occur, the major metabolites of most drugs are detoxication products.

Drug hepatotoxicity is a common cause of acute liver failure, with an incidence of 1 in 10 000–100 000.

Drug biotransformations may occur through one or both of phase I and phase II reactions (Table 6.2). Phase I reactions usually, but not always, precede phase II.

Phase I reactions (also termed non-synthetic reactions) may occur by oxidation, reduction, hydrolysis, cyclisation and decyclisation reactions. Oxidation involves the enzymatic addition of oxygen or removal of hydrogen, carried out by mixed-function oxidases, typically involving a cytochrome P450 (CYP) haemoprotein. If the metabolites of phase I reactions are sufficiently polar, they may be readily excreted at this point. However, many phase I products are not eliminated rapidly and undergo a subsequent reaction in which an endogenous substrate combines with the newly incorporated functional group to form a highly polar conjugate.

Phase II reactions, usually known as conjugation reactions (e.g. with glucuronic acid, sulphonates (sulphation), glutathione or amino acids), are usually detoxifying in nature, and involve the interactions of the polar functional groups of phase I metabolites.

Quantitatively, the smooth endoplasmic reticulum of the hepatocyte is the principal organelle of drug metabolism. Other sites of drug metabolism include epithelial cells of the gastrointestinal tract, the lungs, the kidneys and the skin. These sites are usually responsible for localised toxicity reactions.

The metabolism of paracetamol (Figure 6.6) is an example of potential toxication. Paracetamol is metabolised primarily in the liver, via phase II metabolism, where its major metabolites include inactive sulphate and glucuronide conjugates, which are excreted by the kidneys.

A small yet significant amount is metabolised via the hepatic CYP enzyme system, to a minor alkylating metabolite $N$-acetyl-$p$-benzo-quinone imine (NAPQI). Normally NAPQI is detoxified by conjugation with glutathione. In cases of paracetamol toxicity (overdose), the sulphate and glucuronide pathways become saturated, and more paracetamol is shunted to the CYP system to produce NAPQI. As a result, hepatocellular supplies of glutathione become exhausted and NAPQI is free to react with cellular membrane molecules, resulting in widespread hepatocyte damage and death, and leading to acute hepatic necrosis.

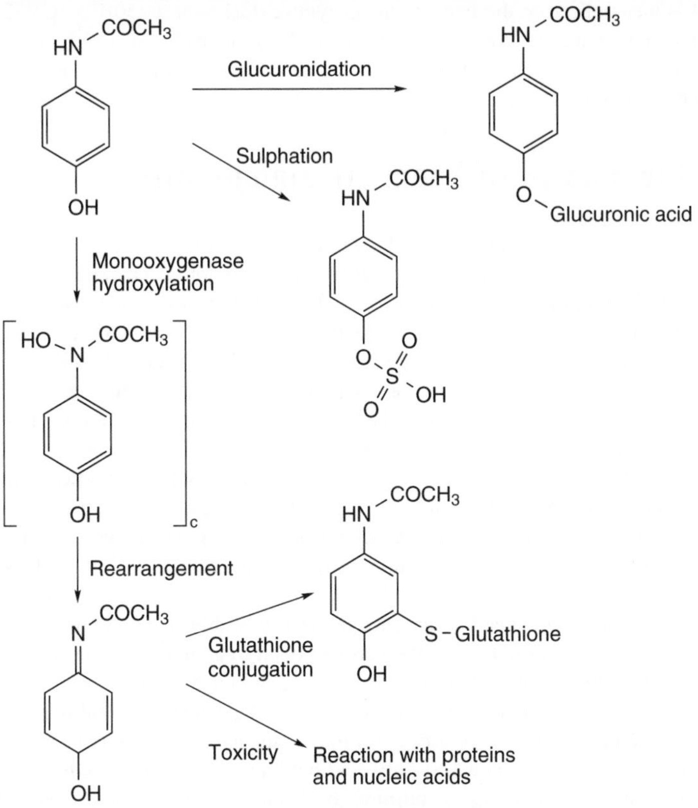

**Figure 6.6** Paracetamol metabolism.

CYP enzymes are a large and diverse superfamily of haemoproteins. Primarily membrane-associated proteins, they are located in the inner membrane of mitochondria and the endoplasmic reticulum of cells, and metabolise thousands of endogenous and exogenous compounds. Most of these enzymes can metabolise multiple substrates, and many can catalyse multiple reactions. While prevalent in the liver, CYP enzymes are also present in most other tissues of the body, and play an important role in hormone synthesis and breakdown (including oestrogen and testosterone synthesis and metabolism), cholesterol synthesis and vitamin D metabolism. Hepatic CYPs are the most widely studied.

The most common reaction catalysed by CYP is a monooxygenase reaction; for example, insertion of one atom of oxygen into an organic substrate (RH):

$$RH + O_2 + 2H^+ + 2e^- \rightarrow ROH + H_2O$$

CYP is the most important element of oxidative metabolism (phase I metabolism). The Human Genome Project has identified 57 human genes coding for various CYP enzymes.

Many drugs may increase or decrease the activity of various CYP isoenzymes in a phenomenon known as enzyme induction and inhibition. This is a major source of adverse drug interaction, since changes in CYP activity may affect the metabolism and clearance of various drugs.

Alcohol is metabolised in the liver by the enzyme cytochrome P450IIE1 (CYP2E1), which may be increased (induced) after chronic drinking. Most alcohol consumed is metabolised in the liver, while the small quantity that remains unmetabolised permits alcohol concentration to be measured in breath and urine.

## 6.12    Breakdown of haem (Haemoglobin)

Red blood cells, the largest repository of haem in the human body, have a life span of about 120 days, representing a turnover of about 6 g/day of haemoglobin. This presents two potential problems: first, the porphyrin haem ring is hydrophobic and must be solubilised to be excreted, and second, iron must be conserved for new haem synthesis. Normally, senescent red blood cells are engulfed (phagocytosed) by cells of the reticuloendothelial system; globin is converted into amino acids, which in turn are recycled or catabolised as required, while haem is oxidised by the haem oxygenase system (a member of the CYP family). Oxidation (introduction of oxygen) splits the porphyrin ring to give the linear tetrapyrrole biliverdin, releasing ferric iron ($Fe^{3+}$) and carbon monoxide (CO). This is the only reaction in the body that is known to produce CO. Most of the CO is excreted through the lungs; the CO content of expired air can be used as a direct measure of the activity of haem oxygenase in an individual. Released iron is recycled. In a reduction reaction, biliverdin reductase converts biliverdin to bilirubin. This process is illustrated in (Figure 6.7).

'Native' bilirubin, as produced, is an unconjugated bilirubin; it is a hydrophobic molecule that must be transported to the liver in the plasma bound to albumin. At the sinusoidal surface of the liver, unconjugated bilirubin detaches from albumin and is transported through the hepatocyte membrane by facilitated diffusion. Within the hepatocyte, bilirubin is bound to two major intracellular proteins, cytosolic Y protein (ligandin or glutathione S-transferase B) and cytosolic Z protein (also known as fatty acid-binding protein). The binding of bilirubin to these proteins decreases the efflux of bilirubin back into the plasma, and therefore increases net bilirubin uptake.

## 6.13    Bilirubin

In adults some 250–400 mg of bilirubin is produced daily; 70–80% is derived from degradation of the haem moiety of haemoglobin, 20–25% is derived from the hepatic turnover of haem proteins, such as myoglobin, cytochromes and catalase.

Bilirubin is a potentially toxic catabolic product of haem metabolism. It is poorly soluble in water at physiologic pH, and conversion to a water-soluble form is essential for elimination by the liver and kidney. This is achieved by hepatic glucuronic acid conjugation of the propionic acid side chains of bilirubin; bilirubin glucuronides are water-soluble and readily excreted in bile. It is primarily excreted as the diglucuronide. Glucuronide addition is catalysed by the hepatic enzyme uridine-diphosphate glucuronosyl transferase (UDP-glucuronyl transferase); addition is through ester linkages; that is, the reaction is an esterification. UDP-glucuronyl transferase is localised to the endoplasmic reticulum. Other compounds, such as xylose and glucose, may also undergo esterification with bilirubin. The conjugation of bilirubin and its subsequent secretion out of hepatocytes and into the bile caniculi is considered a 'linked' process.

Enzyme-catalysed glucuronidation is one of the most important detoxification mechanisms of the body. Of the various isoforms of the UDP-glucuronyl transferase family of enzymes, only one isoform, bilirubin-UDP-glucuronyl transferase UGT-1, is physiologically important in bilirubin glucuronidation.

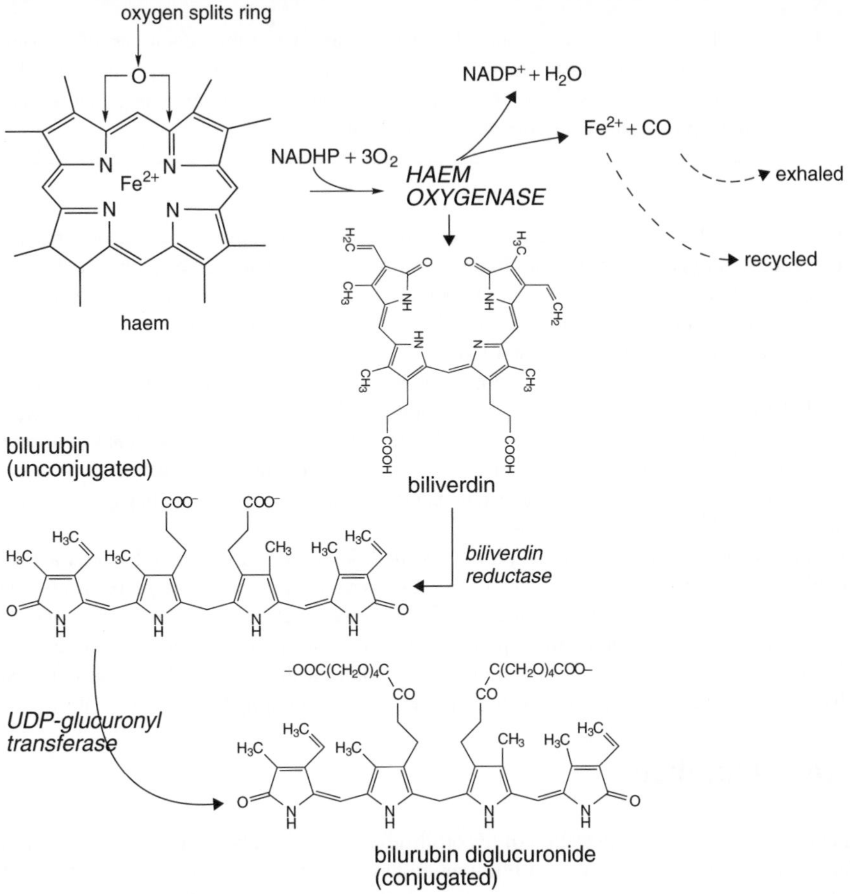

**Figure 6.7** Catabolism of haem to bilirubin.

Glucuronic acid is a carboxylic acid. Its structure is similar to that of glucose, except that carbon-6 is oxidised to a carboxylic acid.

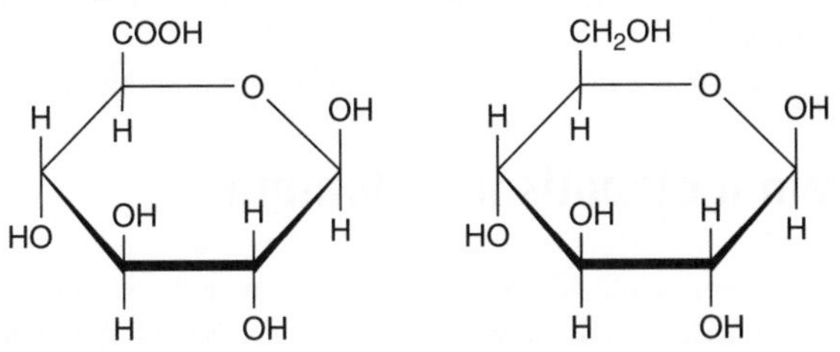

glucuronic acid

beta-D-glucose

Uronic acids have a carboxyl group (-COOH) on the carbon that is not part of the ring. Their names retain the root of the monosaccharides, but the *-ose* sugar suffix is changed to *-uronic acid*. For example, galacturonic acid has the same configuration as galactose, and the structure of glucuronic acid corresponds to glucose.

Deficiency of UGT leads to ineffective esterification of bilirubin, which in turn results in an unconjugated hyperbilirubinaemia. Reduced bilirubin conjugation, as a result of a decreased or absent UGT activity, is found in a number of acquired conditions and inherited diseases, such as Crigler–Najjar syndrome (types I and II) and Gilbert syndrome. Bilirubin-conjugating activity is also very low in the neonatal liver.

UGT activity is modulated by various hormones. Excess thyroid hormone and ethinyl oestradiol (but not other oral contraceptives) inhibit bilirubin glucuronidation. In contrast, the combination of progestational and oestrogenic steroids results in increased enzyme activity. Bilirubin glucuronidation can also be inhibited by certain antibiotics (e.g. novobiocin or gentamicin, at serum concentrations exceeding therapeutic levels) and by chronic hepatitis, advanced cirrhosis and Wilson's disease.

Conjugated bilirubin is excreted with bile as the bile pigments. Intestinal bacteria act upon bilirubin, releasing the glucuronte (which is re-absorbed) and producing the porphyrin products urobilinogens and urobilins. Urobilinogens can be further metabolised to stercobilinogen and the oxidised stercobilin. Stercobilin is partly responsible for the brown colour of faeces. Some urobilinogen is re-absorbed and excreted in the urine along with an oxidised form, urobilin. Normally, a tiny amount of bilirubin is excreted in the urine, accounting for the light yellow colour.

# 6.14  Jaundice

Jaundice is the clinical manifestation of hyperbilirubinaemia. Jaundice may be noticeable in the sclera (white) of the eyes at levels of bilirubin of about 30–50 μmol/l, and in the skin at higher levels. Jaundice is classified, depending upon whether the bilirubin is 'free' or conjugated to glucuronic acid, into:

- conjugated jaundice (direct)

- unconjugated jaundice (indirect).

Total bilirubin measures both direct and indirect; indirect bilirubin is calculated from the total minus the direct bilirubin. Bilirubin levels reflect the balance between production and excretion; there is no 'normal' level of bilirubin (Table 6.3) and levels may be affected by a number of factors (Table 6.4).

Table 6.5 describes a number of disorders associated with bilirubin metabolism.

# 6.15  Protein metabolism – albumin

The liver orchestrates the metabolism of proteins and amino acids. Most blood proteins (except for antibodies) are synthesised and secreted by the liver. One of the most abundant serum proteins is albumin. Impaired liver function that results in decreased amounts of serum albumin may lead to oedema, swelling due to fluid accumulation in the tissues.

**Table 6.3**  Bilirubin levels in adults

| Bilirubin type | Bilirubin level |
|---|---|
| Total bilirubin | 0.3–1.0 mg/dl or 5.1–17.0 mmol/l |
| Direct bilirubin | 0.1–0.3 mg/dl or 1.7–5.1 mmol/l |
| Indirect bilirubin (total bilirubin level minus direct bilirubin level) | 0.2–0.8 mg/dl or 3.4–12.0 mmol/l |

**Table 6.4**  Factors that can affect bilirubin levels in the blood

| Mild rises in bilirubin may be caused by | Moderate rise in bilirubin may be caused by | Very high levels of bilirubin may be caused by |
|---|---|---|
| Haemolysis or increased breakdown of red blood cells. | Drugs (especially anti-psychotic and some sex hormones) | Neonatal hyperbilirubinaemia, where the newborn's liver is not able to properly conjugate the bilirubin |
| Gilbert's syndrome – a genetic disorder of bilirubin metabolism which can result in mild jaundice, found in about 5% of the population. | Hepatitis (levels may be moderate or high) | Unusually large bile duct obstruction, for example gallstone in common bile duct, tumour obstructing common bile duct. Choledocholithiasis (chronic or acute) is the presence of gallstones in the common bile duct |
| | Chemotherapy | Severe liver failure with cirrhosis |
| | Biliary stricture (benign or malignant) | Severe hepatitis. Crigler–Najjar syndrome Dubin–Johnson syndrome |

Albumin is manufactured by the liver at a rate of 9–12 g/day, and catabolised at about the same rate; there is no storage or reserve, and it is not catabolised during starvation. It is a single polypeptide consisting of 585 amino acids, molecular mass approximately 66 248. It is highly soluble, with a strong overall negative charge. Its rate of production is controlled by changes in the colloid osmotic pressure and the osmolarity of the extravascular liver space; production may be increased by a factor of 2 or 3. Synthesis is also increased by insulin, thyroxine and cortisol. It is both an intravascular protein and an extravascular (interstitial) protein. Albumin fulfils a number of functions (Table 6.6).

# 6.16  Protein metabolism – nitrogen metabolism and the urea cycle

The interconversion of amino acids, mainly through transamination reactions, is essential to balancing the requirements for protein synthesis (Figure 6.8), while in protein catabolism the amino nitrogen must be removed in the form of ammonia (ammonium) and converted to urea for excretion by the kidneys (Figure 6.10). Most amino acids are glucogenic, meaning that their carbon skeletons (ketoacid) can be converted to glucose through gluconeogenesis.

**Table 6.5**    Disorders of bilirubin metabolism

**Crigler–Najjar syndrome**, also referred to as congenital non-haemolytic jaundice with glucuronyl transferase deficiency, is a rare, autosomal recessive disorder of bilirubin metabolism. It has been divided into two distinct forms (types I and II), based upon the severity of the disease. The molecular defect in Crigler–Najjar syndrome can be caused by a variety of alterations in the coding sequences of the UDP-glucuronyl transferase (UGT1A1) gene, producing an abnormal protein and resulting in complete loss or very low levels of hepatic UDP-glucuronyl transferase.

**Gilbert's syndrome** results from decreased levels of UDP-glucuronyl transferase. The defect is in the promoter region, the TATAA element, rather than the gene itself. Gilbert's syndrome results in a mild hyperbilirubinaemia, without any clinical sequelae.

**Neonatal hyperbilirubinaemia** is normally a mild unconjugated hyperbilirubinaemia (physiologic jaundice) that affects nearly all newborns and resolves within the first several weeks after birth. It is caused by increased bilirubin production, decreased bilirubin clearance and increased enterohepatic circulation. Bilirubin production in a term newborn is two to three times higher than in adults. This increased production is due to the shorter life span and the greater turnover of neonatal red blood cells. Bilirubin clearance is decreased in newborns, mainly due to the deficiency of the enzyme UDP-glucuronyl transferase; this activity is approximately 1% of that in adults. Furthermore, newborns have fewer intestinal bacteria than adults, resulting in a decreased capacity to reduce bilirubin to urobilinogen, and thus a higher intestinal bilirubin concentration. Additionally, the activity of $\beta$-glucuronidase is also increased, which leads to greater hydrolysis of conjugated to unconjugated bilirubin. The unconjugated bilirubin is re-absorbed from the intestine through the process of enterohepatic circulation, further increasing the bilirubin load in the infant. Higher levels of unconjugated hyperbilirubinaemia are pathologic; this is often the case in premature infants, given the immaturity of the liver and the deficiency in the conjugation system.

**Breast milk jaundice** is a result of an increased enterohepatic circulation. It is thought to result from an unidentified component of human milk that enhances intestinal absorption of bilirubin. One possible mechanism for hyperbilirubinaemia in breast-fed infants compared to formula-fed infants is the increased concentration of $\beta$-glucuronidase in breast milk. $\beta$-glucuronidase deconjugates intestinal bilirubin, increasing its ability to be re-absorbed (i.e. increasing the enterohepatic circulation). Blocking the deconjugation of bilirubin through $\beta$-glucuronidase inhibition may provide a mechanism to reduce intestinal absorption of bilirubin in breast-fed infants; however, this has yet to be proven.

**Kernicterus** (that may result from hyperbilirubinaemia) manifests as various neurological deficits, seizures, abnormal reflexes and eye movements. The blood–brain barrier of the neonate is not fully developed and unconjugated bilirubin can freely pass into the brain interstitium. Some medications, such as the antibiotic co-trimoxazole (a combination of trimethoprim and sulphamethoxazole) may induce this disorder in the infant, either when taken by the mother or if given directly to the infant, due to a displacement of bilirubin from binding sites on serum albumin, thus allowing unconjugated bilirubin to pass across the blood–brain barrier.

**Table 6.6**   Albumin functions

**Binding and transport**. There are four binding sites on the albumin molecule, with varying specificity for different substances (such as fatty acids and bilirubin). Competitive binding of drugs may occur at the same site, with displacement of the originally bound substance.
**Maintenance of colloid osmotic pressure**. Albumin is responsible for 75–80 % of osmotic pressure.
**Free radical scavenging**. Albumin is a major source of sulphydryl groups; 'thiols' scavenge free radicals (nitrogen and oxygen species).
**Platelet function inhibition and antithrombotic effects**. The anticoagulant and antithrombotic effects of albumin are poorly understood, but may be due to the binding of nitric oxide radicals. In diabetes, glycated albumin may increase the incidence of thrombotic events and atherosclerosis.
**Effects on vascular permeability**. In sepsis there is an increased rate of albumin loss into the tissues; this is probably related to an increase in capillary membrane permeability.
**Buffering capacity of blood**. Albumin provides a significant buffering capacity of blood, due mainly to the presence of histidine and the buffering potential of the imidazole side chain.

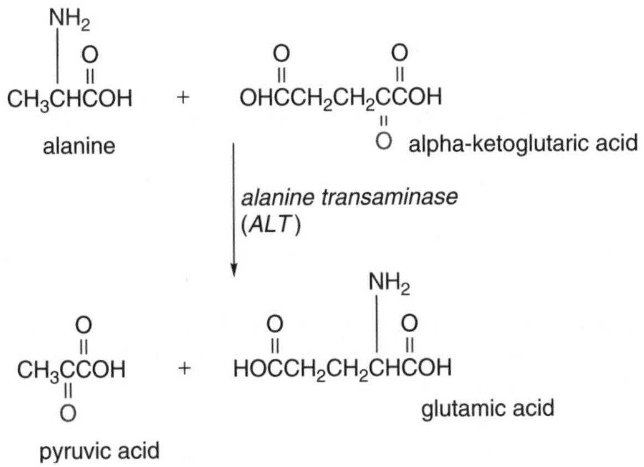

**Figure 6.8**   Interconversion of amino acids occurs through transamination reactions catalysed by aminotransferases.

There are specific aminotransferases for all amino acids, except threonine and lysine, and they are particularly abundant in the liver.

Alanine transaminase (ALT) and aspartate transaminase (AST) are used as clinical markers of tissue damage. ALT has an important function in the delivery of skeletal muscle carbon and nitrogen (in the form of alanine) to the liver. In skeletal muscle, pyruvate is transaminated to alanine, thus affording an additional route of nitrogen transport from muscle to liver. In the liver, ALT transfers the ammonia to $\alpha$-ketoglutarate and regenerates pyruvate. The pyruvate can then be diverted into gluconeogenesis. This process is referred to as the glucose–alanine cycle (Figure 6.9).

In peripheral tissues, two enzymes, namely glutamate dehydrogenase and glutamine synthetase, are important in the removal of reduced nitrogen, and particularly so in the brain, which is highly susceptible to free ammonia.

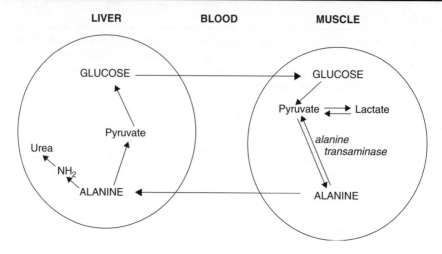

**Figure 6.9** The glucose–alanine cycle.

In the glutamate dehydrogenase reaction:

$$NH_4^+ + alpha\text{-ketoglutarate} + NAD\,(P)\,H + H^+ \rightleftharpoons glutamate + NAD\,(P)^+ + H_2O$$

In the glutamine synthase reaction:

$$glutamate + NH_4^+ + ATP \rightleftharpoons glutamine + ADP + Pi + H^+$$

The glutamine synthase reaction is important in several respects. First, it produces glutamine, one of the 20 major amino acids. Second, in animals, glutamine is the major amino acid found in the circulatory system. Its role is to carry ammonia to and from various tissues, but principally from peripheral tissues to the kidney, where the amide nitrogen is hydrolysed by the enzyme glutaminase (reaction below); this process regenerates glutamate and free ammonium ion, which is excreted in the urine.

$$glutamine + H_2O \rightarrow glutamate + NH_3$$

In this way, ammonia arising in peripheral tissue is carried in a non-ionisable form, with none of the neurotoxic or alkalosis-generating properties of free ammonia.

In the liver, glutamate dehydrogenase converts glutamate to $\alpha$-ketoglutarate and ammonia (the reverse to the reaction occurring in the peripheral tissues), and the ammonia generated enters the urea cycle (Figure 6.10).

$$glutamate + NAD\,(P)^+ + H_2O \rightleftharpoons NH_4^+ + alpha\text{-ketoglutarate} + NAD\,(P)\,H + H^+$$

Liver contains both glutamine synthase and glutaminase, but the enzymes are localised in different cellular locations. This ensures that the liver is neither a net producer nor a net consumer of glutamine; the difference in cellular location of these two enzymes allows the liver to scavenge ammonia that has not been incorporated into urea. The enzymes of the urea cycle are located in the same cells as the glutaminase. The result of this differential distribution

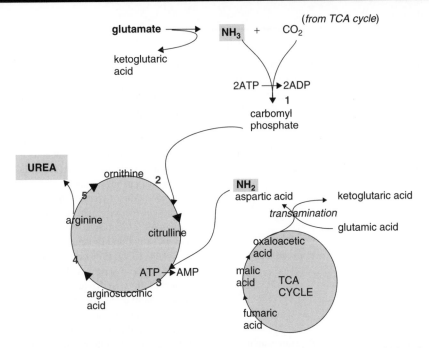

**Figure 6.10**   The urea cycle. The enzymes of the urea cycle include, 1: carbamoyl phosphate synthetase-I, 2: ornithine transcarbamoylase, 3: argininosuccinate synthetase, 4: argininosuccinase, 5: arginase.

is to allow control of ammonia incorporation into either urea or glutamine, the latter leading to excretion of ammonia by the kidney. When acidosis occurs, the body will divert more glutamine from the liver to the kidney. This allows for the conservation of bicarbonate ion, since the incorporation of ammonia into urea requires bicarbonate.

## 6.17   The urea cycle

The urea cycle is responsible for the excretion of some 80% of the body's excreted nitrogen in the form of urea; this is generated in the liver (Figure 6.10).

## 6.18   Regulation of the urea cycle

The urea cycle operates only to eliminate excess nitrogen. On high-protein diets the carbon skeletons of the amino acids (keto acids) are oxidised for energy or stored as fat and glycogen, but the amino nitrogen must be excreted. To facilitate this process, urea-cycle enzymes are closely controlled at the gene level. With long-term changes in the quantity of dietary protein, changes of 20-fold or greater in the concentration of cycle enzymes are observed. Under conditions of starvation, enzyme levels rise as proteins are degraded and amino acid carbon skeletons are used to provide energy, thus increasing the quantity of nitrogen that must be excreted.

**Table 6.7** Urea cycle defects

| UCD | Enzyme deficiency | Symptoms/comments |
|---|---|---|
| Type I hyperammonaemia | Carbamoylphosphate synthetase I (CPS I) | Within 24–72 hours after birth the infant becomes lethargic, needs stimulation to feed. Vomiting, increasing lethargy, hypothermia and hyperventilation. Treatment is with arginine, which activates N-acetylglutamate synthetase. |
| N-acetylglutamate synthetase deficiency | N-acetylglutamate synthetase | Severe to mild hyperammonaemia, associated with deep coma, acidosis, recurrent diarrhoea, ataxia, hypoglycaemia, hyperornithinaemia. Treatment includes administration of carbamoyl glutamate to activate CPS I. |
| Type 2 hyperammonaemia | Ornithine transcarbamoylase | The most commonly occurring, and only X-linked, UCD. Elevated ammonia and amino acids in serum, and increased serum orotic acid. Treatment is a high-carbohydrate, low-protein diet; ammonia detoxification is aided with sodium phenylacetate or sodium benzoate. |
| Classic citrullinaemia | Argininosuccinate synthetase | Episodic hyperammonaemia, vomiting, lethargy, ataxia, seizures, eventual coma. Treatment is with arginine administration to enhance citrulline excretion, also with sodium benzoate for ammonia detoxification. |
| Arginosuccinic aciduria | Argininosuccinate lyase (argininosuccinase) | Episodic symptoms similar to classic citrullinaemia, elevated plasma and cerebral spinal fluid argininosuccinate. Treatment is with arginine and sodium benzoate. |
| Hyperargininaemia | Arginase | A rare UCD, associated with progressive spastic quadriplegia and mental retardation, high ammonia and arginine in cerebral spinal fluid and serum, and high arginine, lysine and ornithine in urine. Treatment includes exclusion of arginine from diet. |

Short-term regulation of the cycle occurs principally at carbamoyl phosphate synthetase-I, which is relatively inactive in the absence of its allosteric activator N-acetylglutamate. The steady-state concentration of N-acetylglutamate is set by the concentration of its components, acetyl-CoA and glutamate, and by arginine, which is a positive allosteric effector of N-acetylglutamate synthetase (glutamate transacylase).

# 6.19   Urea cycle defects

Deficiencies in all enzymes of the urea cycle (UCDs), including N-acetylglutamate synthase, have been identified (Table 6.7), with a common outcome being hyperammonaemia. Blood chemistry will also show elevations in glutamine. In addition to hyperammonaemia, deficiencies will also present with encephalopathy and respiratory alkalosis. In neonates, 24 and 48 hours after birth, there are progressively deteriorating symptoms. Deficiencies in carbamoylphosphate synthetase I (CPS I), ornithine transcarbamoylase, argininosuccinate synthetase and argininosuccinate lyase make up the most common neonatal disorders; deficiency in arginase does not necessarily lead to symptomatic hyperammonaemia. Diagnosis of neonatal UCD, based upon presenting symptoms and observed hyperammonaemia, can often confirm which of the four enzyme deficiencies is the cause.

# 6.20   Neurotoxicity associated with ammonia

Ammonia is highly neurotoxic. Beside its effect on blood pH, ammonia readily traverses the brain–blood barrier; in the brain it is converted to glutamate via glutamate dehydrogenase, depleting the brain of $\alpha$-ketoglutarate. As the $\alpha$-ketoglutarate is depleted, oxaloacetate levels fall and ultimately TCA cycle activity is reduced. In the absence of aerobic oxidative phosphorylation and TCA cycle activity, irreparable cell damage and neural cell death ensue. Additionally, increased glutamate leads to glutamine formation, further reducing ATP levels and depleting glutamate stores, which in neural tissue is both a neurotransmitter and a precursor for the synthesis of $\gamma$-aminobutyrate (GABA), another neurotransmitter. Thus, reductions in brain glutamate affect both energy production and neurotransmission.

# CHAPTER 7
# Alcohol metabolism and cirrhosis

## 7.1  The alcohol dehydrogenase system

Ethanol, $CH_3–CH_2–OH$, is a low-molecular-weight hydrocarbon that is rapidly absorbed across both the gastric mucosa and the small intestines, reaching a blood peak concentration 20–60 minutes after ingestion. Ethanol is a toxin; too high a dose triggers a primary defence mechanism, namely vomiting.

Ethanol absorption is influenced by food intake (which restricts the rate of absorption); the higher the dietary fat content, the slower the time of passage and the longer the process of absorption. 'First-pass' metabolism in the stomach is important. Women absorb and metabolise alcohol differently from men; their lower body water content and a lower activity of alcohol dehydrogenase (ADH) in the stomach result in a more rapid and significant absorption.

## Gastric Effects

At least three different isoenzymes of human stomach ADH exist, plus a more recently recognised class IV, or sigma ADH. A deficiency of sigma ADH has been described in Asians and found to be associated with a lesser first-pass metabolism of ethanol. Chronic ethanol consumption has been shown to produce a significant decrease in levels of gastric ADH, associated with a greater bioavailability (lesser first-pass metabolism) of ethanol. Women have a lower gastric ethanol metabolism than men. Commonly used drugs, such as aspirin, and $H_2$ receptor antagonists, such as cimetidine, have been shown to inhibit gastric

*Essential Biochemistry for Medicine*    Dr Mitchell Fry
© 2010 John Wiley & Sons, Ltd

ADH activity and to enhance the bioavailability of ethanol, with a resulting increase in blood ethanol levels. This 'exaggeration' by certain drugs is particularly important for social drinkers, who commonly take several small drinks, but experience a cumulative effect on blood alcohol levels.

There are three enzyme systems involved in the metabolism of ethanol:

- the ADH system

- the microsomal cytochrome P450 isoenzyme, CYP2E1

- the peroxisome catalase system.

The ADH system catalyses the reactions:

$$CH_3CH_2OH + NAD^+ \rightarrow CH_3CH{=}O + NADH + H^+ \text{ (by alcohol dehydrogenase)}$$

$$CH_3CH{=}O + NAD^+ \rightarrow CH_3COOH + NADH + H^+ \text{ (by acetaldehyde dehydrogenase)}$$

There are two major liver isoenzyme forms of acetaldehyde dehydrogenase, one cytosolic and one mitochondrial. Most Caucasians have both isoenzymes, while approximately 50% of Asians have only the cytosolic isoenzyme. A remarkably higher frequency of acute alcohol intoxication among Asians than among Caucasians could be related to the absence of the mitochondrial isoenzyme. The mitochondrial isoenzyme has a low $K_m$ (high affinity) for acetaldehyde. The mutation in the mitochondrial acetaldehyde dehydrogenase gene, specifically ALDH2, is dominant.

The drug disulfiram (Antabuse) is used in the treatment of chronic alcoholism, producing an acute sensitivity to alcohol. It inhibits the oxidation of acetaldehyde to acetic acid. Metronidazole produces a similar effect, which is why it should never be taken with alcohol.

Acetyl-CoA synthetase catalyses the conversion of acetic acid to acetyl-CoA:

$$ATP + Acetate + CoA \rightarrow AMP + Pyrophosphate + Acetyl\text{-}CoA$$

Acetyl-CoA can enter the TCA cycle or be used in fatty acid synthesis.

Methanol is oxidised to formaldehyde in the liver by ADH, which can lead to blindness or death. An effective treatment to prevent formaldehyde toxicity after methanol ingestion is to administer ethanol. ADH has a higher affinity for ethanol, preventing methanol from binding and acting as a substrate; any remaining methanol will thus have time to be excreted through the kidneys. Remaining formaldehyde will be converted to formic acid and be excreted.

## 7.2 The microsomal cytochrome P450 system

The microsomal cytochrome P450 system consists of a superfamily of haemoproteins that catalyse the oxidative metabolism of a wide variety of exogenous chemicals, including drugs, carcinogens, toxins and endogenous compounds such as steroids, fatty acids and prostaglandins. The principal microsomal enzyme involved in alcohol metabolism is the isoenzyme CYP2E1. This isoenzyme is responsible for about 10–20% ethanol metabolism in a moderate drinker, but represents a major adaptive response to chronic alcohol consumption because of the increased expression of CYP2E1.

### Paracetamol Toxicity

Paracetamol is metabolised primarily in the liver. The main metabolic routes are glucuronidation (about 40%), sulphation (about 20–40%) and N-hydroxylation with glutathione conjugation (less than 15%). This latter route forms a minor yet significant alkylating metabolite known as *N*-acetyl-*p*-benzo-quinone imine (NAPQI). NAPQI is primarily responsible for the toxic effects of paracetamol (an example of toxication). Production of NAPQI is due mainly to two isoenzymes of cytochrome P450: CYP2E1 and CYPIA2. The P450 gene is highly polymorphic however, and individual differences in paracetamol toxicity are believed to be due to a third isoenzyme, CYP2D6.

The paracetamol–alcohol interaction is complex; acute and chronic ethanol intake has opposite effects.

In animals, chronic ethanol causes induction of hepatic microsomal enzymes, and increases paracetamol hepatotoxicity as expected (ethanol primarily induces CYP2E1 and this isoenzyme is important in the oxidative metabolism of paracetamol). This conclusion is not yet fully documented in man.

Acute ethanol inhibits the microsomal oxidation of paracetamol (presumably through competition with the CYP2E1), both in animals and man. This protects against liver damage in animals and there is evidence that it also does so in man.

Alcohol consumption affects the metabolism of a wide variety of other medications.

## 7.3 The peroxisome catalase system

The peroxisome catalase system catalyses the hydrogen peroxide ($H_2O_2$)-dependent oxidation of ethanol to acetaldehyde and water. Normally it contributes little to the oxidation of alcohol because of the limited availability of hydrogen peroxide. However, activation of peroxisomal catalase, by the increased generation of hydrogen peroxide via peroxisomal $\beta$-oxidation, leads to an increased metabolism of alcohol. This state may contribute to an alcohol-related inflammation and necrosis in alcoholic liver disease.

## 7.4 The consequence of alcohol intake

The liver is the only organ that can dispose of significant quantities of alcohol, but its maximum rate of alcohol clearance is fixed. When alcohol is consumed, the liver increases the synthesis of fatty acids (due to increased availability of acetyl-CoA). Fat accumulation has been observed in the liver after just a single bout of heavy drinking, and is the first stage of liver deterioration, interfering with the distribution of nutrients and oxygen to the liver cells. If the condition persists, fibrous scar tissue will result; this is the second stage of liver deterioration, called fibrosis. Fibrosis is reversible, with abstinence from alcohol and good nutrition; the last stage, cirrhosis, is not reversible. The pathological hallmark of cirrhosis is the development of scar tissue that replaces normal parenchyma, blocking the portal flow of blood through the organ and disturbing normal function. Research indicates the pivotal role of stellate cells in the development of cirrhosis (stellate cells normally store vitamin A). Damage to the hepatic parenchyma leads to activation of the stellate cell, which becomes contractile (a myofibroblast), ultimately obstructing blood flow. Additionally it secretes transforming growth factor (TGF) $\beta_1$, which leads to a fibrotic response and proliferation of connective tissue. It also disturbs the balance between matrix metalloproteinases and their naturally occurring inhibitors (tissue inhibitor of metalloproteinases (TIMPs) 1 and 2), leading to matrix breakdown and replacement by connective tissue-secreted matrix. Scar tissue blocks blood flow through the portal vein, producing high blood pressure in that vein (portal hypertension); additionally, scar tissue can block the flow of bile out of the liver.

## 7.5　Short-term metabolic consequences of alcohol intake

Alcohol intake depletes the cellular supply of $NAD^+$ (because of the alcohol and acetaldehyde dehydrogenase reactions) and consequently the $NAD^+$: NADH ratio falls. Reactions that depend on $NAD^+$ will thus be curtailed:

- Oxidation of lactate to pyruvate is inhibited, leading to a reduction in gluconeogenesis and accumulation of lactate (pyruvate + NADH + $H^+$ → lactate + $NAD^+$).

- Hypoglycaemia (low blood glucose) results from the reduction in gluconeogenesis.

- The increased levels of acetyl-CoA are diverted into fatty acid synthesis, and excess NADH promotes the synthesis of glycerol for fat synthesis; this accounts for the fatty liver commonly found in alcoholics. Fatty acid oxidation is also suppressed.

- Pathways that depend upon $NAD^+$, such as the TCA cycle, are reduced; energy production may be compromised.

- Lactate, in high concentration in the blood, competes in the renal mechanism for the excretion of uric acid; if the kidney is unable to fully excrete this end product of nucleic acid metabolism then its accumulation leads to the symptoms of the disease gout.

- High levels of lactate (acidosis) and acetyl-CoA (ketosis) lead to ketoacidosis.

## 7.6　Long-term consequences of chronic alcohol intake

Acetaldehyde may bind to lipids and proteins, principally through the epsilon amino group of lysine, to form unstable acetaldehyde-Schiff base adducts ('adduct', in chemistry: a chemical compound that forms from the addition of two or more substances). Although such adducts are unstable and the reaction is readily reversed, further reduction produces a stable Schiff base that is not easily reversed (Figure 7.1).

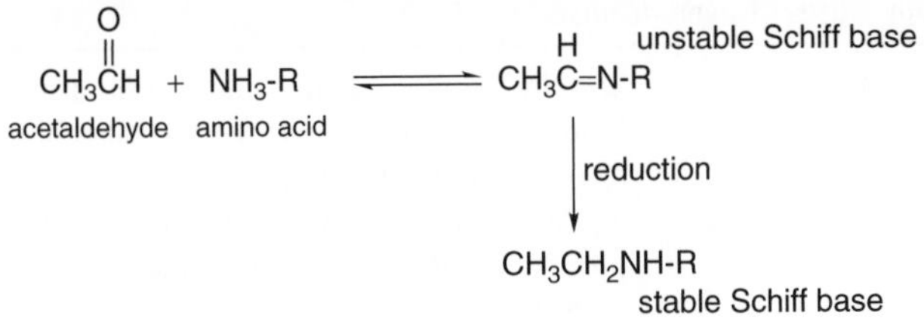

$$\underset{\text{acetaldehyde}}{CH_3\overset{\overset{\textstyle O}{\|}}{C}H} + \underset{\text{amino acid}}{NH_3\text{-}R} \rightleftharpoons CH_3\overset{\overset{\textstyle H}{}}{C}{=}N\text{-}R \quad \text{unstable Schiff base}$$

reduction

$$CH_3CH_2NH\text{-}R$$
stable Schiff base

**Figure 7.1** Schiff base formation between acetaldehyde and an amino acid.

Formation of protein adducts with reactive aldehydic products may provide a general basis for observed pathogenesis.

Formation of hybrid adducts with acetaldehyde and malondialdehyde (MAA adducts) has been shown to act in a synergistic manner and may be involved in the stabilisation of protein adducts *in vivo*. Malondialdehyde ($HOCH{=}CH\text{-}CHO$) is a highly reactive dialdehyde originating from the non-enzymatic lipid peroxidation of a variety of unsaturated fatty acids.

Cell culture studies have shown that aldehydic products derived from ethanol metabolism and lipid peroxidation can increase collagen mRNA levels and enhance the expression of connective tissue proteins. Acetaldehyde is able to increase the production of several extracellular matrix components. Studies also show that hepatic stellate cells, which are the primary source of extracellular matrix, become readily activated under conditions involving enhanced oxidative stress and lipid peroxidation.

Aldehyde-protein adducts and hydroxyl radicals also stimulate immunological responses directed against the specific modifications of proteins. High antibody titres have been observed from patients with severe alcoholic liver disease, particularly IgA and IgG autoantibodies. Such antibodies have considerable specificity towards aldehyde-lysine residues. Alcohol consumption markedly increases the hepatic output of very low-density lipoprotein (VLDL), but decreases the low-density lipoprotein (LDL) levels and apolipoprotein B. Ethylation of apo-B-lysine renders LDL immunogenic and accelerates its clearance. Alcoholics have been shown to develop acetaldehyde adducts in apo-B-containing lipoproteins, particularly VLDL.

# Alcohol and Barbiturates

Alcohol and barbiturates both interact with the $\gamma$-aminobutyrate (GABA)-activated chloride channel. Activation of the chloride channel inhibits neuronal firing, which explains the depressant effects of both these compounds. This drug–alcohol combination is potentially dangerous and normal prescription doses of barbiturates can have lethal consequences in the presence of ethanol. Ethanol inhibits the metabolism of barbiturates, thereby prolonging their effective time in the body, causing increased depression of the CNS. A chronic alcoholic, when sober, has trouble falling asleep even after taking several sleeping pills, because the liver has developed an increased capacity to metabolise barbiturates. In frustration, more pills are consumed, followed by alcohol. Sleep results, but may be followed by respiratory depression and death, because the alcoholic, although less sensitive to barbiturates when sober, remains sensitive to the synergistic effects of alcohol.

**Table 7.1** Some causes of liver cirrhosis

| | |
|---|---|
| Alcoholic liver disease | Develops in 15% of individuals who drink heavily for more than a decade. Patients may also have concurrent alcoholic hepatitis with fever, hepatomegaly, jaundice and anorexia. AST and ALT are both elevated but less than 300 IU/l, with a AST : ALT ratio >2.0, a value rarely seen in other liver diseases. |
| Chronic hepatitis C | Viral infection causes inflammation and low-grade damage that can lead to cirrhosis. Can be diagnosed with serologic assays for hepatitis C antibody or viral RNA. |
| Chronic hepatitis B | Most common worldwide, but less common in the Western world. Inflammation and low-grade damage can lead to cirrhosis. |
| Non-alcohol steatohepatitis | Fat build-up in the liver eventually causes scar tissue; associated with diabetes, protein malnutrition, obesity and coronary artery disease. Treatment with corticosteroid medications. Biopsy is needed for full diagnosis. |
| Primary biliary cirrhosis | May be asymptomatic. Prominent rise in alkaline phosphatase, cholesterol and bilirubin. More common in women. Diagnosis through detection of antimitochondrial antibodies, with biopsy. |
| Primary sclerosing cholangitis | A progressive cholestatic disorder. Strong association with inflammatory bowel disease. |
| Autoimmune hepatitis | Immunologic damage to the liver causing inflammation, scarring and eventually cirrhosis. Elevations in serum globulins, especially gamma globulins. |
| Hereditary haemochromatosis | Usually with family history of cirrhosis, skin hyperpigmentation, diabetes mellitus, pseudo-gout and/or cardiomyopathy, all due to iron overload. |
| Wilson's disease | Autosommal recessive, low serum ceruloplasmin and increased hepatic copper content. |

## 7.7   Cirrhosis of the liver

Cirrhosis of the liver is the third most common cause of death, after heart disorders and cancer, among the 45–65 age group. Cirrhosis has many possible causes (Table 7.1); sometimes more than one cause is present in the same patient. In the Western world, chronic alcoholism and hepatitis C are the most common causes.

## 7.8   Complications of cirrhosis

Complications of cirrhosis can include:

• High blood pressure in the portal veins (due to scarring), causing dilated, twisted veins to form at the lower end of the oesophagus (oesophageal varices), in the stomach (gastric varices) or in the rectum (rectal varices). Vomiting of large amounts of blood may be indicative of the rupture of oesophageal or gastric varices.

- High blood pressure in the portal vein plus impaired liver function, perhaps leading to fluid accumulation in the abdomen. Ascites, also known as peritoneal cavity fluid, is an accumulation of fluid in the peritoneal cavity.

- Poor vitamin D absorption, due to impaired bile excretion, leading to the development of osteoporosis. Poor vitamin K absorption leads to a tendency to bleed easily (lack of clotting factors); an enlarged spleen will reduce platelet numbers in the blood, exasperating this tendency. Bleeding into the gastrointestinal tract will result in anaemia.

- Liver cancer (hepatocellular carcinoma or hepatoma), particularly where the cause is chronic hepatitis B or C, or alcoholism.

# CHAPTER 8

# Protein structures

## 8.1  Protein primary structure

All proteins are polymers of amino acids. The polymers, or polypeptides, consist of a sequence of up to 20 different L-$\alpha$-amino acids (residues). For chains under 40 residues the term peptide is frequently used instead of protein. The term protein is generally used to refer to the complete biological molecule in a stable conformation.

The amino acid sequence in the polypeptide chain is referred to as its primary structure (Table 8.1).

A protein consisting of 100 amino acids could have as many as $20^{100}$ different linear sequences, producing $1.268 \times 10^{130}$ different proteins.

## 8.2  Peptide bonds

Amino acids are combined (linked together) through peptide bonds (-C-N-) (Figure 8.1); the peptide bond formed is planar (flat), due to the delocalisation of electrons that form the partial double bond, restricting rotation about the bond. The rigid peptide dihedral angle, $\omega$ (the bond between C and N), is always close to 180°. The dihedral angles phi $\phi$ (the bond between N and C$\alpha$) and psi $\psi$ (the bond between C$\alpha$ and C) can only have a number of possible values, and so effectively control the protein's three-dimensional structure.

## 8.3  Protein interactions

To be able to perform their biological function, proteins fold into one or more specific spatial conformations, 'driven' by a number of non-covalent interactions, including:

- The hydrogen bond, a relatively strong, electrostatic and directional bond type, that is very dependent upon distance. A hydrogen atom, attached to a relatively electronegative atom, is

**Table 8.1** Amino acids

| Residue | Three-letter code | Single-letter code | $M$r | Side group pK | Charged (C), polar (P), hydrophobic (H) or neutral (N) |
|---|---|---|---|---|---|
| Alanine | ALA | A | 71 | – | H |
| Arginine | ARG | R | 157 | 12.5 | C+ |
| Asparagine | ASN | N | 114 | – | P |
| Aspartate | ASP | D | 114 | 3.9 | C- |
| Cysteine | CYS | C | 103 | – | P |
| Glutamate | GLU | E | 128 | 4.3 | C- |
| Glutamine | GLN | Q | 128 | – | P |
| Glycine | GLY | G | 57 | – | N |
| Histidine | HIS | H | 137 | 6.0 | P, C+ |
| Isoleucine | ILE | I | 113 | – | H |
| Leucine | LEU | L | 113 | – | H |
| Lysine | LYS | K | 129 | 10.5 | C+ |
| Methionine | MET | M | 131 | – | H |
| Phenylalanine | PHE | F | 147 | – | H |
| Proline | PRO | P | 97 | – | H |
| Serine | SER | S | 87 | – | P |
| Threonine | THR | T | 101 | – | P |
| Tyrosine | TRP | Y | 163 | 10.1 | P |
| Valine | VAL | V | 99 | – | H |

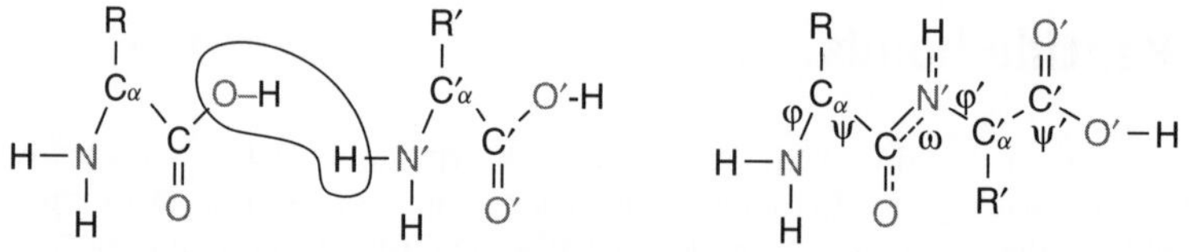

**Figure 8.1** Formation and characteristics of the peptide bond.

a hydrogen bond donor, whereas an electronegative atom such as oxygen or nitrogen is a hydrogen bond acceptor, for example $-O-H^{\delta+}$ ------- $^{\delta-}N-$.

- The charge–charge interaction, consisting of non-directional electrostatic attractions, or repulsions, between charged groups, for example carboxyl $COO^-$, amino $NH_3^+$, phosphate $PO_4^{2-}$.

- The van der Waals force, a very short-range attractive or repulsive force, which can be permanent or transient.

- The hydrophobic group on a protein, for example methyl $(CH_3)$. These 'pack' together to become sequestered away from the polar water solvent.

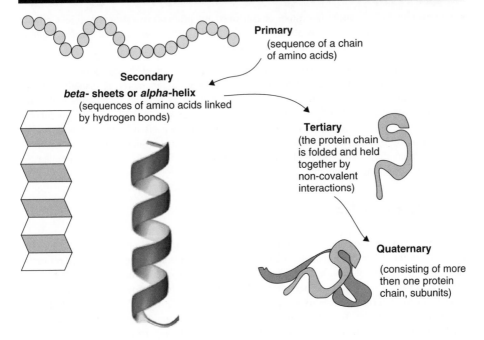

**Primary**
(sequence of a chain
of amino acids)

**Secondary**
*beta*- sheets or *alpha*-helix
(sequences of amino acids linked
by hydrogen bonds)

**Tertiary**
(the protein chain
is folded and held
together by
non-covalent
interactions)

**Quaternary**
(consisting of more
then one protein
chain, subunits)

**Figure 8.2**  Levels of protein structure.

## 8.4  Levels of protein structure

Proteins may adopt different levels of structure (Figure 8.2):

• **Primary**. This structure refers to the amino acid sequence of the peptide chains.

• **Secondary**. This structure refers to the highly regular sub-structures (*alpha helix* and strands of *beta sheets*), which are locally defined.

• **Tertiary**. This structure refers to the three-dimensional structure of a single protein molecule, a spatial arrangement of the secondary structures. It also describes the completely folded and compacted polypeptide chain.

• **Quaternary**. This structure refers to a complex of several protein molecules or polypeptide chains, usually called protein subunits in this context, which function as part of the larger assembly or protein complex.

## 8.5  Types of protein structure

Generally proteins fold into one of three broad structural classes: globular, fibrous or membrane proteins.

• Globular proteins are compactly folded and coiled; almost all are water-soluble and many are enzymes.

- Fibrous proteins are more filamentous or elongated and often serve a structural role. Collagen and elastin are critical components of connective tissue such as cartilage; keratin is found in hard or filamentous structures such as hair and nails.

- Membrane proteins often serve as receptors or provide channels for polar or charged molecules. Proteins that associate with the surfaces of membranes, usually through non-covalent charge–charge interactions, are referred to as peripheral; proteins within the hydrophobic interior of membranes (whose surface is generally made up of hydrophobic amino acids) are referred to as integral. Integral proteins are often transmembrane; that is, they span the membrane; examples are receptors and channels. Such proteins are generally amphipathic; they have both hydrophilic and hydrophobic regions that help orientate the molecule across the membrane.

## 8.6    The $\alpha$-helix

The $\alpha$-helix is a common secondary structure encountered in proteins of the globular class. The formation of the $\alpha$-helix is spontaneous and is stabilised by H-bonding between amide nitrogens and carbonyl carbons of peptide bonds spaced four residues apart. This orientation of H-bonding produces a helical coiling of the peptide backbone such that the R-groups of individual amino acids lie on the exterior of the helix and perpendicular to its axis. Amino acids such as A, D, E, I, L and M favour the formation of $\alpha$-helices, whereas G (glycine) and P (proline) favour disruption of the helix. This is particularly true for proline since it is a pyrrolidine-based amino acid (HN=); it significantly restricts movement about the peptide bond in which it is present, thereby interfering with extension of the $\alpha$-helix. The disruption of the helix is important as it introduces additional folding of the polypeptide backbone to allow the formation of globular proteins.

## 8.7    The $\beta$-sheet

The $\beta$-sheet is composed of two or more different regions of stretches of at least 5–10 amino acids (whereas the $\alpha$-helix is composed of a single linear array of helically disposed amino acids). The folding and alignment of stretches of the polypeptide backbone alongside one another to form $\beta$-sheets is stabilised by H-bonding between amide nitrogens and carbonyl carbons. However, the H-bonding residues are present in adjacently opposed stretches of the polypeptide backbone, as opposed to a linearly contiguous region of the backbone in the $\alpha$-helix. $\beta$-sheets are said to be pleated. This is due to positioning of the $\alpha$-carbons of the peptide bond, which alternates above and below the plane of the sheet. $\beta$-sheets are either parallel or antiparallel. In parallel sheets adjacent peptide chains proceed in the same direction (i.e. the direction of N-terminal to C-terminal ends is the same), whereas in antiparallel sheets adjacent chains are aligned in opposite directions. $\beta$-sheets are often depicted as ribbons in protein formats.

## 8.8    Protein folding

Protein folding is the physical process by which a polypeptide folds into its characteristic and functional three-dimensional structure.

In the compact fold (Figure 8.3) the hydrophobic amino acids (shown as black spheres in the figure) are shielded from the solvent (water).

The amino acid sequence of a protein predisposes it towards its native conformation(s). It will fold spontaneously during or after synthesis. The mechanism of folding will depend on the

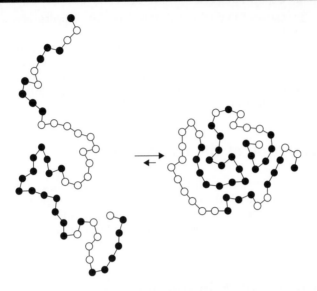

**Figure 8.3**  Protein folding.

characteristics of the cytoplasm, including the nature of the primary solvent (water or lipid), macromolecular crowding, ionic strength, temperature and presence of molecular chaperones.

Most (soluble) folded proteins have a hydrophobic core in which side-chain packing stabilises the folded state, and charged or polar side chains are placed on the solvent-exposed surface, where they interact with surrounding water molecules. It is generally accepted that minimising the number of hydrophobic side chains exposed to water is the principal driving force behind the folding process.

The tertiary structure is fixed only when the parts of a protein are 'locked' into place by structurally specific interactions, involving charge–charge interactions (salt bridges), hydrogen bonds and the tight packing of side chains. The tertiary structure of extracellular proteins can also be stabilised by disulphide bonds; disulfide bonds are extremely rare in cytoplasmic proteins since the cytoplasm is generally a reducing environment. The folded functional state of a protein is known as the native state.

When proteins fold into their tertiary structures, there are often subdivisions within the protein, designated as domains, which are characterised by similar features or motifs. A protein domain is a part of the protein sequence and structure that can evolve, function and exist independently of the rest of the protein chain. Many proteins consist of several structural domains. One domain may appear in a variety of evolutionarily related proteins. Domains vary in length from about 25 up to 500 amino acids. The shortest domains, such as 'zinc fingers', are stabilised by metal ions or disulfide bridges. Domains often form functional units, such as the calcium-binding EF hand domain of calmodulin. As they are self-stable, domains can be 'swapped' by genetic engineering between one protein and another, to make chimera proteins.

## 8.9   Carbohydrates and lipid association with protein

Carbohydrates and lipids are often covalently associated with proteins. These modifications/additions to the protein occur following translation; they are termed post-translational modifications. Proteins covalently associated with lipids or carbohydrates are termed lipoproteins or glycoproteins respectively. A major function of lipoproteins is to aid in the storage and transport of lipid and cholesterol.

Glycoproteins are either N-linked or O-linked, referring to the site of covalent attachment of the sugar moieties. The variability in the composition of the carbohydrate portions of many glycoproteins (and glycolipids) of erythrocytes determines blood group specificities. There are at least 100 blood group determinants, most of which are due to carbohydrate differences, the most common being A, B and O.

## 8.10    Disruption of the native state

Temperatures above (and sometimes below) the normal range will cause thermally unstable proteins to unfold or 'denature'. High concentrations of solutes, extremes of pH, mechanical forces and the presence of chemical denaturants can do the same. A fully denatured protein lacks both tertiary and secondary structure, and exists as a 'random coil'. In most cases denaturation of proteins is irreversible.

## 8.11    Incorrect protein folding and neurodegenerative disease

Failure of a protein to fold into the intended shape usually produces an inactive protein. Several neurodegenerative diseases are believed to result from the accumulation of misfolded (incorrectly folded) proteins. Aggregated proteins are associated with prion-related illnesses such as Creutzfeldt–Jakob disease, bovine spongiform encephalopathy (mad cow disease) and amyloid-related illnesses such as Alzheimer's disease and familial amyloid cardiomyopathy or polyneuropathy, as well as intracytoplasmic aggregation diseases such as Huntington's and Parkinson's disease. These age-onset degenerative diseases are associated with the multimerisation of misfolded proteins into insoluble, extracellular aggregates and/or intracellular inclusions, including cross-$\beta$-sheet myeloid fibrils. It is not clear whether the aggregates are the cause or merely a reflection of the loss of protein homeostasis, the balance between synthesis, folding, aggregation and protein turnover. Misfolding and excessive degradation lead to a number of proteopathy diseases, such as antitrypsin-associated emphysaema, cystic fibrosis and the lysosomal storage diseases, where loss of function is the origin of the disorder.

## 8.12    The study of proteins

The study of proteins and protein purification (Table 8.2) usually begins with tissue homogenisation or cell lysis, to disrupt the cell's membrane and give a crude lysate. The resulting mixture can be ultracentrifuged to fractionate the various cellular components into soluble proteins, membrane lipids and proteins, cellular organelles and nucleic acids. Precipitation (salting out) is often employed to concentrate the proteins from this lysate. Various types of chromatography can be used to isolate the protein(s) on the basis of their size, charge and binding affinity. The level of purification can be monitored using various types of gel electrophoresis, by spectroscopy (if the protein has distinguishable spectroscopic features) or by enzyme assays (if the protein is an enzyme). Additionally, proteins can be isolated according to their charge, using electrofocusing.

**Table 8.2** Purification of proteins

| Fractionation | |
|---|---|
| Crude homogenates | Mortar and pestle, blenders and homogenisers, grinders and sonicators |
| Precipitation | Ammonium sulphate/acetate, 'salting out' |
| Differential centrifugation | Sub-cellular fractionation in a centrifuge, up to 250 000g, supernatant and pellet fractions |
| Column chromatography | |
| Gel exclusion chromatography | Separates by molecular size |
| Ion exchange chromatography | Separates by charge, binding proteins to an anionic or cationic resin |
| Affinity chromatography | Separates by binding a specific protein to a resin with an attached affinity ligand |
| Analyse – identify | |
| Polyacrylamide gel electrophoresis (PAGE) | Proteins migrate in an electric field as a function of their charge |
| Sodium dodecyl sulphate polyacrylamide gel electrophoresis (SDS-PAGE) | As above, but *all* proteins are made −ve by SDS and so separate according to their size |
| Isoelectric focusing | Proteins migrate through gels to their isoelectric point (pH where they have no net charge) |
| 2D electrophoresis | Combines isoelectric focusing and SDS-electrophoresis; samples are run in two directions to increase resolution |

# 8.13  Defects in protein structure and function

Defects in protein structure and function are responsible for numerous disorders, for example:

- **Sickle cell anaemia.** Due to a single nucleotide substitution (adenine to thymine) in the codon for amino acid 6 of globin; this change converts a glutamic acid codon (GAG) to a valine codon (GTG). This form of haemoglobin is referred to as HbS; normal adult haemoglobin is referred to as HbA. Substitution of a hydrophobic (valine) for a polar residue (glutamic acid) results in haemoglobin tetramers that aggregate upon deoxygenation in the tissues. Aggregation results in deformation of the red blood cell into a sickle-like shape, making it relatively inflexible and unable to easily traverse the capillary beds. Sickle cell anaemia is an autosomal recessive disorder. Individuals who are heterozygous are said to have a sickle cell trait. Although heterozygous individuals are clinically normal, their red blood cells can 'sickle' under very low oxygen pressure, for example at high altitudes. Heterozygous individuals exhibit phenotypic dominance, yet are recessive genotypically.

- **$\alpha$ and $\beta$-thalassaemias.** The result of quantitative abnormalities in haemoglobin synthesis, in either the $\alpha$-globin or $\beta$-globin chains. A large number of mutations have been identified leading to decreased ($\alpha^+\beta^+$) or absent ($\alpha^\circ\beta^\circ$) production of globin chains. The primary cause of the $\alpha$-thalassemias is gene deletion, but for the $\beta$-thalassemias the mutations are more subtle, with some 170 different ones identified. These mutations include gene deletions, point mutations in the promoter region, mutations in the coding region leading to defective initiation, insertions and deletions resulting in frameshifts and nonsense mutations, point mutations in the polyadenylation signal and an array of mutations leading to splicing abnormalities.

- **Collagen-related disorders**. Collagens are the most abundant proteins in the body. Mutations that affect the structure and function of type I collagens result in numerous disease states.
  - Ehlers–Danlos syndrome is actually the name associated with at least ten distinct disorders that are biochemically and clinically distinct, yet all manifest structural weakness in connective tissue as a result of defective collagen structure.
  - Osteogenesis imperfecta also encompasses more than one disorder. At least four biochemically and clinically distinguishable maladies have been identified as osteogenesis imperfecta, all of which are characterised by multiple fractures and resultant bone deformities.
  - Marfan syndrome manifests itself as a disorder of the connective tissue and was originally believed to be the result of abnormal collagens. However, recent evidence has shown that Marfan syndrome results from mutations in the extracellular protein, fibrillin, which is an integral constituent of the non-collagenous microfibrils of the extracellular matrix.

- **Blood-clotting factors**. These may be defective or deficient. For example:
  - Afibrinogenaemia is a complete loss of fibrinogen, factor I.
  - Dysfibrinogenaemia is due to a dysfunctional fibrinogen, factor I.
  - Factor V deficiency is caused by production of a labile protein.
  - Haemophilia A is due to a factor VIII deficiency.
  - Haemophilia B is due to a factor IX deficiency.

# 8.14   Glycolipid degradation

The inability to properly degrade membrane-associated gangliosides (glycolipids) can lead to severe psycho-motor developmental disorders, for example:

- **Tay–Sachs disease**. A member of a family of disorders identified as the $G_{M2}$ gangliosidoses. As neural cell membranes are enriched in $G_{M2}$ gangliosides, the inability to degrade this class of sphingolipid results in neural cell death. In addition to Tay–Sachs disease the family includes the Sandhoff diseases and the $G_{M2}$ activator deficiencies. Tay–Sachs disease results from defects in the HEXA gene encoding the $\alpha$-subunit of $\beta$-hexosaminidase.

- **Gauchers disease**. This belongs to a family of disorders identified as lysosomal storage diseases; it is characterised by the lysosomal accumulation of glucosylceramide (glucocerebroside), a normal intermediate in the catabolism of globosides and gangliosides. Gauchers disease results from defects in the gene encoding the lysosomal hydrolase, acid $\beta$-glucosidase (glucocerebrosidase); this gene is located on chromosome 1q21 and spans 7 kb encompassing 11 exons.

- **Niemann–Pick diseases**. These are also classified as lysosomal storage diseases. There are two main distinct sub-families: type A (NP-A) and type B (NP-B), both caused by defects in the acid sphingomyelinase gene. Further type C diseases are caused by defects in a gene involved in LDL–cholesterol homeostasis, identified as the NPC1 gene.

Glycosylation disorders represent a constellation of diseases that result from defects in the synthesis of carbohydrate structures (glycans) and in the attachment of glycans to other compounds. These defective processes involve the $N$-linked and $O$-linked glycosylation pathways, biosynthesis of proteoglycans, as well as lipid glycosylation pathways.

## 8.15    Protein receptor defects

Receptor defects, such as several forms of familial hypercholesterolaemia, are the result of genetic defects in the gene encoding the receptor for LDL. These defects result in the synthesis of abnormal LDL receptors that are incapable of binding to LDLs, or that bind LDLs such that the receptor/LDL complexes are not properly internalised and degraded. The outcome is an elevation in serum cholesterol levels and increased propensity toward the development of atherosclerosis.

## 8.16    Transformation and carcinogenesis

Transformation and carcinogenesis can result from a disruption in protein structure caused by gene mutation. Such genes are termed proto-oncogenes. For some of these proteins, all that is required to convert them to the oncogenic form is a single amino acid substitution. The protein product of the *RAS* gene is observed to sustain single amino acid substitutions at positions 12 or 61 with high frequency in colon carcinomas. Mutations in *RAS* are the most frequently observed genetic alterations in colon cancer.

# CHAPTER 9

# Enzymes and diagnosis

## 9.1 Enzyme nomenclature

Enzymes are biological catalysts, all of which are proteins, except for a class of RNA-modifying catalysts known as ribozymes; ribozymes are molecules of ribonucleic acid that catalyse reactions on the phosphodiester bond of other RNAs. The International Union of Biochemistry and Molecular Biology have developed a nomenclature for enzymes, the EC numbers; each enzyme is described by a sequence of one of four numbers preceded by 'EC'. The first number broadly classifies the enzyme based on its mechanism.

The top-level classification is:

- **EC 1**. *Oxidoreductases* catalyse oxidation/reduction reactions.

- **EC 2**. *Transferases* transfer a functional group (e.g. a methyl or phosphate group).

- **EC 3**. *Hydrolases* catalyse the hydrolysis of various bonds.

- **EC 4**. *Lyases* cleave various bonds by means other than hydrolysis and oxidation.

- **EC 5**. *Isomerases* catalyse isomerisation changes within a single molecule.

- **EC 6**. *Ligases* join two molecules with covalent bonds.

(The full nomenclature can be browsed at http://www.chem.qmul.ac.uk/iubmb/enzyme/.)

Like all proteins, enzymes are made as linear chains of amino acids that fold to produce a three-dimensional product. Each unique amino acid sequence produces a specific structure, which has unique properties. Enzymes can be denatured – that is, unfolded and inactivated – by heating or by chemical denaturants, which disrupt the three-dimensional structure of the protein; denaturation may be reversible or irreversible.

*Essential Biochemistry for Medicine*   Dr Mitchell Fry
© 2010 John Wiley & Sons, Ltd

## 9.2  Catalytic mechanism

Enzymes contain an active (catalytic) site, which three-dimensionally accommodates a particular substrate molecule and determines the catalytic mechanism.

The enzyme (E) binds a substrate (S) and produces a product (P) via a transition complex enzyme–substrate (ES).

$$E + S \rightleftharpoons ES \longrightarrow E + P$$

catalytic step

substrate binding

The favoured model of enzyme–substrate interaction is referred to as the induced fit model (Figure 9.1). An initial interaction between enzyme and substrate induces a conformational change in the protein that strengthens further binding and brings the catalytic site close to substrate bonds to be altered, generating transition-state complexes and reaction products.

## 9.3  Lowering the activation energy

It is the ability of the enzyme to make and 'stabilise' a transition-state complex that underlines its role as a catalyst. Enzymes and other catalysts accelerate reactions by lowering the energy of the transition state (Figure 9.2).

Enzymes can also couple two or more reactions, so that an energetically favourable reaction can be used to 'drive' an energetically unfavourable one. For example, the favourable hydrolysis of ATP is often used to drive other unfavourable chemical reactions.

Enzymes are highly specific for the type of reaction they catalyse and are generally specific for their substrate; they are also specific for a particular steric configuration (optical isomer) of a substrate. Enzymes that act on D sugars (or L amino acids) will not act on the corresponding

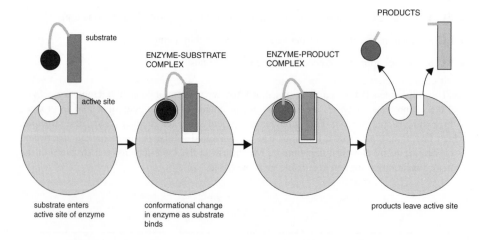

**Figure 9.1**  The induced-fit model of enzyme–substrate interaction.

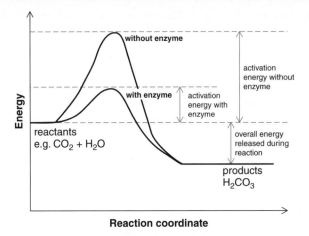

**Figure 9.2** Enzymes lower the activation energy.

L sugar (or D amino acid). Enzymes known as racemases are an exception; indeed the role of racemases is to convert D isomers to L isomers, and vice versa.

## 9.4 Reactions, rates and equilibria

The rate of a chemical reaction is described by the number of molecules of reactant(s) that are converted into product(s) in a specified time. Reaction rate is always dependent on the concentration of the chemicals involved in the process and on rate constants that are character-istic of the reaction.

For example, in the reaction in which A is converted to B (A → B), the rate is expressed algebraically as either a decrease in the concentration of reactant A (the negative sign signifies a decrease), $-[A] = k[B]$, or an increase in the concentration of product B, $[B] = k[A]$. Brackets denote concentration in molarity, $k$ is the rate constant. The rate constant for the forward reaction is defined as $k_{+1}$, and for the reverse reaction as $k_{-1}$.

Rate constants are simply proportionality constants that provide a quantitative connection between chemical concentrations and reaction rates. Each chemical reaction has a characteristic rate constant; these in turn directly relate to the equilibrium constant for that reaction.

At equilibrium, the rate of the forward reaction is equal to the rate of the reverse reaction:

$$K_{eq} = \frac{K_{+1}}{K_{-1}} = \frac{[B]}{[A]}$$

Therefore the equilibrium constant for a chemical reaction is not only equal to the equilibrium ratio of product and reactant concentrations, but is also equal to the ratio of the characteristic rate constants of the reaction.

Catalysts speed up both the forward and reverse reactions proportionately; although the magnitude of the rate constants of the forward and reverse reactions is increased, the ratio of the rate constants remains the same in the presence or absence of enzyme. It is apparent therefore that enzymes and other catalysts have no effect on the equilibrium constant of the reactions they catalyse.

When enzyme and substrate meet, an ES complex is formed; this passes to a transition state (ES*), then to an enzyme product complex (EP), and finally dissociates to yield product and free enzyme.

$$E + S \leftrightarrow ES \leftrightarrow ES^* \leftrightarrow EP \leftrightarrow E + P$$

The kinetics of such reactions was first characterised by Michaelis and Menten. The Michaelis–Menten equation is a quantitative description of the relationship between the rate of an enzyme-catalysed reaction [$v_1$], the concentration of substrate [S], the maximum reaction rate ($V_{max}$) and the Michaelis–Menten constant ($K_m$).

$$V_1 = \frac{V_{max}\,[S]}{\{K_m + [S]\}}$$

## 9.5 Michaelis–menten kinetics

The Michaelis–Menten equation has the same form as the equation for a rectangular hyperbola; graphical analysis of reaction rate (v) versus substrate concentration [S] produces a hyperbolic rate plot (Figure 9.3).

If the Michaelis–Menten plot is extrapolated to infinitely high substrate concentrations, the extrapolated rate is equal to $V_{max}$.

## 9.6 Lineweaver–burk

To avoid dealing with curvilinear plots of enzyme-catalysed reactions, Lineweaver and Burk introduced an analysis of enzyme kinetics based on the following rearrangement of the Michaelis–Menten equation:

$$\frac{1}{V} = \left[ \frac{K_m\,(1)}{V_{max}\,[S]} + \frac{1}{V_{max}} \right]$$

A Lineweaver–Burk plot provides a graphical means of obtaining accurate values of $V_{max}$ and $K_m$. Plots of 1/v versus 1/[S] yield straight lines with a slope of $K_m/V_{max}$, and an intercept on the ordinate at $1/V_{max}$ and on the abscissa of $1/K_m$ (Figure 9.4).

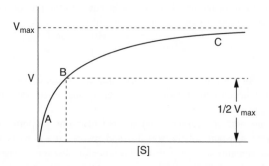

**Figure 9.3**   Plot of substrate concentration versus reaction velocity.

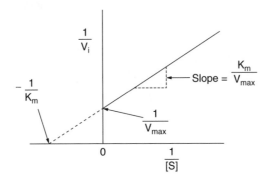

**Figure 9.4**  A Lineweaver–Burk plot.

The $K_m$ of an enzyme is a measure of its affinity for the substrate; the higher the $K_m$ the lower the affinity. Turnover number, related to $V_{max}$, is defined as the maximum number of moles of substrate that can be converted to product per mole of catalytic site per second.

---

Enzyme activity is frequently expressed as the amount of substrate transformed (or product formed) per minute, under standard conditions. A unit (U) of enzyme activity is equivalent to the transformation of 1 μmol of the substrate per minute. This unit was replaced in 1978 by the katal (Kat) (1 U corresponds to 16.67 nKat).

---

## 9.7  Isoenzymes

'Multiple enzyme forms' of one enzyme are known as isoenzymes (or isozymes). Genetically determined differences in primary structure are the basis for the multiplicity in those groups classed as isoenzymes. Isoenzymes of one enzyme group are often expressed to differing extents in different tissues.

Lactate dehydrogenase (LDH) is a tetrameric enzyme composed of all possible arrangements of two different protein subunits; the subunits are known as H (for heart) and M (for skeletal muscle). These subunits come together in various combinations, leading to five distinct isoforms. The all-H isoform is characteristic of that from heart tissue, and the all-M isoform is typically found in skeletal muscle and liver. Isoenzymes all catalyse the same chemical reaction, but with different degrees of efficiency. The detection of specific LDH isoenzymes in the blood is highly diagnostic of tissue damage such as that which occurs during cardiac infarct.

Different isoenzymes often have different $K_m$ values for the same substrate. An often quoted example of this is hexokinase and glucokinase. Both enzymes convert glucose to glucose-6-phosphate, the first reaction in glycolysis, but respond to quite different concentrations of glucose. Glucokinase has a high $K_m$ (low affinity) for glucose, about $2 \times 10^{-2}$ M, and operates within the liver. Since the liver monitors blood glucose, and can both import and export glucose to maintain the blood concentration at about $5 \times 10^{-3}$ M, it is important that the glucokinase activity can change in response to blood glucose levels; an increase or decrease in blood

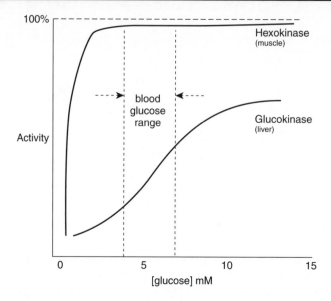

**Figure 9.5**    Response of hexokinase and glucokinase to glucose concentration.

glucose will have a significant effect on its activity. On the other hand, hexokinase has a low $K_m$ (high affinity) for glucose, about $5 \times 10^{-5}$ M. It would be fully saturated at a blood glucose concentration of $5 \times 10^{-3}$ M (5 mM) and its activity would change little with changes in glucose concentration at this level. Hexokinase is located in muscle, where it is responsible for initiating glycolysis to provide energy for muscle contraction; muscle only imports and uses glucose, it does not export it (Figure 9.5).

## 9.8   Enzyme inhibitors

Enzyme inhibitors are broadly classified as irreversible and reversible. Inhibitors of the first class usually cause an inactivating, covalent modification of enzyme structure. Cyanide is a classic example of an irreversible enzyme inhibitor. The kinetic effect of irreversible inhibitors is to decrease the concentration of active enzyme. Irreversible inhibitors are usually considered to be poisons and are generally unsuitable for therapeutic purposes.

Reversible enzyme inhibitors are usually either competitive or non-competitive; a third type, uncompetitive, is rarely encountered (Table 9.1).

Most therapeutic drugs are reversible competitive inhibitors, which bind at the catalytic (active site) of the enzyme. Competitive inhibitors are especially attractive as clinical modulators of enzyme activity because they offer two routes for the reversal of enzyme inhibition, by decreasing the concentration of inhibitor or by raising the concentration of substrate.

Since high concentrations of a substrate can displace its competitive inhibitor, it is apparent that $V_{max}$ should be unchanged by competitive inhibitors. This characteristic of competitive inhibitors is reflected in the identical vertical-axis intercepts of Lineweaver–Burk plots, with and without inhibitor (Figure 9.6).

Competitive inhibitors often structurally resemble the substrate of the enzyme (structural analogues). For example, methotrexate is a competitive inhibitor of the enzyme dihydrofolate

**Table 9.1**  Types of enzyme inhibitor

| Type of inhibitor | Binds to the enzyme at: | Kinetic effect |
| --- | --- | --- |
| Competitive | The catalytic site (usually), competing with substrate for binding in a dynamic equilibrium-like process. Inhibition is reversible by increasing substrate concentration. | $V_{max}$ is unchanged; $K_m$ is increased (the presence of inhibitor effectively decreases the affinity of the enzyme for its substrate). |
| Non-competitive | The E or ES complex, other than at the catalytic site. Substrate binding is unaltered, but the ESI complex cannot form products. Inhibition is not reversed by increasing substrate concentration. | $K_m$ appears unaltered; $V_{max}$ is decreased proportionately to inhibitor concentration. |
| Uncompetitive | The ES complex at locations other than the catalytic site. Substrate binding modifies the enzyme structure, making an inhibitor-binding site available. Inhibition is not reversed by substrate. | Apparent $V_{max}$ decreased; $K_m$ is decreased. |

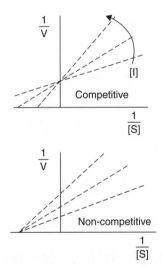

**Figure 9.6**  Lineweaver–Burk plots, with competitive and non-competitive inhibitor.

reductase, which catalyses the reduction of dihydrofolate to tetrahydrofolate. The similarity between the structures of folic acid and methotrexate is shown in Figure 9.7.

## 9.9  The control of enzyme activity

Control of enzyme activity may occur through a variety of strategies involving different time courses (Table 9.2).

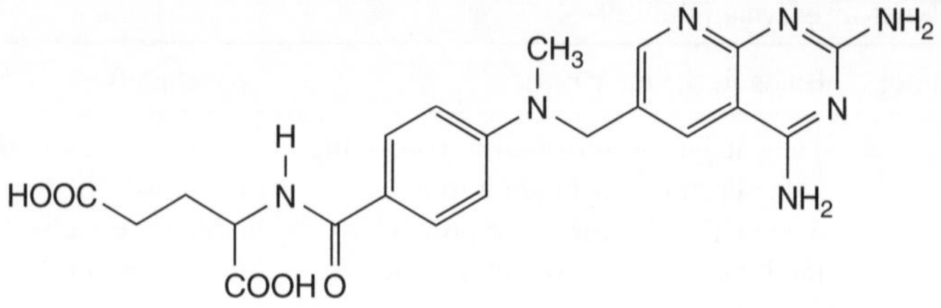

methotrexate

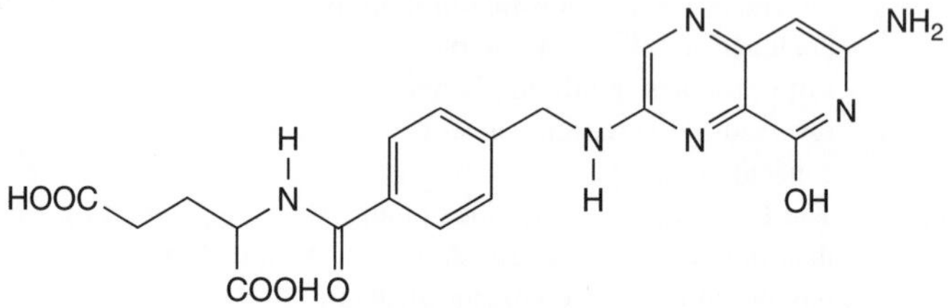

folic acid

**Figure 9.7**  Structures of methotrexate and folic acid.

## 9.10  Allosteric enzymes

Allosteric enzymes bind small regulatory molecules (effectors) that bring about catalytic modification to the enzyme at distinct allosteric sites, well removed from the catalytic site; conformational changes are transmitted through the bulk of the protein to the catalytically active site(s). Effectors may exert a positive or negative action on enzyme activity. Most allosteric enzymes are oligomeric (consisting of multiple subunits); generally they are located at or near branch points in metabolic pathways, where they are influential in directing substrates along one or another of the available metabolic routes. Effectors can range from simple inorganic molecules to complex nucleotides, such as cyclic adenosine monophosphate (cAMP); their single defining feature is that they are not identical to the substrate. Effectors may control enzyme activity by altering the $V_{max}$ or the $K_m$ of the enzyme; $K_m$ effectors work on K-type enzymes, $V_{max}$ effectors work on V-type enzymes.

## 9.11  Covalent modification of enzymes

Covalent modification is a major mechanism for the rapid and transient regulation of enzyme activity. Numerous enzymes of intermediary metabolism are affected by phosphorylation, either positively or negatively. Covalent phosphorylations can be reversed by a separate subclass of enzymes known as phosphatases. The aberrant phosphorylation of growth factor and hormone receptors, as well as of proteins that regulate cell division, often leads to unregulated cell growth or cancer. The usual sites for phosphate addition to proteins are the serine, threonine and tyrosine R-group hydroxyl residues.

**Table 9.2** Controlling enzyme activity

| Control by: | Example | Speed of regulation |
|---|---|---|
| Inhibitors and activators | Negative-feedback mechanism can effectively adjust the rate of synthesis of intermediate metabolites according to the demands of the cell. | Rapid |
| Post-translation modification | Phosphorylation or glycosylation may be used to activate or deactivate an enzyme. In the response to insulin, the phosphorylation of multiple enzymes, including glycogen synthase, helps control the synthesis or degradation of glycogen and allows the cell to respond to changes in blood sugar.<br>Another example of post-translational modification is the cleavage of the polypeptide chain. Chymotrypsin is produced in the inactive form, the proenzyme, as chymotrypsinogen. This type of inactive precursor to an enzyme is known as a zymogen. | Rapid |
| Synthesis | Gene regulation can be induced or inhibited. Bacteria may become resistant to penicillin because enzymes called $\beta$-lactamases are induced which hydrolyse the crucial $\beta$-lactam ring within the penicillin molecule. Cytochrome P450 oxidases in the liver are induced in response to drug metabolism. | Slow |
| Degradation | Key regulatory enzymes are often degraded relatively quickly. | Slow |
| Compartmental-isation | Fatty acids are synthesised by one set of enzymes in the cytosol, and used by a different set of enzymes as a source of energy in the mitochondrion.<br>Some enzymes may become activated when localised to a different environment (e.g. from a reducing (cytoplasm) to an oxidising (periplasm) environment, high pH to low pH, etc.). For example, haemaglutinin, in the influenza virus, is activated by a conformational change caused by the acidic conditions which occur when it is taken up inside its host cell and enters the lysosome. | |

# 9.12 Control proteins

Control proteins, such as calmodulin (a calcium-binding protein), can bind to and regulate various different protein targets, in processes such as inflammation, metabolism, apoptosis, muscle contraction, intracellular movement, short-term and long-term memory, nerve growth and the immune response. Many of the proteins that bind calmodulin are themselves unable to bind calcium, and so use calmodulin as a calcium sensor and signal transducer. Calmodulin undergoes a conformational change upon binding of calcium, which enables it to bind to specific proteins for a specific response.

To achieve overall metabolic control, the body uses a combination of short-term, medium-term and long-term control mechanisms. For example:

- allosteric changes in enzyme activity occur in milliseconds or less

- transmembrane ion channels open or close in milliseconds or less

- G-protein transmembrane signalling operates over a few milliseconds

- protein kinases and phosphatases operate over a few seconds

- a protein switches compartments in a minute or so

- changes in gene expression are evident over about 24 hours

- growth/differentiation occurs over a few days.

## 9.13   Enzymes in medicine

Enzymes have been used for thousands of years, in the production of foodstuffs such as bread, beer and cheese. Today, commercially purified and sometimes immobilised enzymes are used by industry and medicine. Some examples include:

- **Lactose intolerance**. A natural event; the gene for lactase (ß-galactosidase) is 'switched on' at birth and 'switched off' after weaning. An estimated 75% of the world's population is intolerant to lactose in adulthood. As a result, lactose is unabsorbed by the body, ferments in the lower gut and produces intestinal gases (methane), leading to pain and flatulence. In most Europeans, however, the infant condition persists, and the lactase gene remains active (possibly linked with the domestication of cattle and goats in the Near East some 10 000 years ago; the ability to digest lactose throughout life could have conferred some nutritional advantage). The gene encoding lactase in humans is located on chromosome 1; 70% of 'Westerners' have a mutation in the gene such that it fails to 'switch off', thus conferring lactose tolerance. For those lactose-intolerant individuals, lactase may be added to milk or taken as capsules before a meal; it is supplied as a pro-enzyme called prolactazyme. The pro-enzyme is activated by partial digestion in the stomach, so that it has the opportunity to function in the small intestine. So-called 'live yogurts' solve this problem because the lactose (in the yogurt) is digested by the bacteria present. Lactase enzyme is expensive however; nowadays milk can be pre-treated with lactase before distribution. On this industrial scale, the enzyme is 'immobilised', typically on inert resins.

- **Diabetes**. It is useful for diabetics to measure their blood sugar level throughout the day in order to regulate their use of insulin. One test (Clinistix) relies upon a chemical reaction that produces a colour change on a test strip. The test strip contains a chemical indicator called toluidine and the 'immobilised' enzyme glucose oxidase. Glucose oxidase converts the glucose in urine to gluconic acid and hydrogen peroxide; hydrogen peroxide reacts with toluidine, causing the colour change.

- **Enzyme deficiency diseases**. A variety of metabolic diseases are caused by deficiencies or malfunctions of enzymes, due originally to gene mutation. Albinism, for example, may be caused by the absence of tyrosinase, an enzyme essential for the production of cellular pigments. The hereditary lack of phenylalanine hydroxylase results in the disease phenylketonuria (PKU); PKU is usually managed by dietary modifications, but intravenous 'enzyme replacement therapy' can sometimes be employed in some enzyme deficiency diseases. One such example is Gaucher's disease type I, caused by a deficiency in the enzyme glucocerebrosidase, causing lipids to accumulate, swelling the spleen and liver, and triggering anaemia and low blood platelet counts. Such patients often suffer from fatigue, grossly distended abdomens, joint and bone pain, repeated bone fractures and increased bruising and bleeding. This can be treated using intravenous enzyme replacement therapy with a modified version of the enzyme, known generically as alglucerase.

---

Gaucher's disease has three common clinical subtypes. Type I (non-neuropathic type) is the most common; incidence is about 1 in 50 000 live births (particularly common among persons of Ashkenazi Jewish heritage). Type II (acute infantile neuropathic) typically begins within six months of birth; incidence is about 1 in 100 000 live births. Type III (chronic neuropathic form) can begin at any time in childhood or even in adulthood; incidence is about 1 in 100 000 live births.

Alglucerase is a modified form of human $\beta$-glucocerebrosidase. Ceredase is a citrate buffered solution of alglucerase manufactured from human placental tissue. It has largely been replaced by Cerezyme, which is produced by recombinant DNA technology. Both are given intravenously in the treatment of type I Gaucher's disease.

---

- **Heart attacks**. Streptokinase is administered intravenously to patients as soon as possible after the onset of a heart attack, to dissolve clots in the arteries of the heart wall. This minimises the amount of damage to the heart muscle. Streptokinase belongs to a group of drugs known medically as 'fibrinolytics', or colloquially as 'clotbusters'. It works by stimulating production of a naturally produced protease, plasmin, which degrades fibrin, the major constituent of blood clots.

- **Acute childhood leukaemia**. Asparaginase, extracted from bacteria, has proven to be particularly useful for the treatment of acute lymphocytic leukaemia in children, in whom it is administered intravenously. Its action depends upon the fact that tumour cells are deficient in an enzyme called aspartate-ammonia ligase, restricting their ability to synthesise the normally non-essential amino acid L-asparagine. The action of the asparaginase does not affect the functioning of normal cells, which are able to synthesise enough for their own requirements, but reduces the free circulating concentration, thus starving the leukaemic cells. A 60% incidence of complete remission has been reported in a study of almost 6000 cases of acute lymphocytic leukaemia.

- **Cystic fibrosis (CF)**. Patients with CF generally suffer from bacterial infections in the lungs, which is associated with a heavy build-up of thick mucus. Antibiotics are generally prescribed but CF patients also require daily percussion therapy. DNA present in the mucous, which arises from dead white blood cells and bacterial cells, serves to cross link the mucous, changing it from a fluid gel to a semi-solid. The enzyme deoxyribonuclease (DNAase) hydrolyses extracellular DNA, and can be used usefully to alleviate this situation.

- **Enzyme inhibitors**. These can be used, for example to increase the efficacy of the penicillin antibiotics. Bacteria can develop resistance to penicillins by producing enzymes called $\beta$-lactamases, which break down penicillins. It is possible to block the active sites of $\beta$-lactamase using the broad-spectrum inhibitor, Augmentin.

- **Drug manufacture**. Enzymes are particularly useful when it comes to small-molecule pharmaceutical chemicals. This is because they are stereo-specific and are thus able to make single-isomer (chiral) compounds, whereas ordinary chemical methods normally yield racemic mixtures of stereo-isomers. A racemic mixture of thalidomide had tragic consequences in the 1960s; taken by pregnant women as a sedative and to prevent morning sickness, it led in many cases to deformed children. Later research showed that the (+) isomer had the desired effect whilst the (−) isomer had a teratogenic effect. Now optically pure thalidomide is used in the treatment of leprosy, Behcet's syndrome, AIDS and tuberculosis.

## 9.14    Biomarkers and enzymes in diagnosis

Rapid and accurate diagnosis of a patient's condition is an essential part of clinical management. Laboratory tests are used to tailor individual treatment plans according to need, to monitor disease progression, to assess risk, to inform prognosis, and for population screening programs.

Biomarkers may target a disease's aetiology (risk factors for development of the illness), its pathophysiology (abnormalities associated with the illness) or its expression (manifestations of the illness). A biomarker is defined as any characteristic that can be objectively measured and evaluated as an indicator of normal biological processes, pathogenic processes or pharmacological response to a therapeutic intervention.

Biomarkers may cover a range of different approaches, including:

- measurement of physiological function, for example blood pressure, pulse rate, respiratory function

- measurement of individual phenotype, for example height, weight, BMI

- tissue markers, for example blood glucose, serum cholesterol, urine proteins, cell phenotype, prostate specific antigen, enzyme assays

- molecular indicators of environmental exposure, for example human papilloma virus, tobacco exposure (4-(methylnitrosamino)-1-(3-pyridyl)-1-butanone (NNK))

- X-ray, MR or functional imaging data

- genetic polymorphisms

- gene-expression data

- clinical, health, psychological or functional assessment tools.

Any biomarker must generate robust assay performance consistent with the requirements for routine clinical laboratories in the form of analytic validation, and defined disease management value in the form of clinical qualification. Key milestones that must be met for any proposed clinical use of a biomarker would include:

1. **Analytical validity.** That is, the accuracy and precision with which a particular biomarker is identified by the test.

2. **Clinical validity.** That is, the accuracy with which a test identifies or predicts a patient's clinical status.

3. **Clinical utility.** That is, assessment of the risks and benefits, such as cost or patient outcome, resulting from using the test.

4. **Ethical, legal or social implications.**

Point 1 is usually well covered in laboratory quality assurance procedures, point 2 is sometimes assessed in research publications that report sensitivity, specificity, predictive values and related parameters, but points 3 and 4 are often not formally evaluated at all, despite being key to determining whether or not a test actually produces a benefit.

Biomarkers have a key role, in both clinical practice and research, in the monitoring and evaluation of outcomes of interventions, both at individual and at population level. The fundamental need for interdisciplinary collaboration, in order to develop, qualify and properly utilise biomarkers, is widely recognised. Cancer and neurodegenerative diseases are cases in point.

- Identification of biomarkers to identify cancers at an early and curable stage is a priority. For example, the prognosis for patients with lung cancer is strongly dependent on the stage of the disease at the time of diagnosis. Non-small-cell lung cancer, which accounts for 75–80% of cases, has a different clinical presentation, prognosis and response to therapy than small-cell lung cancer (which is less commonly met). Lung cancer is not a result of a sudden transforming event but the end of a multi-step process in which the accrual of genetic and cellular changes results in the formation of an invasive tumour. Patients with early clinical-stage non-small-cell lung cancer have a five-year survival of about 60%, while at later stages the five-year survival may be as low as 5%.

- Current biomarkers for Alzheimer's disease include $\beta$-amyloid, measured in cerebrospinal fluid; Tau protein, measured in cerebrospinal fluid; and neural thread protein/AD7C-NTP, measured in cerebrospinal fluid and in urine. In Alzheimer's patients, cerebrospinal fluid usually contains a reduced level of 42-aminoacid $\beta$-amyloid and an increase in Tau protein. Such biomarkers are however unreliable; they are not accurate for a diagnosis of Alzheimer's, because the same pattern findings are also found in other conditions. At present the costs involved in mass or individual screening would be high; the procedures are also invasive, uncomfortable and not without additional risk.

# 9.15 Enzymes in the diagnosis of pathology

Serum enzyme levels have clinical diagnostic significance; the presence of enzymes in the serum can be indicative of tissue and cellular damage. Commonly assayed biomarker enzymes include:

- the amino transferases (transaminases), alanine transaminase (ALT) and aspartate transaminase (AST)

- lactate dehydrogenase (LDH)

- creatine kinase (CK)

- gamma-glutamyl transpeptidase (GGT).

ALT is particularly diagnostic of liver pathology since it is found predominantly in hepatocytes. The ratio ALT : AST can be diagnostic. In non-viral liver damage the ratio of ALT : AST is less than 1; in viral hepatitis the ALT : AST ratio is greater than 1.

Measurement of AST is also useful in diagnosing heart disease or damage. The level of AST elevation in the serum is directly proportional to the number of cells damaged, as well as to the time following injury at which the AST assay was performed. Following injury, AST levels rise within 8 hours and peak 24–36 hours later. Within 3–7 days the level of AST should return to pre-injury levels, provided a continuous insult is not present. Although measurement of AST alone is not diagnostic for myocardial infarction, taken together with LDH and CK measurements (see below) the level of AST is useful for timing of the infarct.

LDH is especially diagnostic for myocardial infarction. This enzyme exists in five closely related, but slightly different forms (isoenzymes). The five types and their normal distribution and levels in non-disease/injury are listed below:

- **LDH 1**. In heart and red-blood cells, 17–27% of the normal serum total.

- **LDH 2**. In heart and red-blood cells, 27–37% of the normal serum total.

- **LDH 3**. In a variety of organs, 18–25% of the normal serum total.

- **LDH 4**. In a variety of organs, 3–8% of the normal serum total.

- **LDH 5**. In liver and skeletal muscle, 0–5% of the normal serum total.

Following a myocardial infarct, serum levels of LDH rise within 24–48 hours, reaching a peak by 2–3 days, and then returning to normal in 5–10 days. Especially diagnostic is a comparison of the LDH 1 : LDH 2 ratio. This ratio is normally less than 1, but a reversal of the ratio ('flipped LDH') occurs up to 48 hours following an acute myocardial infarct.

CK is found primarily in heart and skeletal muscle, as well as the brain. Measurement of serum CK levels is a good diagnostic for injury to these tissues. Levels of CK rise within 6 hours of injury and peak by around 18 hours, returning to normal within 2–3 days. Like LDH, there are tissue-specific isoenzymes of CK:

- **CK3 (CK-MM)**. The predominant isoenzyme in muscle; 100% of the normal serum total.

- **CK2 (CK-MB)**. About 35% of the CK activity in cardiac muscle, but less than 5% in skeletal muscle; 0% of the normal serum total.

- **CK1 (CK-BB)**. The characteristic isoenzyme in brain and in smooth muscle; 0% of the normal serum total.

Since most of the released CK after a myocardial infarction is CK-MB, an increased ratio of CK-MB to total CK may help in diagnosis of an acute infarction.

GGT is found particularly in hepatocytes and biliary epithelial cells. GGT serum levels may be high in liver disease, but it is particularly a feature of biliary outflow obstruction; more so than hepatocellular damage. GGT serum measurement provides a very sensitive indicator of the presence or absence of hepatobiliary disease. However, raised GGT levels have also been reported in a variety of other clinical conditions, including pancreatic disease, myocardial infarction, chronic obstructive pulmonary disease, renal failure, diabetes, obesity and alcoholism. It is also a sensitive indicator of liver damage through alcohol ingestion.

## 9.16  Liver-function tests

Liver-function tests (LFTs) are commonly undertaken on a blood sample, and include:

- **ALT**. When the liver is injured or inflamed (as in hepatitis), the blood level of ALT usually rises.

- **Alkaline phosphatase (ALP)**. This enzyme occurs mainly in liver and bone. The blood level is raised in some types of liver and bone disease.

- **Albumin**. The main protein made by the liver. The ability to make albumin (and other proteins) is affected in some types of liver disorder.

- **Total protein**. This measures albumin and all other proteins in blood.

- **Bilirubin**. Jaundice is the clinical manifestation of hyperbilirubinaemia. A raised level of 'unconjugated' bilirubin occurs when there is excessive breakdown of red blood cells, for example in haemolytic anaemia, or where the ability of the liver to conjugate bilirubin is compromised, for example in cirrhosis. A raised blood level of 'conjugated' bilirubin occurs in various liver and bile duct conditions. It is particularly high if the flow of bile is blocked, for example by a gallstone in the common bile duct or by a tumour in the pancreas. It can also be raised with hepatitis, liver injury or long-term alcohol abuse.

Liver-function tests often provide the first marker of liver disease. Other blood tests must be used to confirm the diagnosis of a particular disorder and/or to monitor the activity of the disorder and response to treatment. For example:

- **Blood-clotting tests**. Since the liver synthesises many of the blood-clotting proteins, blood-clotting tests may be used as a marker of the severity of certain liver disorders.

- **GGT or 'gamma GT'**. A high level of this enzyme is particularly associated with heavy alcohol drinking.

- **Immunology**. Blood tests can detect viruses and antibodies to viruses, for example hepatitis A/B virus, or auto-antibodies from autoimmune disorders of the liver, for example primary biliary cirrhosis (associated with anti-mitochondrial antibodies), autoimmune hepatitis (associated with smooth muscle antibodies) and primary sclerosing cholangitis (associated with antinuclear cytoplasmic antibodies).

Other types of protein in the blood can identify specific liver diseases, for example ceruloplasmin is reduced in Wilson's disease, lack of 1-antitrypsin is an uncommon cause of cirrhosis and high levels of ferritin is a marker of haemochromatosis.

# CHAPTER 10

# The kidney

## 10.1  Nephron structure

Renal function is an indication of the physiological state of the kidney; glomerular filtration rate (GFR) describes the flow rate of filtered fluid through the kidney, while creatinine clearance rate ($C_{Cr}$) is the volume of blood plasma that is cleared of creatinine per unit time, and is a useful measure for approximating the GFR. Most clinical tests use the plasma concentrations of the waste substances of creatinine and urea, as well as electrolytes, to determine renal function.

The nephron is the functional unit of the kidney (Figure 10.1); it consists of two parts:

- The renal corpuscle, composed of a glomerulus and Bowman's capsule; the glomerulus is the initial filtering unit. Glomerular blood pressure provides the driving force for water and solutes to be filtered out of the blood and into the space made by Bowman's capsule; the resulting glomerular filtrate is further processed along the nephron to form urine.

- The renal tubule, composed of the proximal tubule, loop of Henle (descending and ascending limbs) and distal convulated tubule.

## 10.2  Kidney function

The functions of the kidney can be divided into two groups:

- **Secretion of hormones**. Including erythropoietin, which regulates red blood cell production in the bone marrow, rennin, which is a key part of the rennin–angiotensin–aldosterone system, and the active forms of vitamin D (calcitriol) and prostaglandins.

- **Extracellular homeostasis**. Maintaining a balance of several substances, some of which are summarised in Table 10.1.

---

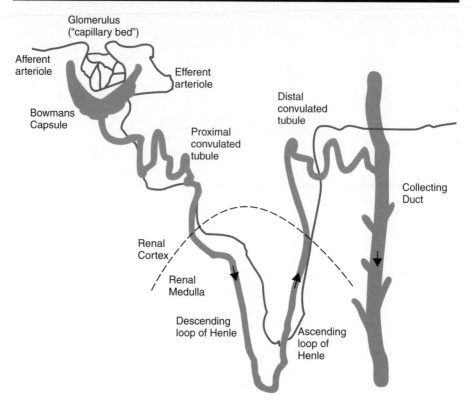

**Figure 10.1**    Nephron structure.

The kidney's ability to perform many of its functions depends on the three fundamental functions of filtration, re-absorption and secretion:

- Filtration is driven by both hydrostatic and oncotic (colloid osmotic pressure) transport.

- Re-absorption is a two-step process beginning with the active or passive extraction of substances from the tubule fluid into the renal interstitium (the connective tissue that surrounds the nephrons), followed by their transport from the interstitium into the bloodstream. These transport processes are driven by hydrostatic, oncotic, diffusion and active transport. Some key regulatory hormones for re-absorption include aldosterone, which stimulates active sodium re-absorption (and water as a result), and antidiuretic hormone, which stimulates passive water re-absorption. Both hormones exert their effects principally on the collecting ducts.

- Secretion is the transfer of materials from peritubular capillaries to the renal tubular lumen. Tubular secretion is mainly an active transport process. Usually only a few substances are secreted, unless they are present in great excess, or are natural poisons.

**Table 10.1** Kidney extracellular homeostasis

| Substance | | Proximal tubule | Loop of Henle | Distal tubule | Collecting duct |
|---|---|---|---|---|---|
| Glucose | Unabsorbed glucose appears in the urine (=glucosuria), associated with diabetes mellitus | Re-absorbed, almost 100% via $Na^+$-glucose transporters (apical) and GLUT (basolateral) | – | – | – |
| Peptides/ amino acids | Re-absorbed nearly completely | Re-absorbed | – | – | – |
| Urea | Regulation of osmolarity | Re-absorbed, 50% by passive transport | Secretion | – | Re-absorbed |
| Sodium | $Na^+$-$H^+$ antiport and $Na^+$-glucose symport | Re-absorbed, 65% | Re-absorbed, 25%, ascending loop, $Na^+$-$K^+$ symport | Re-absorbed, 5%, $Na^+$-$Cl^-$ symport | Re-absorbed, 5% |
| Chloride | Usually follows sodium | Re-absorbed | Re-absorbed, $Na^+$-$K^+$-$2Cl^-$ symport | Re-absorbed, $Na^+$-$Cl^-$ symport | – |
| Water | Uses aquaporin water channels | Absorbed osmotically along with solutes | Re-absorbed, descending loop | – | Re-absorbed |
| Bicarbonate | Helps maintain acid–base balance | Re-absorbed (80–90%) | Re-absorbed, ascending loop | – | Re-absorbed |
| Protons | Uses vacuolar $H^+$ATPase | – | – | – | Secretion |
| Potassium | – | Re-absorbed, 65% | Re-absorbed, 20%, ascending loop, $Na^+$-$K^+$-$2Cl^-$ symport | – | Secretion via $Na^+$-$K^+$ ATPase |
| Calcium | Uses $Ca^{2+}$ ATPase and $Na^+$–$Ca^{2+}$ exchanger | Re-absorbed | Re-absorbed, ascending loop via passive transport | – | – |
| Magnesium | Calcium and magnesium compete; an excess of one can lead to excretion of the other | Re-absorbed | Re-absorbed, ascending loop | Re-absorbed | – |
| Phosphate | – | Re-absorbed, 85%, via $Na^+$-phosphate cotransporter | – | – | – |

# 10.3   Diuretics

Diuretics elevate the rate of urination and thus provide a means of forced dieresis. There are several categories of diuretics, all of which increase the excretion of water, but in a distinct way:

1. **High-ceiling loop diuretics.** Diuretics that cause a substantial diuresis, up to 20% of the filtered load of NaCl and water. Loop diuretics, such as furosemide, inhibit the re-absorbtion of sodium at the ascending loop, which leads to a retention of water in the urine. Other examples of high-ceiling loop diuretics include ethacrynic acid, torsemide and bumetanide.

2. **Thiazides.** Diuretics that act on the distal convoluted tubule and inhibit the sodium chloride symporter, leading to retention of water in the urine. An example is hydrochlorothiazide.

3. **Potassium-sparing diuretics.** Diuretics that do not promote the secretion of potassium into the urine. Two specific classes include:

   (a) Aldosterone antagonists. Spironolactone is a competitive antagonist of aldosterone. Aldosterone normally acts to add sodium channels in the principal cells of the collecting duct and late distal tubule of the nephron. Spironolactone prevents aldosterone from entering the principal cells, preventing sodium re-absorption. A similar agent is potassium canreonate.

   (b) Epithelial sodium channel blockers, including amiloride and trimterene.

4. **Calcium-sparing diuretics.** Diuretics that result in a relatively low rate of excretion of calcium. The sparing effect on calcium can be beneficial in hypocalcaemia. The thiazides and potassium-sparing diuretics are considered to be calcium-sparing diuretics. The thiazides cause a net decrease in calcium lost in urine; the potassium-sparing diuretics cause a net increase in calcium lost in urine, but the increase is much smaller than that associated with other diuretic classes. By contrast, loop diuretics promote a significant increase in calcium excretion. This can increase risk of reduced bone density.

5. **Osmotic diuretics.** Diuretics that are filtered in the glomerulus but cannot be re-absorbed. An example is mannitol. Their presence leads to an increase in the osmolarity of the filtrate; to maintain osmotic balance, water is retained in the urine. Glucose, like mannitol, can also behave as an osmotic diuretic. In diabetes mellitus, the concentration of glucose in the blood exceeds the maximum resorption capacity of the kidney; glucose remains in the filtrate, leading to the osmotic retention of water in the urine.

6. **Low-ceiling diuretics.** Diuretics that have a rapidly flattening dose effect curve (in contrast to 'high ceiling', where the relationship is close to linear). 'Low ceiling' refers to a pharmacological profile, not a chemical structure. However, there are certain classes of diuretic which usually fall into this category, such as the thiazides.

Diuretics are used to treat oedema in heart failure, liver cirrhosis, hypertension and certain kidney diseases. Some diuretics, such as acetazolamide, make the urine more alkaline and are helpful in increasing excretion of substances such as aspirin in cases of overdose. Diuretics are often abused by sufferers of eating disorders (e.g. bulimia) in an attempt at weight loss.

Xanthines, including caffeine and theophylline, are diuretics; they inhibit re-absorption of $Na^+$, increasing the GFR. Alcohol is a diuretic; it reduces production of anti-diuretic hormone.

## 10.4  Anti-diuretic hormone

Anti-diuretic hormone (arginine vasopressin) is a nine-amino-acid peptide secreted from the posterior pituitary. Within hypothalamic neurons, the hormone is packaged in secretory vesicles together with a carrier protein called neurophysin; both are released upon secretion.

The single most important effect of anti-diuretic hormone is to conserve body water, by reducing the loss of water in urine. In the absence of anti-diuretic hormone, the collecting ducts of the kidney are virtually impermeable to water. Anti-diuretic hormone stimulates water re-absorbtion through the insertion of 'water channels', or aquaporins (see Section 10.5), into the membranes of kidney tubules. Aquaporins transport solute-free water through tubular cells and back into blood, leading to a decrease in plasma osmolarity and an increased osmolarity of urine.

Regulation of anti-diuretic hormone secretion is primarily through the plasma osmolarity. Osmolarity is sensed in the hypothalamus by neurons known as osmoreceptors, which in turn stimulate secretion from those neurons that produce anti-diuretic hormone. Secretion of anti-diuretic hormone is also simulated by decreases in blood pressure and volume, conditions sensed by stretch receptors in the heart and large arteries. Changes in blood pressure and volume are not nearly as sensitive a stimulator as increased osmolarity, but are nonetheless potent in severe conditions. For example, loss of 15–20% of blood volume by haemorrhage results in a massive secretion of anti-diuretic hormone. Another potent stimulus of anti-diuretic hormone is nausea and vomiting, both of which are controlled by regions in the brain with links to the hypothalamus.

The most common disease state related to anti-diuretic hormone is diabetes insipidus. This condition can arise from either of two situations:

- **Hypothalamic ('central') diabetes insipidus**. This results from a deficiency in secretion of antidiuretic hormone from the posterior pituitary. Causes may include head trauma, infections or tumours involving the hypothalamus.

- **Nephrogenic diabetes insipidus**. This occurs when the kidney is unable to respond to anti-diuretic hormone. Most commonly this is the result of renal disease, but mutations in the ADH receptor gene or in the gene encoding aquaporin-2 have also been demonstrated in affected humans.

The major indication of either type of diabetes insipidus is excessive urine production; as much as 16 l of urine per day. If adequate water is available for consumption, the disease is rarely life-threatening. Hypothalamic diabetes insipidus can be treated with exogenous anti-diuretic hormone.

## 10.5  Aquaporins

Aquaporins selectively conduct water molecules in and out of the cell; also known as 'water channels', aquaporins are integral membrane pore proteins. Some, known as aquaglyceroporins,

**Table 10.2**　Kidney aquaporins

| Type | Location | Function |
|------|----------|----------|
| Aquaporin 1 | Kidney (apically), proximal tubule, descending loop of Henle | Water re-absorption |
| Aquaporin 2 | Kidney (apically) | Water re-absorption in response to anti-diuretic hormone |
| Aquaporin 3 | Kidney (basolateral) | Water re-absorption |
| Aquaporin 4 | Kidney (basolateral) | Water re-absorption |

can also transport other small uncharged solutes, such as glycerol, $CO_2$, ammonia and urea. Water pores are however completely impermeable to charged species, such as protons.

Aquaporins comprise six transmembrane $\alpha$-helices, and five interhelical loop regions (A–E) that form the extracellular and cytoplasmic vestibules. Loops B and E are hydrophobic loops which contain the highly, although not completely, conserved Asn-Pro-Ala (NPA) motif, which overlaps the middle of the lipid bilayer of the membrane, forming a 3D 'hourglass' structure through which water flows. Two 'constrictions' in the channel act as selectivity filters.

There are 13 known types of aquaporin in mammals; six of these are located in the kidney. The most studied aquaporins are compared in Table 10.2.

# CHAPTER 11

# Haemostasis

The haemostatic system must preserve intravascular integrity by achieving a balance between haemorrhage and thrombosis. Haemostasis provides several important functions: it maintains blood in a fluid state while circulating within the vascular system; it arrests bleeding at the site of injury by formation of a haemostatic plug (clot); and it ensures the removal of the haemostatic plug once healing is complete.

## 11.1 Blood vessel trauma

Blood vessel trauma is followed by:

- **Vascular spasm**. In which the blood vessels contract as a result of neurological reflexes and local myogenic (muscle) spasm. The degree of constriction is directly proportional to the degree of trauma. The spasm may last up to 30 minutes.

- **Platelet phase (primary haemostasis)**. In which a 'platelet plug', a loose collection of platelets, forms and acts as a base for the formation of a stable clot. Exposure of blood vessel sub-endothelium proteins, most notably collagen, binds circulating platelets through surface collagen-specific glycoprotein Ia/IIa-receptors; adhesion is strengthened further by the large, multimeric circulating protein, von Willebrand factor (vWF), which forms links between the platelets glycoprotein Ib/IX/V and the collagen fibrils. Adhesion activates the platelets, which release the contents of stored granules into the blood plasma, including adenosine diphosphate (ADP), serotonin (a powerful vasoconstrictor), platelet activating factor (PAF), vWF, platelet factor 4 and thromboxane $A_2$, which in turn activates additional platelets. In the platelet a $G_q$-linked protein receptor cascade results in an increased cytosolic calcium concentration; calcium activates protein kinase C, which in turn activates phospholipase $A_2$, which modifies the integrin membrane glycoprotein IIb/IIIa, increasing its affinity to bind fibrinogen. Activated platelets change shape from spherical to stellate and fibrinogen cross-links with glycoprotein IIb/IIIa to aid in aggregation of adjacent platelets. The platelet plug is only effective in stopping blood loss through small rents in vessel walls, which are continually

*Essential Biochemistry for Medicine*   Dr Mitchell Fry
© 2010 John Wiley & Sons, Ltd

occurring naturally. Absent or defective platelets are noted in thrombocytopenic patients, who develop petechiae (small pinpoint haemorrhage).

Platelets, derived from the fragmentation of megakaryocytes, are essential both in maintaining the integrity of the adherens junctions, which provide a tight seal between the endothelial cells that line the blood vessels, and in forming a clot where blood vessels have been damaged. The role of thromboxane $A_2$ in platelet activation accounts for the beneficial effect of low doses of aspirin, a cyclooxygenase inhibitor, in preventing inappropriate blood clotting (recovery after surgery, prevention of deep-vein thrombosis, avoiding heart attack). ReoPro is a monoclonal antibody directed against platelet receptors. It inhibits platelet aggregation and appears to reduce the risk that 'reamed out' coronary arteries (after coronary angioplasty) will plug up again.

- **Plasma phase (secondary haemostasis – clotting)**. In which blood clots develop within a minute or two. Traumatised vessels and platelets liberate activating factors, which initiate the clotting process.

# 11.2   Blood coagulation

Initiation of blood coagulation (clotting) occurs through the contact activation pathway (intrinsic pathway) and the tissue factor (TF) pathway (extrinsic pathway). The contact activation pathway is quantitatively the most important, but is much slower to initiate; the TF pathway is considered to be the primary pathway for the initiation of blood coagulation and affords a more rapid response (the so-called thrombin burst), which augments the contact activation pathway. Both pathways share a common pathway that converges at factor X with the production of thrombin (Figure 11.1).

# 11.3   The coagulation cascade

The coagulation cascade is a series of reactions in which a zymogen (an inactive enzyme precursor) and its glycoprotein co-factor are activated, catalysing the next reaction in the cascade, and ultimately resulting in cross-linked fibrin. The coagulation factors are generally serine proteases, although there are some exceptions; factors VIII and V are glycoproteins and factor XIII is a transglutaminase. Serine proteases act by cleaving other proteins at specific sites.

# 11.4   The tissue factor (TF) pathway

The TF pathway (extrinsic) generates the thrombin burst, a process by which thrombin, the most important constituent of the coagulation cascade in terms of its feedback activation roles, is generated almost instantaneously. Following blood vessel damage, endothelium TF is released, forming a complex with and activating factor VII (TF-VIIa); factor VIIa circulates in a higher amount than any other activated coagulation factor. In turn, TF-VIIa activates factors IX and X (VII is itself activated by thrombin, factor XIa, and by plasmin, factors XII and Xa). The

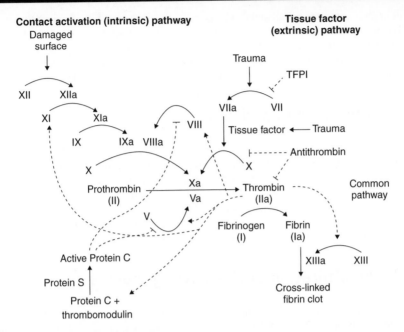

**Figure 11.1**  Pathways of blood coagulation. Nomenclature: the use of roman numerals rather than eponyms or systematic names was agreed upon during annual conferences, starting in 1955, and a consensus achieved in 1962 on the numbering of factors I–XII. Assignment of numerals ceased in 1963 after the naming of factor XIII. Fletcher factor and Fitzgerald factor were given to further coagulation-related proteins, namely prekallikrein and high-molecular weight kininogen respectively. A lowercase 'a' is appended to indicate an active form of the factor.

activation of factor X by TF-VIIa is almost immediately inhibited by tissue factor pathway inhibitor (TFPI).

Factor Xa and its co-factor Va form the prothrombinase complex, which activates prothrombin to thrombin. Thrombin then activates other components of the coagulation cascade, including factor V and factor VIII (which activates factor XI, in turn activating factor IX), and activates factor VIII and releases it from being bound to vWF.

Factor VIIIa is the co-factor of factor IXa; together they form the 'tenase' complex which activates factor X, and so the cycle continues.

> The tenase complex is formed by the activated forms of the blood coagulation factors VIII and IX. It forms on a phospholipid surface in the presence of calcium and is responsible for the activation of factor X. 'Tenase' is a contraction of 'ten' and the suffix '-ase', used for enzymes

## 11.5  The contact activation pathway

The contact activation pathway (intrinsic) begins with formation of the primary complex on collagen by high-molecular weight kininogen, prekallikrein and factor XII (Hageman factor).

Prekallikrein is converted to kallikrein and factor XII is activated to XIIa. Factor XIIa activates XI–XIa, which activates factor IX, which together with its co-factor VIIIa forms the tenase complex, resulting in activation of factor X. The minor role that the contact activation pathway has in initiating clot formation can be illustrated by the fact that patients with severe deficiencies of factor XII, kininogen and prekallikrein do not exhibit a bleeding disorder.

## 11.6   The common pathway

Finally, the common pathway involves thrombin in a number of functions. Its primary role is the conversion of fibrinogen to fibrin, the building block of a haemostatic plug. In addition, it activates factors VIII and V and their inhibitor protein C (in the presence of thrombomodulin), and it activates factor XIII, which cross-links the fibrin polymers.

Following activation by the contact factor or TF pathways, the coagulation cascade is maintained in a prothrombotic state by the continued activation of factors VIII and IX to form the tenase complex, until it is down-regulated by the anticoagulant pathways (see Section 11.9).

## 11.7   Amplification of the clotting process

The clotting process has several positive-feedback loops which quickly magnify a tiny initial event into a 'cascade'. The TF-VII complex (which started the process) also activates factor IX. Factor IX binds to factor VIII, a protein that circulates in the blood, stabilised by another protein, vWF. This complex activates more factor X. As thrombin is generated, it activates more factor V, factor VIII and factor IX.

Factor XI amplifies the production of activated Factor IX. Thus what may have begun as a tiny, localised event rapidly expands into a coagulation cascade.

## 11.8   Co-factors in coagulation

Various co-factors are required for the proper functioning of the coagulation cascade:

- **Calcium and phospholipid**. Required for the tenase and prothrombinase complexes to function. Calcium mediates the binding of the complexes via the terminal $\gamma$-carboxy residues on factor Xa and factor IXa to the phospholipid surfaces expressed by platelets, as well as those procoagulant microparticles or microvesicles shed from them.

- **Vitamin K**. An essential factor for hepatic $\gamma$-glutamyl carboxylase, which adds a carboxyl group to glutamic acid residues on factors II, VII, IX and X, as well as protein S, protein C and protein Z. In adding the $\gamma$-carboxyl group to glutamate residues on the immature clotting factors, vitamin K is itself oxidised. Another enzyme, vitamin K epoxide reductase, reduces vitamin K back to its active form. Vitamin K epoxide reductase is pharmacologically important as a target for the anticoagulant drugs warfarin and related coumarins (acenocoumarol, phenprocoumon and dicumarol). These drugs create a deficiency of reduced vitamin K by blocking the epoxide reductase, thereby inhibiting maturation of clotting factors. Other deficiencies of vitamin K (e.g. in malabsorption or disease, e.g. hepatocellular carcinoma) also impair the function of this enzyme.

Warfarin is sometimes prescribed as a 'blood thinner' because it is an effective vitamin K antagonist. Warfarin is also used as a rat poison, causing death by lethal (internal) bleeding.

## 11.9 Regulators of coagulation

Five known mechanisms regulate the coagulation cascade. Abnormalities can lead to an increased tendency towards thrombosis:

- **Protein C.** A major physiological anticoagulant, this is a vitamin K-dependent serine protease enzyme that is activated by thrombin. Protein C is activated in a sequence that begins with it binding, together with thrombin, to the cell-surface protein thrombomodulin. Thrombomodulin activates the protein C. The activated form, along with protein S and phospholipid as co-factors, degrades factors Va and VIIIa. Quantitative or qualitative deficiency of either may lead to thrombophilia (a tendency to develop thrombosis). Impaired action of protein C (activated protein C resistance), for example through having the 'Leiden' variant of factor V or high levels of factor VIII, can lead to a thrombotic tendency. Recombinant protein C is now available to treat people threatened with inappropriate clotting, as a result of widespread infection (sepsis) for example.

- **Antithrombin.** A serine protease inhibitor (serpin) that degrades the serine proteases of thrombin, factors IXa, Xa, XIa and XIIa. It is constantly active, but its adhesion to these factors is increased by the presence of heparin sulphate (a glycosaminoglycan) or the administration of heparins (different heparinoids increase affinity to factor Xa, thrombin, or both). Deficiency of antithrombin (inborn or acquired, e.g. in proteinuria) leads to thrombophilia.

Serpins were first identified as a set of proteins able to inhibit proteases. The name 'serpin' is derived from this activity: 'serine protease inhibitors'. Heparin is a mixture of polysaccharides that bind to antithrombin III, inducing an allosteric change that greatly enhances its inhibition of thrombin synthesis. Some surgical patients, especially those receiving hip or heart valve replacements and those at risk of ischemic stroke (clots in the brain), are given heparin.

- **Tissue factor pathway inhibitor.** This limits the action of TF. It also inhibits excessive TF-mediated activation of factor IX and X.

- **Plasmin.** Generated by proteolytic cleavage of plasminogen, a plasma protein synthesised in the liver. Cleavage is catalysed by tissue plasminogen activator (t-PA), which is synthesised and secreted by endothelium. Plasmin proteolytically cleaves fibrin into fibrin degradation products, which inhibits excessive fibrin formation.

- **Prostacyclin (PGI$_2$).** Released by endothelium. Activates platelet G$_s$ protein-linked receptors, which in turn activate adenyl cyclase; increased levels of cyclic AMP inhibit platelet activation by counteracting the actions that result from increased cytosolic levels of calcium. This inhibits the release of granules that would lead to activation of additional platelets and the coagulation cascade.

# 11.10    Breaking down the clot

Fibrinolysis is the solubilisation and removal of a clot, a process orchestrated by plasmin (Figure 11.2).

- **Plasmin**. This is produced in an inactive form, plasminogen, in the liver. Although plasminogen cannot cleave fibrin, it has an affinity for it and is incorporated into the clot when it is formed. Plasminogen contains secondary structure motifs ('kringles') which specifically bind lysine and arginine residues on fibrin. When converted from plasminogen into plasmin, it functions as a serine protease, cutting C-terminal to these lysine and arginine residues. Fibrin monomers, when polymerised, form protofibrils, each consisting of two strands arranged anti-parallel. Within a single strand the fibrin monomers are covalently linked through the actions of coagulation factor XIII. Thus, plasmin action on a clot initially creates nicks in the fibrin that lead to digestion and solubilisation.

- **t-PA and urokinase**. These convert plasminogen to the active plasmin, thus allowing fibrinolysis to occur. t-PA is released into the blood very slowly by the damaged endothelium of blood vessels, such that after several days (when the bleeding has stopped) the clot is broken down. t-PA and urokinase are themselves inhibited by plasminogen activator inhibitor-1 and plasminogen activator inhibitor-2. Plasminogen activator inhibitor-1 is the principal inhibitor of t-PA and urokinase; it is a serine protease protein (serpin 1). Plasminogen activator inhibitor-2 is secreted by the placenta and is only present in significant amounts during

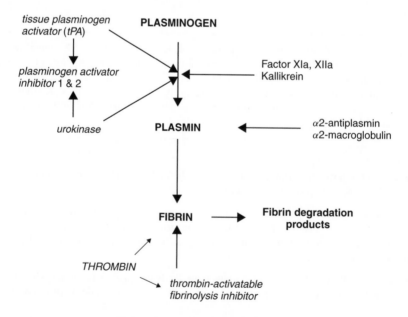

**Figure 11.2**   The plasmin pathway.

pregnancy. Plasmin further stimulates plasmin generation by producing more active forms of both t-PA and urokinase.

- **Alpha 2-antiplasmin and alpha 2-macroglobulin**. These inactivate plasmin. Plasmin activity is also reduced by thrombin-activatable fibrinolysis inhibitor, which modifies fibrin to make a less potent co-factor for the t-PA-mediated plasminogen.

> Recombinant human t-PA is now produced by recombinant DNA technology. Injected within the first hours after a heart attack, it dissolves the clot blocking the coronary artery, restoring blood flow before the heart muscle becomes irreversibly damaged. It is also used in treatment of ischaemic stroke.
>
> Female sheep, transgenic for the human factor IX gene, have been cloned. The human gene is coupled to the promoter for the ovine (sheep) milk protein $\beta$-lactoglobulin; thus they secrete large amounts of human factor IX in their milk, which can be purified for human therapy.

# 11.11   Disorders of haemostasis

Disorders of haemostasis can be roughly divided into:

- platelet disorders
- disorders of coagulation.

## 11.11.1   Platelet disorders

Platelet disorders may be inborn or acquired. Inborn platelet pathologies include Glanzmann's thrombasthenia (platelets lack glycoprotein IIb/IIIa), Bernard–Soulier syndrome (abnormal glycoprotein Ib-IX-V complex), gray platelet syndrome (deficient alpha granules) and delta storage pool deficiency (deficient dense granules); these are rare conditions that predispose to haemorrhage. von Willebrand disease, a deficiency or abnormal function of vWF, manifests in a similar bleeding pattern; its milder forms are relatively common. Decrease in platelet numbers may be due to various causes, including insufficient production (e.g. in myelodysplastic syndrome or other bone marrow disorders), destruction by the immune system (immune thrombocytopenic purpura) and consumption due to various causes, for example thrombotic thrombocytopenic purpura. Most consumptive conditions lead to platelet activation, and some are associated with thrombosis.

## 11.11.2   Disorders of coagulation

The best-known coagulation factor disorders are the haemophilias:

- haemophilia A (factor VIII deficiency)
- haemophilia B ('Christmas disease': factor IX deficiency)
- haemophilia C (factor XI deficiency; mild bleeding tendency).

Haemophilia A and B are X-linked recessive disorders. Haemophilia C is a much more rare autosomal recessive disorder, most commonly seen in Ashkenazi Jews (Ashkenazi Jews are those that are descended from the medieval Jewish communities of the Rhineland in the west of Germany).

- **von Willebrand disease (vWF)**. This behaves more like a platelet disorder, except in severe cases. It is the most common hereditary bleeding disorder, inherited autosomal recessive or dominant. A defect in vWF prevents the normal mediation of the binding of glycoprotein Ib to collagen.

- **Bernard–Soulier syndrome**. Caused by a defect or deficiency in glycoprotein Ib, the receptor for vWF. It is an autosomal recessive inherited disorder.

- **Glanzmann's thrombasthenia**. An extremely rare disorder in which the platelets lack glycoprotein IIb/IIIa, hence no fibrinogen bridging can occur, and bleeding time is significantly prolonged. This is an autosomal recessive inherited disorder.

- **Deficiency of vitamin K**. Contributes to bleeding disorders as clotting factor maturation is dependent upon this vitamin.

- **Thrombosis**. This is the pathological development of blood clots. Clots formed may detach and become mobile, forming an embolus, or grow to such a size that they occlude the vessel in which they developed. An embolism is said to occur when the embolus migrates to another part of the body, interfering with blood circulation and hence impairing organ function downstream of the occlusion. This can cause ischaemia, leading to ischaemic necrosis of the tissue. Most cases of thrombosis are due to acquired extrinsic problems (surgery, cancer, immobility, obesity, 'economy class syndrome'), but some are due to predisposing conditions, known collectively as thrombophilia (e.g. antiphospholipid syndrome, factor V Leiden; these are rare genetic disorders).

# 11.12   Pharmacology of haemostasis

Pharmacology of haemostasis involves the use of procoagulants and anticoagulants.
   Procoagulants include:

- The use of adsorbent chemicals, such as zeolites and other haemostatic agents, to seal severe injuries quickly. Thrombin and fibrin glue are used surgically to treat bleeding and to thrombose aneurysms.

- Desmopressin, used to improve platelet function by activating arginine vasopressin receptor 1A.

- Coagulation factor concentrates, used to treat haemophilia, reverse the effects of anticoagulants and treat bleeding in patients with impaired coagulation factor synthesis or increased consumption.

- Prothrombin complex concentrate, cryoprecipitate and fresh frozen plasma, commonly-used coagulation factor products.

- Recombinant activated human factor VII, increasingly popular in the treatment of major bleeding.

- Tranexamic acid and aminocaproic acid, used to inhibit fibrinolysis. They lead to a *de facto* reduced bleeding rate.

Anticoagulants and anti-platelet agents are amongst the most commonly used medicines. Warfarin (and related coumarins) and heparin are the most commonly used anticoagulants. Warfarin affects the vitamin K-dependent clotting factors (II, VII, IX, X), while heparin and related compounds increase the action of antithrombin on thrombin and factor Xa. Aspirin, clopidogrel, dipyridamole and ticlopidine are among the anti-platelet agents. Parenteral glycoprotein IIb/IIIa inhibitors are used during angioplasty.

A newer class of drug, the direct thrombin inhibitors, are under development; some members (such as lepirudin) are already in clinical use.

# Focus on: anaemia

Anaemia is defined as a qualitative or quantitative deficiency of haemoglobin, which may lead to hypoxia (lack of oxygen) in organs. The three main ways in which anaemia may arise are:

- excessive blood loss (acutely, such as a haemorrhage, or chronically through low-volume loss, e.g. ulcer)

- excessive blood cell destruction (haemolysis)

- deficient red blood cell production (ineffective haematopoiesis).

Anaemia is the most common disorder of the blood; there are several kinds, produced by a variety of underlying causes. Anaemia may be classified by a 'kinetic' approach, which involves evaluating the production, destruction and loss of red blood cells, or a 'morphologic' approach, based on red blood cell size. The morphologic approach uses a quickly available and cheap lab test as its starting point (the mean corpuscular volume (MCV)). MCV is a measure of the average red blood cell volume.

Anaemia often goes undetected; signs and symptoms can be related to the anaemia itself or to the underlying cause.

## Types of anaemia

Microcytic anaemia is primarily a result of a failure or deficiency of haemoglobin synthesis, which may be caused by several aetiologies:

- Haem synthesis defect, for example iron-deficiency anaemia and anaemia of chronic disease (more commonly presenting as normocytic anaemia).

- Globin synthesis defect, for example $\alpha$- or $\beta$-thalassaemia, HbE syndrome, HbC syndrome and various other unstable haemoglobin diseases.

- Sideroblastic defect (abnormal production of red blood cells), for example hereditary sideroblastic anaemia, acquired sideroblastic anaemia, including lead poisoning, and reversible sideroblastic anaemia.

Iron-deficiency anaemia is the most common type of anaemia. Red blood cells often appear hypochromic (paler than usual) and microcytic (smaller than usual). Iron-deficiency anaemia is caused by insufficient dietary intake or absorption of iron, or by loss of blood, for example bleeding lesions of the gastrointestinal tract. Worldwide the most common cause of iron-deficiency anaemia is parasitic infestation (hookworm, amoebiasis, schistosomiasis and whipworm).

Macrocytic anaemia is defined as one in which the red blood cells are larger than normal. Megaloblastic anaemia is the most common cause of macrocytic anaemia, caused by a deficiency of either vitamin B12 or folic acid (or both). Deficiency in folate and/or vitamin B12 may result from either inadequate intake or malabsorption. Pernicious anaemia is caused by a lack of intrinsic factor, which is required to absorb vitamin B12 from food (see below).

Macrocytic anaemia can also be caused by removal of the functional portion of the stomach, such as during gastric bypass surgery, leading to reduced vitamin B12 and folate absorption.

Alcoholism commonly causes a macrocytosis, although not specifically anaemia. Other types of liver disease can also cause macrocytosis.

Macrocytic anaemia can be further divided into 'megaloblastic anaemia' and 'non-megaloblastic macrocytic anaemia'. Megaloblastic anaemia is primarily a failure of DNA synthesis with preserved RNA synthesis, which results in restricted cell division of the progenitor cells. Non-megaloblastic macrocytic anaemias have different aetiologies (i.e. an unimpaired DNA globin synthesis).

---

Pernicious anaemia is a megaloblastic anaemia, caused by a deficiency of vitamin B12; it is associated with both haematopoietic and neurological disorder. In the stomach, vitamin B12 is bound to one of two B12 binding proteins present in gastric juice; in the less acidic environment of the small intestine, these proteins dissociate from the vitamin. It is then bound by intrinsic factor, produced by the parietal cells of the gastric mucosa; the B12–intrinsic factor complex is specifically bound by epithelial receptors in the ileum, where the vitamin B12 is then absorbed. Intrinsic factor is a glycoprotein. The most common cause for impaired binding of vitamin B12 by intrinsic factor is autoimmune atrophic gastritis. Autoantibodies are directed against parietal cells, which atrophy and cannot make intrinsic factor, and consequently cannot transport vitamin B12. Less frequently, loss of parietal cells may simply be part of a widespread atrophic gastritis of non-autoimmune origin, such as that frequently occurring in elderly people affected with long-standing chronic gastritis (including *H. pylori* infection).

Treatment of pernicious anaemia has traditionally involved the parenteral delivery of vitamin B12 to ensure absorption. Oral replacement is now an accepted route, using large doses of vitamin B12, 1–2 mg daily.

---

Normocytic anaemia occurs when the overall haemoglobin levels are decreased, although red blood cell volume remains normal. Causes include:

• acute blood loss

• anaemia of chronic disease

- aplastic anaemia (bone marrow failure)

- haemolytic anaemia.

## Treating anaemia

- Mild to moderate iron-deficiency anaemia is treated by iron supplementation with ferrous sulphate or ferrous gluconate. Vitamin C can aid iron absorption. A diagnosis of iron deficiency may indicate other potential sources of iron loss, such as gastrointestinal bleeding from ulcers or colon cancer.

- Vitamin supplements given orally (folic acid) or subcutaneously (vitamin B12) will replace specific deficiencies.

- In anaemia of chronic disease, anaemia associated with chemotherapy or anaemia associated with renal disease, recombinant erythropoietin can stimulate red cell production.

- In severe cases of anaemia, or with ongoing blood loss, a blood transfusion may be necessary. Blood transfusions for anaemia are generally to be avoided due to adverse clinical outcomes, but in severe, acute bleeding, transfusions of donated blood are often lifesaving.

- Hyperbaric oxygenation (HBO) can be a treatment for exceptional loss of blood. HBO may be used on medical or religious grounds: medically, if there is a threat of blood product incompatibility or concern of transmissible disease, religiously where belief prohibits the use of transfused blood.

# Focus on: angiogenesis

Angiogenesis is the physiological growth of new blood vessels from pre-existing vessels. It is a normal process in growth and development, as well as in wound healing, but is also a fundamental step in the transition of tumours from a dormant to a malignant state.

A variety of angiogenic factors have been implicated in the stimulation and control of angiogenesis:

- **FGF (Fibroblast growth factor).** The FGF family consists of at least 22 members. Most are 16–18 kDa single-chain peptides, which stimulate a variety of cellular functions by binding to cell-surface FGF (tyrosine kinase)-receptors. Receptor activation gives rise to a signal transduction cascade that leads to gene activation and diverse biological responses, including cell differentiation, proliferation and matrix dissolution, thus initiating a process of mitogenic activity critical for the growth of endothelial cells, fibroblasts and smooth-muscle cells.

- FGF-1 is the broadest acting member of the FGF family, and a potent mitogen for the diverse cell types needed to mount an angiogenic response in damaged (hypoxic) tissues, where upregulation of FGF-receptors occurs. FGF-1 stimulates the proliferation and differentiation of all cell types necessary for building an arterial vessel, including endothelial cells and smooth-muscle cells; this fact distinguishes FGF-1 from other pro-angiogenic growth factors, such as vascular endothelial growth factor (VEGF), which primarily drives the formation of new capillaries.

- **VEGF.** This has been demonstrated to be a major contributor to angiogenesis, increasing the number of capillaries in a given network. Upregulation of VEGF is a major component of the physiological response to exercise, and its role in angiogenesis is suspected to be a possible treatment in vascular injuries. *In vitro* studies demonstrate VEGF to be a potent stimulator of angiogenesis, causing endothelial cells to proliferate and migrate, eventually forming tube structures resembling capillaries. Binding to vascular endothelial growth factor receptor-2 (VEGFR-2) initiates a tyrosine kinase signalling cascade which stimulates the production of factors that variously stimulate vessel permeability, proliferation, migration and differentiation into mature blood vessels. Mechanically, VEGF is upregulated with muscle contractions as a result of increased blood flow to affected areas.

- **Angiopoietins.** These are protein growth factors that promote angiogenesis (as demonstrated by mouse knock-out studies). These growth factors bind to tyrosine kinase receptors. There are four identified angiopoietins, Ang1–4; in addition there are a number of proteins that are closely related to angiopoietins.

- **MMPs (Matrix Metalloproteinases).** These act to degrade those proteins that keep the vessel walls solid. This proteolysis allows the endothelial cells to escape into the interstitial matrix, as seen in sprouting angiogenesis. Inhibition of MMPs prevents the formation of new capillaries. The MMPs play an important role in tissue remodelling associated with various physiological and pathological processes, such as morphogenesis, angiogenesis, tissue repair, cirrhosis, arthritis and metastasis. MMP-2 and MMP-9 are thought to be important in metastasis; MMP-1 is thought to be important in rheumatoid and osteoarthritis.

- **NO (nitric oxide).** Widely considered to be a major contributor to the angiogenic response; reduction of NO significantly diminishes the effects of angiogenic growth factors.

## Angiogenesis as a therapeutic target

Angiogenesis may be a target for combating diseases in which there is either poor vascularisation or abnormal vasculature.

- **Tumour angiogenesis.** Tumours induce blood vessel growth by secreting various growth factors, such as FGF and VEGF; furthermore it has been shown that cancerous colon cells stop producing PKG. PKG, in normal cells, inhibits the action of VEGF, but in cancerous cells the lack of PKG would solicit angiogenesis.

- An increase in $\beta$-catenin production has been noted in individuals with basal cell carcinoma. $\beta$-catenin is known to play an important role in various aspects of liver biology, including liver development (both embryonic and postnatal), liver regeneration following partial hepatectomy, and the pathogenesis of liver cancer.

- Angiogenesis-based tumour therapy relies on natural and synthetic angiogenesis inhibitors, such as angiostatin, endostatin and tumstatin. These proteins mainly originate as specific fragments to pre-existing structural proteins such as collagen or plasminogen.

- A therapy recently approved for angiogenesis in cancer in the USA is a monoclonal antibody (Avastin) directed against an isoform of VEGF; Avastin has been approved for use in colorectal cancer.

- **Angiogenesis for cardiovascular disease**. Angiogenesis represents an excellent therapeutic target for the treatment of cardiovascular disease, namely the production of new collateral vessels to overcome the ischaemic insult. However, despite the large number of pre-clinical trials in animal models of cardiac ischaemia, no therapy designed to stimulate angiogenesis in underperfused tissue has yet become viable in man.

- **Fat loss**. White adipose tissue is vascularised, much like a tumour, and growth of adipose tissue is highly dependent on the building of new blood vessels (angiogenesis). Recent studies with obese mice models have shown that proapoptotic peptide, directed against blood vessels, results in decreased food intake and significant fat loss.

Bone is a relatively dynamic organ that undergoes significant turnover; that is, bone resorption and deposition; it is broken down by osteoclasts and rebuilt by osteoblasts. Besides an adequate supply of calcium, a close cooperation is required between these two types of cell. Complex signalling pathways achieve proper rates of growth and differentiation; these pathways include the action of several hormones, including parathyroid hormone (PTH), vitamin D, growth hormone, steroids and calcitonin, as well as several cytokines.

## 12.1 Mineral density test

A bone mineral density (BMD) test measures the density of minerals (such as calcium) using X-ray, computed tomography (CT) or ultrasound. Dual-energy X-ray absorptiometry (DEXA) is the most accurate way to measure BMD. DEXA can measure as little as 2% of bone loss per year; it is fast, uses low doses of radiation, but is more expensive than ultrasound.

An individual's 'T-score' is a standard deviation (SD) from the mean of a large group of healthy 30 year olds (young adult reference range) (Table 12.1).

## 12.2 Osteoblasts

Osteoblasts are mature, metabolically active, bone-forming cells that secrete osteoid (see Section 12.5). Mature osteoblasts synthesise type 1 collagen, osteocalcin, cell attachment proteins (thrombospondin, fibronectin, bone sialoprotein, osteopontin), proteoglycans and growth-related proteins. PTH binds to specific receptors on osteoblasts, increasing intracellular cAMP and stimulating ion and amino acid transport and collagen synthesis. Vitamin D stimulates synthesis of alkaline phosphatase, matrix and bone-specific proteins. The growth

**Table 12.1** T-score ranges

| Bone mineral density (BMD) | T-score |
|---|---|
| Normal | Less than 1 SD below reference range |
| Low (osteopenia) | 1–2.5 SD below range |
| Low (osteoporosis) | 2.5 SD or more below range |

factors transforming growth factor (TGF-$\beta$), insulin-like growth factors (IGFs) -I and -II and platelet-derived growth factor (PDGF) serve as autocrine regulators of osteoblast function.

---

Serum biochemical markers that reflect osteoblast function include:

- bone-specific alkaline phosphatase

- osteocalcin

- markers of collagen formation, carboxy-terminal procollagen extension peptide (ICTP) and amino-terminal procollagen extension peptide (PINP).

---

## 12.3 Osteoclasts

Osteoclasts are multinucleated cells found on the endosteal surface of bone, in Haversian systems and periosteal surfaces. PTH activates osteoclasts (indirectly via osteoblasts that possess PTH receptors). Calcitonin is a potent inhibitor of osteoclast activity. Local cytokine factors, including interleukin-1 (IL-1), tumour-necrosis factor (TNF), TGF-$\beta$ and interferon-$\gamma$ (INF-$\gamma$), are important regulators. Osteoclast resorption of bone releases collagen peptides, pyridinoline cross-links and calcium from the bone matrix, through the action of lysosomal enzymes (collagenases and cathepsins). The collagen breakdown products in serum and urine (e.g. hydroxyproline) can be used as biochemical markers.

## 12.4 Bone structure

There are three primary types of bone:

- woven

- cortical

- trabecular (cancellous).

Woven bone is formed during embryonic development, during fracture healing (callus formation) and in some pathological states, such as hyperparathyroidism and Paget's disease. It is composed of randomly arranged collagen bundles and irregularly shaped vascular spaces lined with osteoblasts. Woven bone is normally remodelled and replaced with cortical or trabecular bone.

Cortical bone, also called compact or lamellar bone, is remodelled from woven bone by means of vascular channels that invade the embryonic bone from its periosteal and endosteal surfaces. It forms the internal and external tables of flat bones and the external surfaces of long bones. The primary structural unit is an osteon, also known as a Haversian system, a cylindrical shaped lamellar bone surrounding longitudinally oriented vascular channels (the Haversian canals). Horizontally oriented canals (Volkmann canals) connect adjacent osteons. The mechanical strength of cortical bone results from the tight packing of the osteons.

Trabecular (cancellous) bone lies between cortical bone surfaces and consists of a network of honeycombed interstices containing haematopoietic elements and bony trabeculae. The trabeculae are predominantly oriented perpendicular to external forces to provide structural support. Trabecular bone continually undergoes remodelling on the internal endosteal surfaces.

## 12.5   Composition of bone

Bone comprises water, inorganic calcium phosphate and an organic matrix of fibrous protein and collagen. Osteoid is the unmineralised organic matrix secreted by osteoblasts, comprising 90% type I collagen and 10% ground substance, which consists of non-collagenous proteins, glycoproteins, proteoglycans, peptides, carbohydrates and lipids. Hydroxylapatite (hydroxyapatite) is a naturally occurring form of calcium apatite ($Ca_5(PO_4)_3(OH)$) that constitutes some 70% of bone mineralisation. Mineralisation of bone involves crystals of calcium phosphate that are laid down in precise amounts within the bone's fibrous matrix. Regulation of this process relies largely on a substance called inorganic pyrophosphate, which inhibits abnormal calcification; levels of this important bone regulator are controlled at least in part by alkaline phosphatase. The initial calcification of osteoid typically occurs within a few days of secretion but is completed over the course of several months.

## 12.6   Collagen

Collagen is the major insoluble fibrous protein in the extracellular matrix and in connective tissue; 80–90% of the collagen in the body consists of types I, II and III (Table 12.2). The collagen superfamily consists of at least 20 collagen types, with as many as 38 distinct polypeptide chains and more than 15 additional proteins that have collagen-like domains.

The basic structural unit of collagen is tropocollagen, which consists of three polypeptide chains ($\alpha$-peptides) that form left-handed helices, twisted together into a right-handed triple helix (the collagen microfibril). Microfibrils are arranged to form the collagen fibril. The characteristic right-handed triple-helical structure of collagen is a result of the three amino acids glycine, proline and hydroxyproline, which make up the characteristic repeating motif Gly-Pro-X, where X can be any amino acid. Each amino acid has a precise function. The side chain of glycine, an H atom, is the only one that can fit into the crowded centre of a three-stranded helix.

**Table 12.2**   Main collagen types

| Type | Composition | Tissue |
|------|-------------|--------|
| I | $[\alpha1(I)]2[\alpha2(I)]$ | Skin, tendon, bone, ligaments, dentin, interstitial tissues |
| II | $[\alpha1(II)]3$ | Cartilage, vitreous humour |
| III | $[\alpha1(III)]3$ | Skin, muscle, blood vessels |

Hydrogen bonds linking the peptide bond NH of a glycine residue with a peptide carbonyl (C=O) group in an adjacent polypeptide help hold the three chains together. The fixed angle of the C–N peptidyl–proline or peptidyl–hydroxyproline bond enables each polypeptide chain to fold into a helix with a geometry such that three polypeptide chains can twist together to form a three-stranded helix. Short segments at either end of the collagen chains are of particular importance in the formation of collagen fibrils. These segments do not assume the triple-helical conformation and contain the unusual amino acid hydroxylysine. Covalent aldol cross-links form between two lysine or hydroxylysine residues at the C-terminus of one collagen molecule, with two similar residues at the N-terminus of an adjacent molecule, thereby stabilising the side-by-side packing of collagen molecules and generating a strong fibril.

A procollagen triple helix is assembled in the endoplasmic reticulum; helix formation is aided by disulfide bonds between N- and C-terminal propeptides, which align the polypeptide chains. Post-translational modification of procollagen is crucial in allowing for collagen fibril formation. For example, in cells deprived of ascorbate, as in the disease scurvy, the procollagen chains are not hydroxylated sufficiently to form stable triple helices at normal body temperature (hydroxylation is through the activity of prolyl hydroxylase, which requires the cofactor ascorbic acid).

Collagen has specific structural requirements and is very susceptible to mutation, especially in glycine residues. As mutant collagen chains can affect the function of wild-type ones, such mutations have a dominant phenotype.

## 12.7   Bone disorders

Osteoporosis (fragile bone tissue) leads to an increased risk of fracture; it is most common in women after menopause (postmenopausal osteoporosis), and is the most common bone disorder.
   The main mechanisms underlying osteoporosis are:

- an inadequate peak bone mass (the skeleton develops insufficient mass and strength during growth)

- excessive bone resorption

- inadequate formation of new bone during remodelling.

It can occur in the presence of particular hormonal disorders and chronic diseases, or as a result of medications, specifically glucocorticoids (steroid- or glucocorticoid-induced osteoporosis). A major risk factor for osteoporosis in both men and women is advanced age . In females, oestrogen deficiency following menopause is correlated with a rapid reduction in BMD, while in men a decrease in testosterone has a comparable (but less pronounced) effect. European or Asian ancestry predisposes for osteoporosis.

Paget's disease is a chronic disorder that typically results in enlarged and deformed bones; breakdown and formation of bone tissue is excessive. As a result, bone can weaken, leading to bone pain, arthritis, deformities and fractures. Paget's disease is rarely diagnosed below 40 years of age; men and women are affected equally. The underlying problem resides with the osteoclasts; in affected areas of bone, abnormal osteoclasts (larger than normal) resorb more bone than normal. In response to this, osteoblasts increase in activity to make new bone material, but the increase in bone turnover leads to badly structured areas that are wrongly 'woven'.

Osteomalacia is a general term that describes the softening of the bone due to defective bone mineralisation. Osteomalacia in children is known as rickets. It is caused by a lack of vitamin D, calcium and phosphate, which may be a result of dietary or genetic factors or malabsorption disorders; kidney disease may cause a deficiency of vitamins and minerals, resulting in rickets. Vitamin D is vital for the growth and health of bone; without it, bones become soft, malformed and unable to repair themselves normally.

Osteogenesis imperfecta (sometimes known as brittle bone disease) is an autosomal dominant genetic bone disorder. There are eight different types, type I being the most common. Type I results from cysteine substitution of glycine in collagen; the larger cysteine molecule creates a steric hindrance that prevents correct formation of the collagen triple helix. Glycosylation of the cysteine molecule promotes further interference within the structure. Individuals with osteogenesis imperfecta either have less collagen than normal, or poorer quality collagen than normal.

Bone metastasis is one of the most frequent causes of pain in patients with cancer. Bone is a common site for circulating cancer cells to settle and grow, either near or far from the primary tumour site. Metastatic bone disease is not the same as primary bone cancer.

Kidney disease can be associated with a decreased ability to clear phosphorus; calcium levels in the blood become low and can lead to a loss of calcium from the bones.

Hyperparathyroidism leads to an increased release of PTH. Normally the same level of calcium leaves the bones and enters the blood as is absorbed by the bones from the blood; with an increase in PTH, more calcium leaves the bones than is absorbed, resulting in a net loss of calcium and leading to weakened bones.

# 12.8  Contributing factors to bone disorders

Some of the more common contributing factors are listed below:

- Chronic heavy drinking (alcohol intake greater than 2 units/day), especially at a younger age, increases risk significantly.

- Vitamin D deficiency is common among the elderly worldwide.

- Tobacco smoking inhibits the activity of osteoblasts and is an independent risk factor for osteoporosis. Smoking also results in increased breakdown of exogenous oestrogen, lower body weight and earlier menopause, all of which contribute to lower BMD.

- Malnutrition (low dietary intake of calcium and vitamins K and C) is associated with lower peak bone mass during adolescence.

- Physical inactivity. Bone remodelling occurs in response to physical stress, and weight-bearing exercise can increase peak bone mass in adolescence. In adults, physical activity helps maintain bone mass.

- Excess physical activity can lead to constant damage to the bones. There are numerous examples of marathon runners who develop severe osteoporosis later in life. In women, heavy exercise can lead to decreased oestrogen levels, which predisposes to osteoporosis.

- There is a strong association between cadmium, lead and bone disease. Low-level exposure to cadmium is associated with an increased loss of BMD in both genders, leading to pain and increased risk of fractures, especially in the elderly and in females. Higher cadmium exposure results in osteomalacia (softening of the bone).

- High body mass index may protect against osteoporosis, either by increasing load, or through the effects of the hormone leptin.

- Certain medications have been associated with an increase in osteoporosis risk, particularly steroids and anticonvulsants. Steroid-induced osteoporosis arises due to use of glucocorticoids, analogous to Cushing's syndrome, and involving mainly the axial skeleton; prednisone is a main candidate after prolonged intake. Barbiturates, phenytoin and some other antiepileptics probably accelerate the metabolism of vitamin D. Anticoagulants, for example heparin and warfarin, have been associated with a decreased bone density after prolonged use. Proton-pump inhibitors may interfere with calcium absorption.

## 12.9    Treatments of bone disorders

Lifestyle changes are frequently an aspect of treatment. Medications to treat osteoporosis, depending on gender, include:

- Bisphosphonates, a first-line treatment in women; they inhibit osteoclast resorptive activity.

- Teriparatide (a recombinant PTH, residues 1–34), which has been shown to be effective in osteoporosis. It acts like PTH and stimulates osteoblasts. It is used mostly for patients with established osteoporosis (who have already fractured), who have particularly low BMD or several risk factors for fracture, or who cannot tolerate the oral bisphosphonates.

- Strontium citrate, an alternative oral treatment with proven efficacy. In laboratory experiments, strontium ranelate was shown to stimulate the proliferation of osteoblasts, as well as inhibit the proliferation of osteoclasts.

- Oestrogen replacement therapy, a good treatment for prevention of osteoporosis. It is not recommended unless there are other indications for its use as well. There is uncertainty and controversy about whether oestrogen should be recommended in women in the first decade after the menopause.

## 12.10    Calcium homeostasis

Calcium is the most abundant mineral in the body, about 1 kg, 99% of which is found in the skeleton in the form of calcium phosphate salts. There is a significant exchange of calcium between the bone and extracellular fluid (ECF); serum levels of calcium are tightly regulated (it can vary with the level of serum albumin, to which calcium is bound).

The kidney excretes and reabsorbs calcium, leading to a small overall loss. The kidney also processes vitamin D into calcitrol, which assists in intestinal absorption of calcium; both processes are stimulated by PTH, which also regulates calcium release from bone.

Tight control of $[Ca^{2+}]_{ECF}$ is essential because calcium ions have a stabilising effect on voltage-gated ion channels. When $[Ca^{2+}]_{ECF}$ is too low (hypocalcaemia), voltage-gated ion channels start opening spontaneously, causing nerve and muscle cells to become hyperactive; this is called hypocalcaemic tetany. Conversely, when $[Ca^{2+}]_{ECF}$ is too high (hypercalcaemia), voltage-gated ion channels don't open as easily, and there is depressed nervous system function. In hypercalcaemia calcium will combine with phosphate ions, forming deposits of calcium phosphate (stones) in blood vessels and in the kidneys.

## 12.11    Endocrine regulation of $[Ca^{2+}]_{ECF}$

Endocrine regulation of $[Ca^{2+}]_{ECF}$ is dependent on PTH and $1,25(OH)_2D_3$ (the active form of vitamin D); the major regulator is PTH, which is part of a negative-feedback loop to maintain

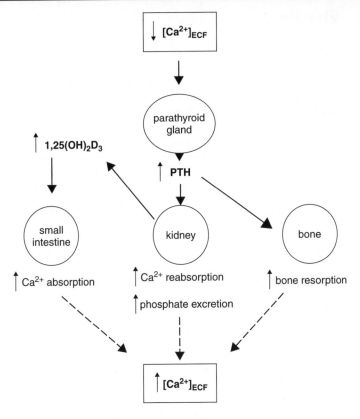

**Figure 12.1**   Endocrine regulation of $[Ca^{2+}]_{ECF}$.

$[Ca^{2+}]_{ECF}$. PTH secretion is stimulated by hypocalcaemia; it works through three mechanisms to increase $Ca^{2+}$ levels (Figure 12.1):

1. It stimulates the release of $Ca^{2+}$ from bone, stimulating bone resorption.

2. It decreases urinary loss of $Ca^{2+}$ by stimulating $Ca^{2+}$ reabsorption.

3. It indirectly stimulates $Ca^{2+}$ absorption in the small intestine by stimulating synthesis of $1,25(OH)_2D_3$ in the kidney.

## 12.12   Parathyroid hormone

PTH has a rapid effect (occurring within minutes), stimulating osteoblasts to pump $Ca^{2+}$ ions out of the fluid surrounding the bone (which has a higher $Ca^{2+}$ concentration) and into the ECF. Over a longer period, PTH stimulates bone resorption via its action on osteoblasts (osteoblasts express a signalling molecule that activates osteoclasts).

PTH has two important effects on the kidney that work to increase $[Ca^{2+}]_{ECF}$. First, it decreases the loss of $Ca^{2+}$ ions in the urine by stimulating $Ca^{2+}$ re-absorption (from the urine back into the ECF); second, it inhibits phosphate re-absorption.

## 12.13   Vitamin D

Vitamin D refers to a group of compounds that are formed in different parts of the body (Figure 12.2). Vitamin $D_3$ is synthesised in a photochemical reaction in the skin, in response to sunlight. A constitutively active enzyme in the liver (25-hydroxylase) produces 25-$(OH)D_3$, whereas PTH stimulates a kidney enzyme (1-hydroxylase), resulting in the production of $1,25(OH)_2D_3$; $1,25(OH)_2D_3$ acts in the small intestine to promote $Ca^{2+}$ absorption. In kidney disease, inadequate amounts of $1,25(OH)_2D_3$ are synthesised and $Ca^{2+}$ homeostasis must be maintained at the expense of that in bone. Hypocalcaemia stimulates high levels of PTH secretion; this is termed secondary hyperparathyroidism, since the problem that causes hyperparathyroidism is in the kidney, not in the parathyroid gland. Secondary hyperparathyroidism is treated by administering vitamin D and $Ca^{2+}$ supplements. The drug Cinacalcet has recently been approved for the treatment of secondary hyperparathyroidism. Cinacalcet is a calcimimetic drug that binds to the $Ca^{2+}$ receptor on cells in the parathyroid gland, inhibiting the secretion of PTH.

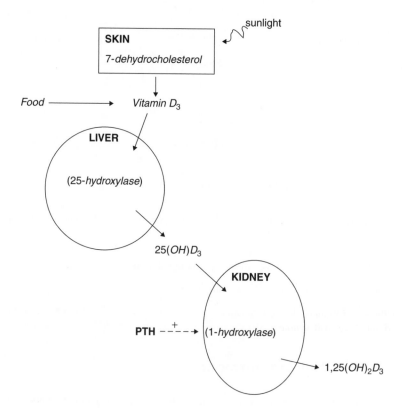

**Figure 12.2**   Synthesis of vitamin D forms.

# CHAPTER 13

# Intracellular signalling

The body uses a diverse range of chemical messengers to mediate and control growth, function and metabolism. These include hormones, eicosanoids and cytokines. Their effects may be evident throughout the body, or on the cell that synthesises them alone. They may affect one target or many targets about the body, in the same or sin different ways.

A chemical messenger may be defined as autocrine (affects the same cell that synthesises it), paracrine (affects a nearby target cell), intracrine (acts within the cell (e.g. steroid hormones)) or juxtacrine (signals are transmitted along cell membranes via protein or lipid components integral to the membrane and are capable of affecting either the emitting cell or cells immediately adjacent).

## 13.1 Hormones

Hormones are those messengers produced by cells in the endocrine glands that serve as regulators at the overall organism level; however, they may also exert their effects solely within the tissue in which they are produced. General features of endocrine glands are their ductless nature, their vascularity and the intracellular secretory granules that store their hormones. Figure 13.1 shows the major human endocrine glands.

Hormones are also synthesised by exocrine glands and secreted via ducts. These are the counterparts to the endocrine glands, which secrete their products directly into the bloodstream. Typical exocrine glands include sweat, salivary, mammary and gastrointestinal glands, as well as the liver and the pancreas.

Several hormones may control one process (e.g. regulation of plasma glucose), or one hormone may control several processes.

*Essential Biochemistry for Medicine*   Dr Mitchell Fry
© 2010 John Wiley & Sons, Ltd

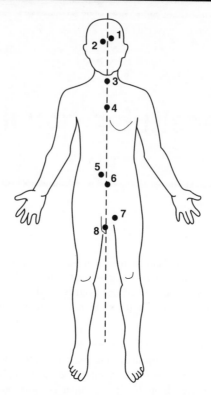

**Figure 13.1** Major human endocrine glands: (1) pineal gland, (2) pituitary gland, (3) thyroid gland, (4) thymus, (5) adrenal gland, (6) pancreas, (7) ovary and (8) testes.

Hormone immunoassay is the most widely applied technique for detecting and quantitating hormones in biological samples. Most immunoassays employ monoclonal antibodies, produced by fusion of spleen cells from an immunised mouse with a mouse myeloma cell line. The resulting hybridoma secretes a single antibody. Measuring hormones is a matter of timing. Most hormones are released in bursts, from single bursts to sustained release; they also conform to strict biological rhythms, for example occurring once an hour (e.g. luteinizing hormone (LH)), a day (e.g. cortisol) or a month (e.g. progesterone).

## 13.2 The hierarchical nature of hormonal control

The master coordinator of hormonal activity in mammals is the hypothalamus, which acts on input that it receives from the central nervous system. Hormone secretion from the anterior pituitary gland is regulated by 'releasing' hormones secreted by the hypothalamus. Neuroendocrine neurons in the hypothalamus project axons to the median eminence, at the base of the brain; these neurons release substances into small blood vessels that travel directly to the anterior pituitary gland (the hypothalamo–hypophysial portal vessels). While the anterior pituitary

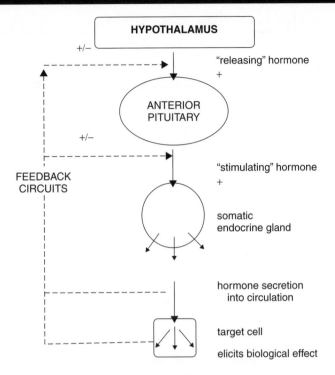

**Figure 13.2** The hierarchical control of hormone release. Releasing hormones from the hypothalamus travel directly to the anterior pituitary; hormones secreted by the anterior pituitary control most somatic endocrine glands.

controls hormone release in most somatic endocrine glands, the pituitary itself is controlled by the hypothalamus (Figure 13.2).

Releasing hormones, from the hypothalamus, control the release of other hormones. The main releasing hormones secreted by the hypothalamus are:

- thyrotropin-releasing hormone (TRH)

- corticotropin-releasing hormone (CRH)

- gonadotropin-releasing hormone (GnRH)

- growth hormone-releasing hormone (GHRH).

Two others are also classed as releasing hormones, although they in fact inhibit pituitary hormone release, namely somatostatin and dopamine.

Hormones synthesised by the anterior pituitary that regulate the activity of various endocrine glands are referred to as tropic hormones (e.g. thyroid-stimulating hormone (TSH), adrenocorticotropic hormone (ACTH), luteinizing hormone (LH), follicle-stimulating hormone (FSH)).

The major hormones secreted by the pituitary gland are listed in Table 13.1.

A number of endocrine glands that signal one another in sequence are usually referred to as an axis. The hypothalamo–pituitary axes are listed in Table 13.2.

**Table 13.1**    Major hormones secreted by the pituitary gland

| Hormone | | Target | Effect |
|---|---|---|---|
| **Anterior pituitary** | | | |
| Adrenocorticotropic hormone | ACTH | Adrenal gland | Secretion of glucocorticoids |
| Endorphins | | Opioid receptor | Inhibit perception of pain |
| Follicle-stimulating hormone | FSH | Ovaries, testes | Growth of reproductive system |
| Human growth hormone | hGH | Liver, adipose tissue | Promotes growth; lipid and carbohydrate metabolism |
| Luteinizing hormone | LH | Ovaries, testes | Sex hormone production |
| Prolactin | | Ovaries, mammary glands | Secretion of oestrogens, progesterone, milk production |
| Thyroid-stimulating hormone | TSH | Thyroid gland | Secretion of thyroid hormones |
| **Posterior pituitary** | | | |
| Oxytocin | | Uterus, mammary gland ducts | Contraction of smooth muscle of uterus and myoepithelial cells lining mammary ducts |
| Vasopressin (antidiuretic hormone) | ADH | Kidney, constriction of arterioles | Re-absorption of water, increase in blood pressure |

**Table 13.2**    The hypothalamo–pituitary axes

| Hypothalamic hormone | +TRH −Somatostatin | +CRH −AVP | +GnRH | +GHRH −Somatostatin | −Dopamine |
|---|---|---|---|---|---|
| Anterior pituitary hormone | TSH | ACTH | FSH/LH | GH | Prolactin |
| Target organ | Thyroid | Adrenal cortex | Ovary/testes | Liver tissues | Breast |
| Target organ hormone | $T_4/T_3$ | Cortisol | Oestradiol Testosterone | IGF-1 | |

TRH = thyrotropin-releasing hormone, TSH = thyroid-stimulating hormone, CRH = corticotrophin-releasing hormone, AVP = arginine vasopressin, ACTH = adrenocorticotropic hormone, GnRH = gonadotropin-releasing hormone, FSH = follicle-stimulating hormone, LH = luteinizing hormone, GHRH = growth hormone-releasing hormone, GH = growth hormone, IGF-1 = insulin-like growth factor.

## 13.3   Hormone synthesis and secretion

Protein and peptide hormones are often synthesised as an extended peptide, a pro-hormone (inactive) or even a pre-pro-hormone (e.g. insulin); these are stored within the cell in secretory granules, which subsequently fuse with the plasma membrane to release their contents (by exocytosis). The pro-hormone must be cleaved to produce the active molecule. In contrast, steroid hormones are fat-soluble and readily cross membranes, so they cannot be stored, but are synthesised as needed. Their release is therefore slower than that of peptide hormones. Peptide hormones are both degraded quickly and excreted in the urine; steroid and thyroid hormones are transported in plasma bound to proteins and may remain in the plasma for days.

---

Protein and peptide hormones must be administered more frequently if used therapeutically. They cannot be administered by mouth since they would be degraded in the digestive tract. Hence, diabetics must regularly inject with insulin.

---

## 13.4   Hormonal control

The physiological effect of hormones, and of other chemical messengers, depends largely upon their concentrations in the blood and extracellular fluids; this concentration is determined primarily by:

1. **Rate of production.** Synthesis and secretion of hormones is highly regulated by both positive- and negative-feedback circuits.

2. **Rate of delivery.** Dependent on, for example, the blood flow to a target organ or group of target cells.

3. **Rate of degradation and elimination.** Like all biomolecules, chemical messengers have characteristic rates of decay and biological half-lives, and are metabolised and excreted from the body through several routes.

Hierarchical control and feedback control, both positive and negative, are a fundamental feature of endocrine systems (Figure 13.2). Each of the major hypothalamic–pituitary–hormone axes is governed by negative feedback:

- thyroid hormones, by negative feedback on the TRH–TSH axis

- cortisol, by negative feedback on the CRH–ACTH axis

- gonadal steroids, by negative feedback on the GnRH–LH/FSH axis

- insulin-like growth factor (IGF-1), by negative feedback on the GHRH–GH axis.

---

Such regulatory circuits include both positive (e.g. TRH, TSH) and negative (e.g. $T_4$, $T_3$) components. A fall in $T_4/T_3$ levels causes a positive feedback and increase in TRH and TSH synthesis; once $T_4/T_3$ levels are restored to a new steady state, negative feedback suppresses TRH and TSH.

# Focus on: Cushing's syndrome

Cushing's syndrome (hypercorticism) is an endocrine hormone disorder caused by high levels of cortisol in the blood. Cushing's disease refers to one specific cause, a non-cancerous tumour (adenoma) in the pituitary gland that produces a large amount of ACTH, which in turn elevates cortisol. It can usually be cured by surgery.

The paraventricular nucleus of the hypothalamus releases corticotrophin-releasing hormone, which stimulates the pituitary gland to release adrenocorticotropin (ACTH). ACTH travels via the blood to the adrenal gland, where it stimulates the release of cortisol. Cortisol is secreted by the cortex of the adrenal gland from a region called the zona fasciculata in response to ACTH. Elevated levels of cortisol exert negative feedback on the pituitary, which decreases the amount of ACTH released from the pituitary gland.

Cushing's syndrome may be sub-classified according to whether or not the excess cortisol is dependent on increased ACTH:

- ACTH-dependent Cushing's syndrome is driven by increased ACTH and includes exogenous ACTH administration as well as pituitary and ectopic Cushing's.

- ACTH-independent Cushing's syndrome shows increased cortisol, but the ACTH is not elevated but rather decreased due to negative feedback. It can be caused by exogenous administration of glucocorticoids or by adrenal adenoma, carcinoma or nodular hyperplasia.

Symptoms may include:

- rapid weight gain, particularly of the trunk and face

- excess sweating, dilation of capillaries, thinning of the skin (which causes easy bruising and dryness, particularly the hands)

- insomnia, reduced libido, impotence and infertility, caused by the effect of excess cortisol on other endocrine systems

- various psychological disturbances, ranging from euphoria to psychosis, depression and anxiety

- polyuria, persistent hypertension (due to cortisol's enhancement of adrenaline's vasoconstrictive effect) and insulin resistance (especially common in ectopic ACTH production), leading to hyperglycaemia, which in turn can result in diabetes mellitus

- hyperpigmentation, due to melanocyte-stimulating hormone, produced as a by-product of ACTH synthesis from pro-opiomelanocortin (POMC)

- gastrointestinal disturbances, opportunistic infections and impaired wound healing. Cortisol is a stress hormone, so depresses the immune and inflammatory responses

- osteoporosis, caused by the stress-like response; the body's maintenance of bone (and other tissues) becomes secondary to maintenance of the false stress response

- hypokalaemia (low plasma potassium), caused by mineralcorticoid activity

# 13.5 Types of chemical messenger

As knowledge is gained about the variety and types of chemical messengers, a definitive classification becomes more challenging; some terminology has already become redundant.

- **Hormones** are released by cells that (usually) affect other distant cells in other parts of the body. Hormone biosynthetic cells are typically of a specialised cell type, residing within a particular endocrine or exocrine gland. Hormones in animals are often transported in the blood. Endocrine hormones are secreted (released) directly into the bloodstream, while exocrine hormones are secreted directly into a duct, and from the duct they either flow into the bloodstream or flow from cell to cell by diffusion in a process known as paracrine signalling. Vertebrate hormones fall into three chemical classes: amine-derived hormones (derivatives of the amino acids tyrosine and tryptophan), for example catacholamines, thyroxine; peptide/protein hormones, for example insulin and growth hormone, including glycoproteins such as LH, FSH and TSH; and lipid and phospholipid-derived hormones, for example steroid hormones (glucocorticoids, mineralocorticoids, androgens, oestrogens and progestagens).

- **Eicosanoids** are lipids derived from arachidonic acid; they include the prostaglandins, prostacyclins, thromboxanes and leukotrienes. In humans, eicosanoids are local hormones that are released by most cells, act on those same cells or nearby cells (i.e. they are autocrine and paracrine mediators), and then are rapidly inactivated. Eicosanoids have a short half-life, ranging from seconds to minutes. They exert complex control over many bodily systems, mainly in inflammation or immunity, and act as messengers in the central nervous system. Most eicosanoid receptors are members of the G-protein-coupled receptor superfamily.

- **Cytokines** are proteins, peptides or glycoproteins. The action of cytokines may be autocrine, paracrine or endocrine. The term 'cytokine' encompasses a large and diverse family of polypeptide regulators that are produced widely throughout the body by cells of diverse embryological origin. Historically, the term 'cytokine' has been used to refer to the immunomodulating agents (interleukins, interferons, etc.). The distinction between hormone and cytokine is muddled. Classic protein hormones circulate in nanomolar ($10^{-9}$) concentrations that usually vary by less than one order of magnitude. In contrast, some cytokines (such as IL-6) circulate in picomolar ($10^{-12}$) concentrations that can increase up to 1000-fold during trauma or infection. The widespread distribution of cellular sources for cytokines may be a feature that differentiates them from hormones. Virtually all nucleated cells, but especially endo/epithelial cells and resident macrophages, are potent producers of certain cytokines. In contrast, classic hormones, such as insulin, are secreted from discrete glands (e.g. the pancreas). As of 2008, the current terminology refers to cytokines as 'immunomodulating agents'.

- **Chemokines** are a family of small (8–10 kDa) protein cytokines, with four cysteine residues in conserved locations that are key to forming their three-dimensional shape. Their name is derived from their ability to induce directed chemotaxis in nearby responsive cells; they are *chemo*tactic cyto*kines*.

- **Lymphokines** are cytokines secreted by helper T-cells, in response to stimulation by antigens, which act on other cells of the immune system.

## 13.6    Intracellular signalling and signal transduction

For any chemical messenger to elicit an effect it must first be recognised by the cell. Peptide messengers interact with cell-surface receptors (peptides do not cross the membrane), whereas steroid- and lipid-derived messengers diffuse across the cell membrane and interact with intracellular receptors. Ligand-gated ion-channel receptors are a class of receptor that may occur either at the cell surface or within the cell.

Recognition of the initial signal initiates signal transduction, an ordered sequence of biochemical reactions. Such processes are usually rapid, lasting in the order of milliseconds in the case of ion flux, minutes for the activation of protein- and lipid-mediated kinase cascades, or hours and even days for gene expression. The number of proteins and other molecules participating in the signal transduction event increases as the process emanates from the initial stimulus, resulting in a signal cascade; thus a small stimulus elicits a large response, referred to as signal amplification.

## 13.7    Cell-surface and intracellular receptors

### 13.7.1    Cell-surface receptors

Cell-surface receptors are integral transmembrane proteins that span the plasma membrane of a cell, with both an extracellular and an intracellular domain. Signal transduction is initiated when the chemical messenger, the ligand, binds to its specific receptor at the cell surface. Transmission of the signal across the plasma membrane is induced by a change in the shape (the conformation) of the intracellular domain of the receptor. Changes in conformation result in the activation of enzyme activity, either within the receptor, or through the exposure of a binding site for other signalling proteins within the cell. Most of the intracellular proteins that are activated by a ligand–receptor interaction possess enzymic activity.

- **Transmembrane receptors** include G-protein-coupled receptors, receptor tyrosine kinases, integrins, pattern-recognition receptors and ligand-activated ion channels.

- **G-protein-coupled receptors, guanine nucleotide-binding proteins,** are a family of proteins involved in second messenger cascades; they alternate between an inactive guanosine diphosphate (GDP) and active guanosine triphosphate (GTP) bound state, ultimately regulating downstream cell processes. G-proteins belong to a large group of enzymes called GTPases.

    G-proteins can refer to two distinct families of proteins:

- **Heterotrimeric G-proteins,** sometimes referred to as the 'large' G-proteins, which are activated by G-protein-coupled receptors and made up of alpha ($\alpha$), beta ($\beta$) and gamma ($\gamma$) subunits.

- **'Small' G-proteins** (20–25 kDa), which belong to the Ras superfamily of small GTPases. These proteins are homologous to the alpha ($\alpha$) subunit found in heterotrimers, and are in fact monomeric. However, they also bind GTP and GDP and are involved in signal transduction.

    Different types of heterotrimeric G-proteins share a common mechanism. They are activated in response to a conformation change in the G-protein-coupled receptor, exchange GDP for

GTP, and dissociate to activate other proteins in the signal transduction pathway. The specific mechanisms, however, differ among the types.

Receptor-activated G-proteins are bound to the inside surface of the cell membrane. They consist of the $G_\alpha$ and the tightly associated $G_{\beta\gamma}$ subunits. There are four classes of $G_\alpha$ subunit: $G_{\alpha s}$, $G_{\alpha i}$, $G_{\alpha q/11}$ and $G_{\alpha 12/13}$. They behave differently in the recognition of the effector, but share a similar mechanism of activation.

When a ligand activates the G-protein-coupled receptor, it induces a conformational change in the receptor that allows the receptor to function as a guanine nucleotide exchange factor (GEF), which exchanges GDP for GTP on the $G_\alpha$ subunit. In the traditional view of heterotrimeric protein activation, this exchange triggers the dissociation of the $G_\alpha$ subunit, bound to GTP, from the $G_{\beta\gamma}$ dimer and the receptor. Both $G_\alpha$-GTP and $G_{\beta\gamma}$ can then activate different signalling cascades (or second messenger pathways) and effector proteins, while the receptor is able to activate the next G-protein.

The $G_\alpha$ subunit will eventually hydrolyse the attached GTP to GDP by its own inherent enzymic activity, allowing it to re-associate with $G_{\beta\gamma}$ and begin a new cycle. A group of proteins known as GTPase-activating proteins (GAPs), which are specific for $G_\alpha$ subunits, act to accelerate hydrolysis and terminate the transduced signal. In some cases the effector itself may possess intrinsic GAP activity, which helps deactivate the pathway. This is true in the case of phospholipase $C_\beta$, which possesses GAP activity within its C-terminal region. This is an alternate form of regulation for the $G_\alpha$ subunit.

The activated G-protein subunits detach from the receptor and initiate signalling from many downstream effector proteins, including phosphodiesterases, adenyl cyclase, phospholipases and ion channels, which permit the release of second messenger molecules such as cyclic adenosine monophosphate (cAMP), cyclic guanosine monophosphate (cGMP), inositol triphosphate, diacylglycerol and calcium ions.

---

Adenyl cyclase, also referred to as adenylate cyclase or adenylyl cyclase, is a lyase enzyme (in biochemistry, a lyase is an enzyme that catalyses the breaking of various chemical bonds by means other than hydrolysis and oxidation, often forming a new double bond or a new ring structure. Lyases differ from other enzymes in that they only require one substrate for the reaction in one direction, but two substrates for the reverse reaction). There are 10 known adenyl cyclases in mammals, ADCY1 through ADCY10.

Adenyl cyclase catalyses the conversion of ATP to $3',5'$-cAMP and pyrophosphate (Figure 13.3):

$$ATP \rightarrow cAMP + PPi$$

Adenyl cyclase is a transmembrane protein, consisting of 12 membrane loops. Its functional domains are located in the cytoplasm, sub-divided into the N-terminus, C1a, C1b, C2a and C2b. The C1a and C2a domains form a catalytic dimer where ATP binds and is converted to cAMP.

Adenyl cyclase is stimulated by G-proteins and by forskolin, as well as other class-specific substrates.

- Isoforms I, III and VIII are also stimulated by $Ca^{2+}$/calmodulin.

- Isoforms V and VI are inhibited by $Ca^{2+}$ in a calmodulin-independent manner.

In neurons, adenyl cyclases are located next to calcium ion channels for faster reaction to $Ca^{2+}$ influx; they are suspected to play an important role in learning processes.

Adenyl cyclase can be activated or inhibited by G-proteins, which are coupled to membrane receptors and thus can respond to hormonal or other stimuli. Following activation of adenyl cyclase, the resulting cAMP acts as a second messenger by interacting with and regulating other proteins, such as protein kinase A and cyclic nucleotide-gated ion channels.

- **Receptor tyrosine kinases**. These are transmembrane receptor proteins with an intracellular kinase domain and an extracellular domain that binds the ligand. Examples include many growth factor receptors, such as the insulin receptor and the insulin-like growth factor receptors. Upon ligand binding, the tyrosine kinase receptor dimerises, causing autophosphorylation of tyrosines within the cytoplasmic tyrosine kinase, resulting in their conformational change. The kinase domain of the receptors is subsequently activated, initiating signalling cascades of phosphorylation of downstream cytoplasmic molecules (Figure 13.4).

The mutation of certain receptor tyrosine kinase genes can result in the expression of receptors that exist in a constitutively active state (i.e. they are always produced). Such mutated genes may act as oncogenes, genes that contribute to the initiation or progression of cancer.

- **Integrins**. These are produced by a wide variety of cell types and play a role in the attachment of a cell to the extracellular matrix, as well as to other cells.

- **Pattern recognition receptors**. These are proteins expressed by cells of the immune system that identify molecules associated with microbial pathogens or cellular stress. They include the so-called Toll-like receptors.

- **Ligand-activated ion channels**. These recognise a specific ligand and then undergo a structural change that opens a gap (channel) in the plasma membrane through which ions can pass. These ions will then relay the signal. An example of this mechanism is found in the receiving cell, the post-synaptic cell, of a neural synapse. By contrast, other ion channels open in response to a change in cell-membrane potential.

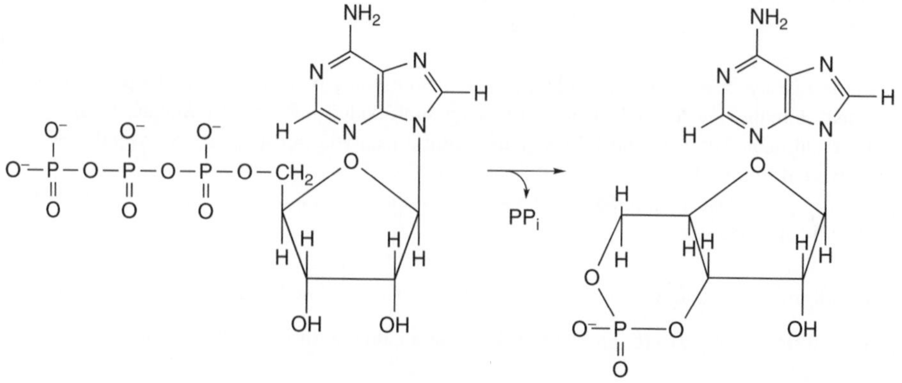

**Figure 13.3**  Conversion of ATP to 3′,5′-cAMP and pyrophosphate.

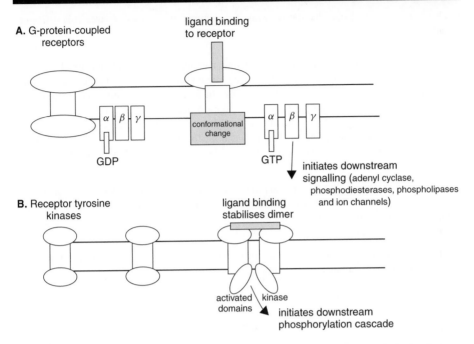

**Figure 13.4** Transmembrane receptors must bind their ligand and propagate their signal across the plasma membrane, usually by way of a protein conformational change. On the cytosolic face of the plasma membrane, that change initiates an enzymic sequence (e.g. G-protein or tyrosine kinase), leading to 'downstream' events.

## 13.7.2 Intracellular receptors

Intracellular receptors include both nuclear receptors and cytoplasmic receptors, soluble proteins that are localised within the nucleoplasm and the cytoplasm, respectively. Typical ligands include the steroid hormones (e.g. testosterone, progesterone, cortisol) and derivatives of vitamins A and D, which freely diffuse through the plasma membrane.

Nuclear receptors are ligand-activated transcription activators. On binding the ligand (the hormone), the activated receptor passes through the nuclear membrane to enable the transcription of a certain gene, and thus the production of a protein. The activated nuclear receptors attach to DNA at receptor-specific hormone-responsive elements (HREs), DNA sequences that are located in the promoter region of the genes and activated by the hormone–receptor complex. The activation of gene transcription is much slower than signals that directly affect existing proteins. As a consequence, the effects of hormones that use nucleic receptors are usually long-term. The steroid receptors are a subclass of nuclear receptors, located primarily within the cytosol. In the absence of steroid hormone, the receptors cling together in a complex called an aporeceptor complex, which also contains chaperone proteins (also known as heat shock proteins). Chaperone proteins are necessary to activate the receptor by assisting the protein with folding in such a way that the signal sequence that enables its passage into the nucleus is accessible. Steroid receptors can also have a repressive effect on gene expression, when their transactivation domain is hidden and cannot activate transcription. Furthermore, steroid receptor activity can be enhanced by phosphorylation of serine residues at their N-terminal end, as a result of another signal transduction pathway, for example by a growth factor. This behaviour is called crosstalk.

- **The glucocorticoid receptor protein** contains a number of different functional domains, including a DNA-binding domain. The glucocorticoid receptor resides in the cytosol, complexed with a variety of proteins including so-called heat shock proteins plus a number of other binding proteins. Upon diffusion of the glucocorticoid hormone cortisol across the cell membrane into the cytoplasm, binding to the glucocorticoid receptor occurs, resulting in release of the heat shock proteins. The activated glucocorticoid receptor homodimerises and is actively translocated into the nucleus, where it binds to specific DNA-responsive elements and activates gene transcription.

- **Retinoid X receptor (RXR) and orphan receptors** are not accompanied by chaperone proteins. In the absence of hormone, they bind to their specific DNA sequence, repressing the gene. Upon activation by the hormone, they activate the transcription of the gene that they were repressing.

- **NOD-like receptors** belong to the pattern recognition receptors group and are involved in the regulation of inflammatory and apoptotic processes.

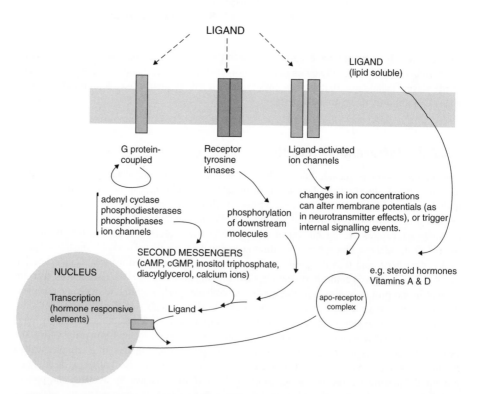

**Figure 13.5**  General scheme of intracellular signalling. Different strategies for communicating signals into the cell and propagating them within the cell are invariably directed to the nucleus and the control of transcription.

## 13.8   Second messengers

Second messengers are signalling molecules that are rapidly generated/released and can then go on to activate effector proteins within the cell to exert a cellular response (see Figure 13.5). There are three basic types of secondary messenger molecule:

- Hydrophobic molecules like diacylglycerol, inositol triphosphate (IP3) and phosphatidylinositols, which are membrane-associated and diffuse from the plasma membrane into the cell, where they can reach and regulate membrane-associated effector proteins.

- Hydrophilic molecules like cAMP, cGMP and calcium, which are located within the cytosol.

- Gases, such as nitric oxide and carbon monoxide, which can diffuse both through the cytosol and across cellular membranes.

## Focus on: the insulin receptor

The insulin receptor is a transmembrane receptor belonging to the tyrosine kinase receptor class (Figure 13.6). Two $\alpha$ and two $\beta$ subunits make up the insulin receptor. The $\beta$ subunits pass through the cell membrane and are linked by disulphide bonds. The $\alpha$ and $\beta$ subunits are encoded by a single gene (*INSR*).

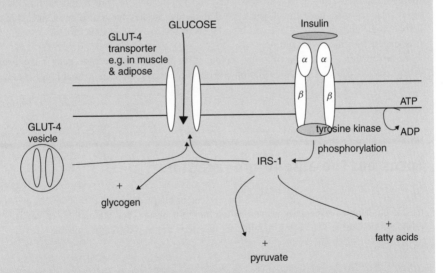

**Figure 13.6**   Insulin-receptor signalling. Insulin binding to its receptor causes the phosphorylation of 'substrate proteins', in this case IRS-1; IRS-1 mediates the recruitment of GLUT-4 transporters to the plasma membrane. IRS-1 also has effects on glycogen synthesis, glycolysis and fatty acid production.

Binding of insulin to its receptor can initiate a number of signal cascades, affecting translocation of GLUT-4 transporters to the plasma membrane and influx of glucose, glycogen synthesis, glycolysis and fatty acid synthesis. Activation of the tyrosine kinase receptor leads to phosphorylation of 'substrate' proteins and their activation. IRS-1 (insulin receptor protein) is one such substrate protein that leads to the movement of high-affinity GLUT-4 glucose transporters to the plasma membranes of insulin-responsive tissues (e.g. muscle and adipose).

Glycogen synthesis is also stimulated by the insulin receptor via IRS-1. The activated receptor kinase IRS-1 complex converts membrane phosphatidylinositol, which indirectly activates a protein kinase via phosphorylation. The activated kinase phosphorylates several target proteins, including glycogen synthase kinase. Glycogen synthase kinase is responsible for phosphorylating (and thus deactivating) glycogen synthase. When glycogen synthase kinase is phosphorylated, it is deactivated and prevented from deactivating glycogen synthase. In this roundabout manner, insulin increases glycogen synthesis.

Insulin insensitivity, or a decrease in insulin-receptor signalling, leads to diabetes mellitus type 2; the cells are unable to take up glucose and the result is hyperglycaemia (an increase in circulating glucose). The nature of insulin insensitivity has been difficult to ascertain; in some patients the insulin receptor is abnormal, in others one or more aspect of insulin signalling is defective. Hyperinsulinaemia, excessive insulin secretion, is most commonly a consequence of insulin resistance, associated with type 2 diabetes. More rarely, hyperinsulinaemia results from an insulin-secreting tumour (insulinoma). At the cellular level, down-regulation of insulin receptors occurs due to high circulating insulin levels, apparently independently of insulin resistance. There is clearly an inherited component; sharply increased rates of insulin resistance and type 2 diabetes are found in those with close relatives who have developed type 2 diabetes. Studies have also implicated high-carbohydrate and -fructose diets, and high levels of fatty acids and inflammatory cytokines (associated with the obese state).

A few patients with homozygous mutations in the insulin-receptor gene have been described; this causes Donohue syndrome or leprechaunism. This autosomal recessive disorder results in a totally non-functional insulin receptor.

## Focus on: the adrenergic receptors

The adrenergic receptors are a class of G-protein-coupled receptor that are targets of the catecholamines, especially noradrenaline (norepinephrine) and adrenaline (epinephrine) (although dopamine is a catecholamine, its receptors are in a different category).

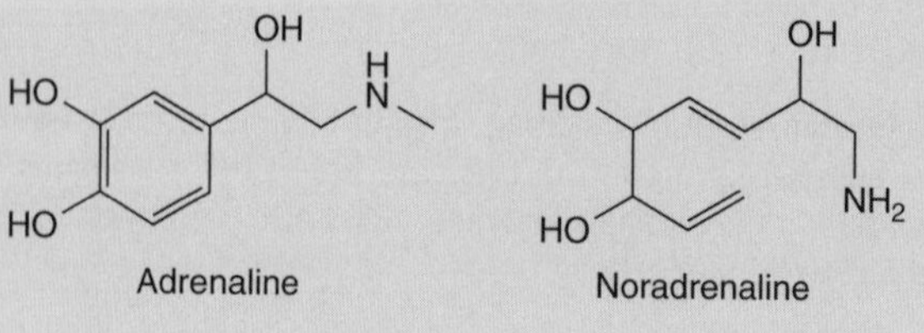

Adrenaline                    Noradrenaline

**Figure 13.7** Adrenaline and noradrenaline.

**Table 13.3**   $\alpha$-$\beta$-adrenergic receptors

| Receptor type | Selected action | Mechanism | Antagonists |
|---|---|---|---|
| $\alpha_1$ | Smooth-muscle contraction. Vasoconstriction of coronary artery, vasoconstriction of veins, decreased motility of GI smooth muscle | $G_q$: phospholipase C is activated, cleaves phosphatidylinositol, increases inositol trisphosphate (IP3) and diacylglycerol; the former interacts with $Ca^{2+}$ channels and calcium increased | $\alpha$-1 blockers, for example alfuzosin, doxazosin, phentoxybenzamine, phentolamine |
| $\alpha_2$ | Smooth-muscle contraction and neurotransmitter inhibition | $G_i$: adenyl cyclase is inactivated, cAMP is reduced | $\alpha$-2 blockers, for example yohimbine, idazoxan, atipamezole |
| $\beta_1$ | Heart-muscle contraction increases cardiac output by raising heart rate (positive chronic tropic) and increasing impulse conduction and contraction, thus increasing the volume expelled with each beat (increased ejection fraction) | $G_s$: adenyl cyclase is activated, cAMP is increased | $\beta$ blockers, for example metoprolol, atenolol |
| $\beta_2$ | Smooth-muscle relaxation, for example in bronchi, non-pregnant uterus, detrusor urinae muscle of bladder wall. Vasodilation of arteries to skeletal muscle. Contract sphincters of GI tract. Increases lipolysis in adipose tissue, increases anabolism in skeletal muscle, increase glycogenolysis and gluconeogenesis. Promotes insulin release from pancreatic $\beta$ cells. Inhibits histamine release from mast cells. | $G_s$: adenyl cyclase is activated, cAMP is increased | $\beta$ blockers, for example butxamine, propranolol |

There are two main groups of adrenergic receptor, $\alpha$ and $\beta$, with several subtypes:

- $\alpha$-receptors have the subtypes $\alpha_1$ (a $G_q$-coupled receptor) and $\alpha_2$ (a $G_i$-coupled receptor). Phenylephrine is a selective agonist of the $\alpha$-receptor.

- $\beta$-receptors have the subtypes $\beta_1$, $\beta_2$ and $\beta_3$. All three are linked to $G_s$-proteins, which in turn are linked to adenyl cyclase. Adrenaline binds its receptor, which associates with a heterotrimeric G-protein. The G-protein associates with adenyl cyclase, which converts ATP to cAMP, spreading the signal.

Adrenaline and noradrenaline (Figure 13.7) are receptor ligands to $\alpha_1$-, $\alpha_2$- or $\beta$-adrenergic receptors.

- $\alpha_1$ couples to $G_q$, resulting in an increased intracellular $Ca^{2+}$, leading to smooth-muscle contraction.

- $\alpha_2$ couples to $G_i$, resulting in a decrease of cAMP activity, leading to smooth-muscle relaxation.

- $\beta$-receptors couple to $G_s$, resulting in an increased intracellular cAMP, leading to heart-muscle contraction, smooth-muscle relaxation and glycogenolysis.

Adrenaline reacts with both $\alpha$- and $\beta$-adrenergic receptors, causing vasoconstriction and vasodilation, respectively. Although $\alpha$-receptors are less sensitive to adrenaline, when activated they override the vasodilation mediated by $\beta$-adrenergic receptors. The result is that high levels of circulating adrenaline cause vasoconstriction. At lower levels of circulating adrenaline, $\beta$-adrenergic-receptor stimulation dominates, producing an overall vasodilation.

The actions and mechanisms of different receptor types are summarised in Table 13.3.

# 13.9   The glucagon receptor

The glucagon receptor is a member of the G-protein-coupled family of receptors.

On binding glucagon, the receptor undergoes a conformational change, activating a $G_s$-protein. On substitution of GDP for GTP, the $\alpha$-subunit of the heterotrimeric $G_s$-protein activates adenyl cyclase. Adenyl cyclase manufactures cAMP, which activates protein kinase A, activating phosphorylase B kinase, which phosphorylates phosphorylase $b$, converting it to the active form phosphorylase $a$. Phosphorylase $a$ is the enzyme responsible for the release of glucose-1-phosphate from glycogen.

Glucagon receptors are mainly expressed in the liver and kidney, with lesser amounts in other tissues. Glucagon binding to its receptor on hepatocytes causes the liver to release glucose from glycogen (glycogenolysis), as well as synthesise additional glucose by gluconeogenesis.

# 13.10   The gastrin receptor

The gastrin receptor is one of the receptors that bind cholecystokinin (Table 13.4), and is known as the CCK-B receptor; it is another member of the G-protein-coupled receptor family.

**Table 13.4** Cholecystokinin receptors

| Protein receptor | Main tissue distribution | Function |
|---|---|---|
| CCKA (gastrin) | Gastrointestinal tract | Stimulation of bicarbonate secretion, gall-bladder emptying and inhibition of gut motility |
| CCKB | CNS | Regulation of anxiety, memory and hunger |

Binding of gastrin stimulates an increase in intracellular $Ca^{2+}$, activation of protein kinase C and production of inositol phosphate.

The five C-terminal amino acids of gastrin and cholecystokinin are identical, which explains their overlapping biological effects.

Proglumide (Milid) acts as a cholecystokinin antagonist, blocking both the $CCK_A$ and $CCK_B$ subtypes. It was used mainly in the treatment of stomach ulcers but has now been largely replaced by newer drugs.

The histamine receptor is another example of a G-protein-coupled receptor.

# CHAPTER 14

# Inflammation

## 14.1 The acute inflammatory response

The acute inflammatory response (see Table 14.1) leads to inflammation, redness, swelling, heat and pain, due to extravasation (leakage) of plasma and infiltration of leukocytes into the site of inflammation. Mechanisms that allow cells and proteins to gain access to damaged or infected extravascular sites include:

- vasodilation

- increased vascular permeability

- cellular infiltration.

Vasodilation occurs first at the arteriole level, progressing to the capillary level, with a net increase in the amount of blood present, causing the redness and heat of inflammation.

Increased vascular permeability and the slowing of blood flow are induced by cells already present in all tissues, mainly fixed macrophages (dendritic cells, endothelial cells and mastocytes); these release various inflammatory mediators (for example bradykinin increases the sensitivity to pain). Increased permeability of the vessels results in the movement of plasma into the tissues (this may lead to oedema (swelling)); the subsequent increase in cellular concentration of the blood leads to stasis.

Stasis allows leukocytes to marginate along the endothelium, a process critical to their recruitment into the damaged tissues (in normal flowing blood the shearing force along the periphery of the vessels moves cells into the middle). After margination, cells will cross the blood vessel wall (diapedesis) into the extravascular tissue, along a chemotactic gradient. Cellular infiltration occurs over a few hours, with the appearance of granulocytes, particularly neutrophils, in the tissue. Erythrocytes may also leak into the tissues and a haemorrhage can occur (i.e. a blood blister). If the vessel is damaged, fibrinogen and fibronectin are deposited, platelets aggregate and red cells stack together to aid clot formation. The dead and dying cells contribute to pus formation.

*Essential Biochemistry for Medicine*   Dr Mitchell Fry
© 2010 John Wiley & Sons, Ltd

**Table 14.1** Comparison between acute and chronic inflammation

|  | Acute | Chronic |
|---|---|---|
| Causative agent | Pathogens, injured tissue | Non-degradable pathogens, foreign bodies, autoimmune reaction |
| Major cells involved | Neutrophils, mononuclear cells (monocytes, macrophages) | Mononuclear cells (monocytes, macrophages, lymphocytes, plasma cells), fibroblasts |
| Primary mediators | Vasoactive amines, eicosanoids | IFN-$\gamma$ and other cytokines, growth factors, reactive oxygen species, hydrolytic enzymes |
| Onset | Immediate | Delayed |
| Duration | A few days | Can be months or years |
| Outcome | Resolution, abscess formation, chronic inflammation | Tissue destruction, fibrosis |

In addition to cell-derived mediators, several acellular biochemical cascade systems act in parallel to initiate and propagate the inflammatory response. These include:

- the complement system, activated by bacteria (Section 15.3)

- the coagulation system (Section 11.2)

- the fibrinolysis systems, activated by necrosis (e.g. a burn or a trauma); fibrinolysis is the breakdown of fibrin by plasmin (Section 11.10).

The acute inflammatory response requires constant stimulation if it is to be sustained. Inflammatory mediators have short half-lives and are quickly degraded in the tissue. Hence, inflammation ceases once the stimulus has been removed.

Recruitment of leukocytes is receptor-mediated. The products of inflammation, such as histamine, promote the immediate expression of P-selectin on endothelial cell surfaces. This receptor binds weakly to carbohydrate ligands on leukocyte surfaces and causes them to 'roll' along the endothelial surface as bonds are made and broken. Cytokines from injured cells induce the expression of E-selectin on endothelial cells, which functions similarly to P-selectin. Cytokines also induce the expression of integrin ligands on endothelial cells, which further slow the movement of leukocytes. These weakly bound leukocytes are free to detach if not activated by chemokines produced in injured tissue. Activation increases the affinity of bound integrin receptors for ligands on the endothelial cell surface, firmly binding the leukocytes to the endothelium.

# 14.2    Leukocyte transmigration

Migration across the endothelium, known as transmigration, occurs via the process of diapedesis. Chemokine gradients stimulate the adhered leukocytes to move between endothelial cells and pass the basement membrane into the tissues.

Movement of leukocytes within the tissue occurs via chemotaxis. Leukocytes reaching the tissue interstitium bind to extracellular matrix proteins via expressed integrins and CD44 to

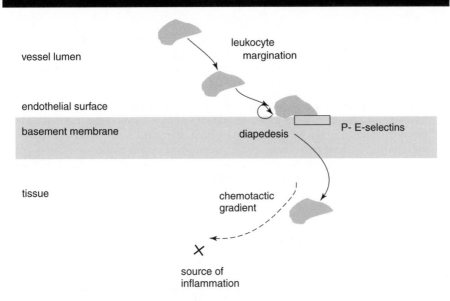

**Figure 14.1** Leukocyte transmigration. Expression of P- and E-selectins (and integrins) on the endothelial surface of damaged cells weakly binds leukocyte carbohydrate ligands, slowing and causing the leukocyte to 'roll' over the endothelial surface, eventually holding the leukocyte in place.

prevent their loss from the site. Chemoattractants cause the leukocytes to move along a chemotactic gradient towards the source of inflammation (Figure 14.1).

- **P-selectin** plays an essential role in the initial recruitment of leukocytes. When endothelial cells are activated by molecules such as histamine or thrombin during inflammation, P-selectin moves from an internal cell location to the endothelial cell surface. Thrombin is one trigger of endothelial-cell release of P-selectin. Ligands for P-selectin on eosinophils and neutrophils are similar; they are sialylated carbohydrates, but clearly different from those reported for E-selectin. P-selectin attaches to the actin cytoskeleton through anchor proteins.

- **E-selectin** recognises and binds to sialylated carbohydrates present on the surface proteins of certain leukocytes. The local release of cytokines IL-1 and TNF induces the over-expression of E-selectin on the endothelial cells of nearby blood vessels. Leukocytes with a low affinity for E-selectin will slow and 'roll' along the internal surface of the vessel.

- **Integrins** are receptors that mediate attachment between a cell (and the tissues surrounding it) and other cells or the extracellular matrix. They also play a role in cell signalling and help define cellular shape and mobility, and regulate the cell cycle. Integrins transduce information 'outside-to-in', as well as 'in-to-outside'.

- **Chemotaxis** is the phenomenon in which cells and bacteria direct their movements according to certain chemicals in their environment. In multicellular organisms, chemotaxis is critical to early (e.g. movement of sperm towards the egg during fertilisation) and subsequent (e.g. migration of neurons or lymphocytes) phases of development, as well as to normal function. It is now recognised that mechanisms that allow chemotaxis in animals can be subverted during cancer metastasis.

- **The extracellular matrix** is the extracellular part of animal tissues that provides structural support; it is the defining feature of connective tissue. Extracellular matrix includes the interstitial matrix and the basement membrane. Interstitial matrix is present between the cells (that is, in the intercellular spaces); it consists of polysaccharides and fibrous proteins, which act as a compression buffer. Basement membranes are sheet-like depositions of extracellular matrix on which various epithelial cells rest.

# 14.3   Chronic inflammation

Chronic inflammation (see Table 14.1) is a pathological condition characterised by concurrent active inflammation, tissue destruction and attempts at tissue repair. Chronically inflamed tissue is characterised by the infiltration of mononuclear immune cells (including monocytes, macrophages, lymphocytes and plasma cells), tissue destruction and attempts at tissue repair (including angiogenesis and fibrosis). Endogenous causes include persistent acute inflammation. Exogenous causes include bacterial infection, especially by *Mycobacterium tuberculosis*, prolonged exposure to chemical agents such as silica or tobacco smoke, and autoimmune reactions as seen in rheumatoid arthritis. In chronically inflamed tissue the stimulus is persistent; therefore recruitment of monocytes is maintained, existing macrophages are tethered in place and proliferation of macrophages is stimulated (especially in atheromatous plaques).

Mast cells appear to be key players in the initiation of inflammation. Cytoplasmic granules contain mediators of inflammation, and their surface is coated with a variety of receptors which, when engaged by the appropriate ligand, trigger exocytosis of the granules. Toll-like receptors trigger exocytosis when they interact with the lipopolysaccharide of Gram-negative bacteria or the peptidoglycan of Gram-positive bacteria. Complement receptors trigger exocytosis when they bind C3a and C5a (Table 14.2). Activated mast cells release dozens of potent mediators, leading to:

- Cell recruitment to the site of damage (monocytes that infiltrate tissues to become macrophages, neutrophils, antigen-presenting dendritic cells, lymphocytes including B- and T-cells (leading to an adaptive immune response), natural killer cells (an effector cell in innate immunity) and eosinophils).

- Activation of many of these recruited cells, producing their own mediators of inflammation.

Resolution may occur with restoration of normal tissue architecture; blood clots are removed by fibrinolysis. If it is not possible to return the tissue to its original form, scarring results from in-filling with fibroblasts, collagen and new endothelial cells. Infectious agents may be walled off from the surrounding tissue in a granuloma; this is a ball of cells formed when macrophages

**Table 14.2** Plasma-derived mediators

| | |
|---|---|
| Bradykinin (produced by the kinin[a] system) | A vasoactive protein which is able to induce vasodilation, increase vascular permeability, cause smooth-muscle contraction and induce pain. |
| C3 (produced by the complement[b] system) | Cleaves to produce C3a and C3b. C3a stimulates histamine release by mast cells, thereby producing vasodilation. C3b is able to bind to bacterial cell walls and act as an opsonin, which marks the invader as a target for phagocytosis. |
| C5a (produced by the complement[b] system) | Stimulates histamine release by mast cells, thereby producing vasodilation. It is also able to act as a chemoattractant to direct cells via chemotaxis to the site of inflammation. |
| Factor XII, Hageman factor (produced in the liver) | A protein which circulates inactively, until activated by collagen, platelets or exposed basement membranes via conformational change. When activated, it in turn is able to activate three plasma systems involved in inflammation: the kinin system, fibrinolysis system and coagulation system. |
| Membrane-attack system (produced by the complement system[b]) | A complex of the complement proteins C5b, C6, C7, C8 and multiple units of C9. The combination and activation of this range of complement proteins forms the membrane-attack complex, which is able to insert into bacterial cell walls and causes cell lysis with ensuing death. |
| Plasmin (produced by the fibrinolysis system) | Able to break down fibrin clots, cleave complement protein C3 and activate factor XII. |
| Thrombin (produced by the coagulation system) | Cleaves the soluble plasma protein fibrinogen to produce insoluble fibrin, which aggregates to form a blood clot. Thrombin can also bind to cells via the PAR1 receptor to trigger several other inflammatory responses, such as production of chemokines and nitric oxide. |

[a] The kinin-kallikrein system (kinin system) is a poorly delineated system of blood proteins that plays a role in inflammation, blood-pressure control, coagulation and pain. Kinins are small peptides produced from kininogen by kallikrein, which are subsequently degraded by kininases. They act on phospholipase and increase arachidonic acid release and thus prostaglandin ($PGE_2$) production.
[b] The complement system is discussed in Chapter 15.

and lymphocytes accumulate together with epitheloid cells and gigant cells (perhaps derived from macrophages).

## 14.4 Mediators of inflammation

There are many mediators of inflammation, some plasma derived and some cell derived. Examples of some are given in Tables 14.2 and 14.3.

## 14.5 Acute-phase proteins

These are proteins whose plasma concentrations increase (positive acute-phase proteins) or decrease (negative acute-phase proteins) in response to inflammation. This response is called the

**Table 14.3**  Cell-derived mediators

| | |
|---|---|
| Lysosome granules, containing enzymes from granulocytes | Contain a large variety of enzymes, which perform a number of functions. |
| Histamine, vasoactive amine from mast cells, basophils and platelets | Stored in preformed granules and released in response to a number of stimuli; causes arteriole dilation and increased venous permeability. |
| IFN-$\gamma$, cytokine from T-cells and natural killer cells | Antiviral, immunoregulatory and anti-tumour properties. Originally called macrophage-activating factor. Especially important in the maintenance of chronic inflammation. |
| IL-8, chemokine, primarily from macrophages | Activation and chemoattraction of neutrophils, with a weak effect on monocytes and eosinophils. |
| Leukotriene B4, eicosanoid from leukocytes | Mediates leukocyte adhesion and activation, allowing them to bind to the endothelium and migrate across it. In neutrophils, a potent chemoattractant and able to induce the formation of reactive oxygen species and the release of lysosomal enzymes by these cells. |
| Nitric oxide, soluble gas from macrophages, endothelial cells | Potent vasodilator, relaxes smooth muscle, reduces platelet aggregation, aids in leukocyte recruitment, has direct antimicrobial activity in high concentrations. |
| Prostaglandins, eicosanoid from mast cells | A group of lipids which can cause vasodilation, fever and pain. |
| TNF-$\alpha$ and IL-1, cytokines, primarily from macrophages | Both affect a wide variety of cells to induce many similar inflammatory reactions: fever, production of cytokines, endothelial gene regulation, chemotaxis, leukocyte adherence, activation of fibroblasts. Responsible for the systemic effects of inflammation, such as loss of appetite and increased heart rate. |

acute-phase reaction (or acute-phase response). On injury, local inflammatory cells (neutrophils, granulocytes and macrophages) secrete a number of cytokines into the blood stream, most notable of which are the interleukins IL-1, -6 and -8, and TNF-$\alpha$. The liver responds by producing a large number of acute-phase reactants. At the same time, the production of a number of other proteins is reduced (negative acute-phase reactants).

Positive acute-phase proteins serve different physiological functions for the immune system. Some act to destroy or inhibit growth of microbes, for example C-reactive protein, mannose-binding protein, complement factors, ferritin, ceruloplasmin, serum amyloid A and haptoglobulin. Others give negative feedback on the inflammatory response, for example serpins; $\alpha$2-macroglobulin and coagulation factors affect coagulation.

Measurement of acute-phase proteins, especially C-reactive protein, is a useful marker of inflammation in both medical and veterinary clinical pathology. Their presence may also indicate liver failure.

**Table 14.4**  *Morphological patterns of inflammation*

| | |
|---|---|
| Granulomatous inflammation | Formation of granulomas is the result of a limited but diverse number of diseases, including tuberculosis, leprosy and syphilis. |
| Fibrinous inflammation | Formation of a fibrinous exudate, which can be converted to a scar. |
| Purulent inflammation | Inflammation which results in a large amount of pus (pyogenic bacteria such as staphylococci is characteristic of this kind of inflammation). Large, localised collections of pus enclosed by surrounding tissues are called abscesses. |
| Serous inflammation | Effusion of non-viscous serous fluid; skin blisters exemplify this pattern of inflammation. |
| Ulcerative inflammation | Inflammation occurring near an epithelium can result in the necrotic loss of tissue from the surface, exposing lower layers. The subsequent excavation in the epithelium is known as an ulcer. |

# 14.6   Patterns of acute and chronic inflammation

Morphological patterns of acute and chronic inflammation are seen during particular situations that arise in the body, such as when inflammation occurs on an epithelial surface, or when pyogenic bacteria (bacterial infections that make pus) are involved (see Table 14.4).

# 14.7   Inflammatory disorders

These form a large, unrelated group of disorders which underlie a variety of human diseases. The immune system is often involved with inflammatory disorders, demonstrated in both allergic reactions and some myopathies. Non-immune diseases with aetiological origins in inflammatory processes are thought to include cancer, atherosclerosis (Chapter 5) and ischaemic heart disease.

## Focus on: asthma

Asthma is an example of a type 1 hypersensitivity (Chapter 15). The incidence of asthma in the United States (as well as in many other developed countries) has reached epidemic proportions.

In an asthmatic attack, the bronchi become constricted, making it difficult to breathe, especially to breathe out. Severe attacks can be life-threatening. An attack of asthma begins when an allergen is inhaled; the allergen binds to IgE antibodies on mast cells in the lungs, triggering degranulation (exocytosis) of the mast cells and the release of a variety of substances, but in particular histamine and leukotrienes, leading to:

• Smooth-muscle-cell contraction of the bronchi, narrowing the lumen of the bronchi. This is referred to as the 'early phase'.

- Attraction and accumulation of inflammatory cells, especially eosinophils, with the production of mucus. This is referred to as the 'late phase'. With repeated attacks, the lining of the bronchi becomes damaged.

Although asthma begins as an allergic response, subsequent attacks can be triggered by a variety of nonspecific factors (cold air, exercise, tobacco smoke).

For reasons not fully understood, some individuals have a predisposition to respond to antigens by making antibodies of the IgE class; the trait tends to run in families, suggesting a genetic component. Such individuals are said to suffer from 'atopy'.

T-helper cells of atopic people are largely of the Th2 type rather than the Th1; that is, they will induce IgE class switching in B-cells (Th2) rather than IgG class switching (Th1). Th2 helper cells synthesise IL-4 and IL-13, which promote class switching, and release IL-5, which attracts eosinophils and other inflammatory cells to the site, producing the late phase of the response.

'Knock-out' mice, for a transcription factor T-bet (required for Th1-cell synthesis), make fewer Th1 and more Th2 cells, and suffer the lung changes typical of human asthma, even though they are not exposed to any particular allergen.

Asthma is mostly a disease of developed countries. The reason for this is uncertain, but it might be that sanitation and widespread childhood immunisation have enabled children to avoid the infections, especially viral, that stimulate the immune system to respond with Th1 helper cells, rather than Th2 cells. Children in Europe who give a positive response to tuberculin (the tuberculin test), a response mediated by Th1 cells, have lower rates of asthma than children who are negative in the tuberculin test. However, children in tropical, undeveloped countries, who are often infected with parasitic worms, have high levels of Th2 cells and IgE, but a very low incidence of asthma. Perhaps a variety of chronic infections in childhood activate mechanisms (e.g. production of regulatory T-cells) that suppress inflammatory immune responses, both Th1- and Th2-mediated. It is certainly not something as simple as air pollution; although air pollution can trigger asthma, some regions with heavily polluted air have a much lower incidence of asthma than regions with relatively clean air.

---

Atopy, or atopic syndrome, is an allergic hypersensitivity affecting parts of the body not in direct contact with the allergen. It can involve eczema (atopic dermatitis), allergic conjunctivitis, allergic rhinitis and asthma. There appears to be a strong hereditary component. Atopic syndrome can be fatal for those who experience serious allergic reactions such as anaphylaxis. Although atopy has various definitions, it is most consistently defined by the presence of elevated levels of total and allergen-specific IgE in the serum, leading to positive skin-prick tests to common allergens.

---

Treatments of asthma include:

- **$\beta$2-adrenergic agonists**. Drugs such as salbutamol (albuterol) mimic the action of adrenaline (epinephrine), relaxing the smooth muscle of the bronchi. They may be inhaled or given by mouth. While useful in the early phase of an attack, they provide no protection against the long-term damage produced during the late phase.

- **Corticosteroids**. These reduce the inflammation of the late phase of the response. They may be given in an inhaler (e.g. beclomethasone) or by mouth (e.g. prednisone).

- **Cromolyn sodium** (sodium cromoglycate) inhibits exocytosis of mast cells, thus blocking the release of histamine and leukotrienes. It is used mainly to prevent attacks (e.g. those triggered by exercise).

- **Leukotriene inhibitors**. one leukotriene inhibitor blocks leukotriene synthesis by inhibiting the action of 5-lipooxygenase (Zileuton; Zyflo). Another blocks leukotriene receptors on the surface of smooth-muscle cells and eosinophils (Montelukast; Singulair).

# Focus on: urticaria

Urticaria (or hives) is a skin rash notable for dark red, raised, itchy bumps. It is caused by the release of histamine and other mediators of inflammation (cytokines) from cells in the skin. This process can be the result of an allergic or non-allergic reaction, differing in the eliciting mechanism of histamine release.

The skin lesions of urticarial disease are caused by an inflammatory reaction in the skin, causing leakage of capillaries in the dermis and resulting in an oedema which persists until the interstitial fluid is absorbed into the surrounding cells. Types of urticaria include:

- **Allergic urticaria**. Histamine and other pro-inflammatory substances are released from mast cells in the skin and tissues in response to the binding of allergen-bound IgE antibodies to high-affinity cell-surface receptors. Basophils and other inflammatory cells release histamine and other mediators, and are thought to play an important role, especially in chronic urticarial diseases.

- **Autoimmune urticaria**. This seems to be an underlying basis of many cases of chronic idiopathic urticaria. About one third of patients with chronic urticaria spontaneously develop auto-antibodies directed at the receptor FcεRI, located on skin mast cells. Chronic stimulation of this receptor leads to chronic hives. Patients often have other autoimmune conditions, such as autoimmune thyroiditis.

- **Infectious urticaria**. This is commonly seen with viral illnesses, such as the common cold. It usually appears three to five days after the cold has started, and sometimes even a few days after the cold has resolved.

- **Non-allergic urticaria**. Many drugs, for example morphine, can induce direct histamine release without an immune component. A diverse group of signalling substances called neuropeptides have been found to be involved in emotionally induced urticaria. Dominantly inherited cutaneous and neurocutaneous porphyrias (porphyria cutanea tarda, hereditary coproporphyria, variegate porphyria and erythropoietic porphyria) have been associated with solar urticaria. The occurrence of drug-induced solar urticaria may be associated with porphyrias. This may be caused by IgG binding rather than IgE binding.

- **Stress and chronic idiopathic urticaria**. These forms have been anecdotally linked to both poor emotional well-being and a reduced health-related quality of life. Evidence has been found for a link between stressful life events (e.g. bereavement, divorce, etc.) and chronic idiopathic urticaria, and preliminary evidence shows a link with post-traumatic stress.

## Treatment and management

Chronic urticaria can be difficult to treat. Drug treatment is typically in the form of antihistamines, such as diphenhydramine, hydroxyzine, cetirizine and other H1-receptor antagonists. The H2-receptor antagonists, such as cimetidine and ranitidine, may help control symptoms either prophylactically or by lessening their effects during an attack. When taken in combination with an H1 antagonist they have been shown to have a synergistic effect. Treatment of urticaria with ranitidine or other H2 antagonists is considered an off-label use, since these drugs are primarily used in the treatment of peptic ulcer disease and gastroesophageal reflux disease (Chapter 4). Tricyclic antidepressants, such as doxepin, are also potent H1 and H2 antagonists, and may have a role in therapy, although side effects limit their use. For very severe outbreaks, an oral corticosteroid such as prednisone is sometimes prescribed. An analogue of $\alpha$-melanocyte-stimulating hormone, called afamelanotide, is in clinical trial for the treatment of solar urticaria.

# Focus on: rheumatoid arthritis

Rheumatoid arthritis is a chronic, systemic inflammatory disorder that may affect many tissues and organs, but principally attacks the joints, producing an inflammatory synovitis; it often progresses to destruction of the articular cartilage and ankylosis (from the Greek, meaning 'bent', 'crooked') of the joints. With time it nearly always affects multiple joints; it is a polyarthritis. Rheumatoid arthritis can also produce diffuse inflammation in the lungs, pericardium, pleura and sclera. About 1% of the world's population is estimated to be afflicted by rheumatoid arthritis, women three times more often than men. Onset is most frequent at 40–50 years. The arthritis of rheumatoid arthritis is due to synovitis, inflammation of the synovial membrane, which lines joints and tendon sheaths. Joints become swollen, tender and warm, and stiffness limits their movement. Most commonly, small joints of the hands, feet and cervical spine are affected, but larger joints like the shoulder and knee can also be involved; this varies from individual to individual. Synovitis can lead to tethering of tissue, with loss of movement and erosion of the joint surface, causing deformity and loss of function.

Although the exact cause of rheumatoid arthritis is unknown, autoimmunity seems to play a pivotal role in its chronicity and progression.

Key pieces of evidence relating to the pathogenesis are:

- A genetic link with human leukocyte antigen (HLA)-DR4 and related allotypes of MHC class II and the T-cell-associated protein protein tyrosine phosphatase N22 (PTPN22).

- A link with cigarette smoking that appears to be causal.

- A dramatic response in many cases to blockade of the cytokine TNF ($\alpha$).

- A similar dramatic response in many cases to depletion of B-lymphocytes, but no comparable response to depletion of T-lymphocytes.

- A more or less random pattern of whether and when predisposed individuals are affected.

- The presence of autoantibodies to IgGFc, known as rheumatoid factors (RFs), and antibodies to citrullinated peptides (ACPAs).

---

The best known genes in the MHC region are the subset that encode antigen-presenting proteins on the cell surface. In humans, these genes are referred to as human leukocyte antigen (HLA) genes. The human immune response, d-related antigen, encoded by the d locus on chromosome 6 and found on lymphoid cells, is strongly associated with rheumatoid arthritis and juvenile diabetes. A missense polymorphism in the PTPN22 gene has also been shown to be associated with risk of developing rheumatoid arthritis, as well as some other autoimmune diseases, such as lupus and type 1 diabetes.

---

Data suggests that the disease involves abnormal B-cell–T-cell interaction, with presentation of antigens by B-cells to T-cells via HLA-DR eliciting T-cell help and consequent production of RFs and ACPAs. Inflammation is then driven by B-cell or T-cell products stimulating release of TNF and other cytokines. The process may be facilitated by an effect of smoking on citrullination.

---

Citrullination is the term used for the post-translational modification of the amino acid arginine to citrulline. The reaction is performed by enzymes called peptidylarginine deiminases. The conversion of arginine into citrulline can have important consequences for the structure and function of proteins, since arginine is positively charged at a neutral pH, whereas citrulline is uncharged. This increases the hydrophobicity of the protein, leading to protein unfolding. Proteins such as fibrin and vimentin become citrullinated during cell death and tissue inflammation. Fibrin and fibrinogen may be favoured sites for citrullination within rheumatoid joints. Tests for the presence of anti-citrullinated protein antibodies are highly specific (88–96%) for rheumatoid arthritis, and are detectable even before the onset of clinical disease.

---

Various treatments are available. Non-pharmacological treatment includes physical therapy and occupational therapy. Analgesics (painkillers) and anti-inflammatory drugs, as well as steroids, are used to suppress the symptoms, while disease-modifying antirheumatic drugs (DMARDs) are often required to inhibit or halt the underlying immune process and prevent long-term damage.

---

DMARDs are an otherwise unrelated group of drugs defined by their use in rheumatoid arthritis. The term 'disease-modifying antirheumatic drug' is often used in contrast to non-steroidal anti-inflammatory drugs (NSAIDs), which refers to agents that treat the inflammation but not the underlying cause. Although the use of DMARDs was first demonstrated in rheumatoid arthritis (hence their name), they have come to be used in the treatement of many other diseases, such as Crohn's disease, lupus erythematosus (SLE), idiopathic thrombocytopenic purpura and myasthenia gravis. Many of these are autoimmune disorders, but others, such as ulcerative colitis, are not. Some DMARDs are mild chemotherapeutics, but they are also immunosuppressants.

# Focus on: aspirin

Aspirin (acetylsalicylic acid; Figure 14.2; see also Table 14.5) is synthesised by the treatment of salicylic acid with acetic anhydride.

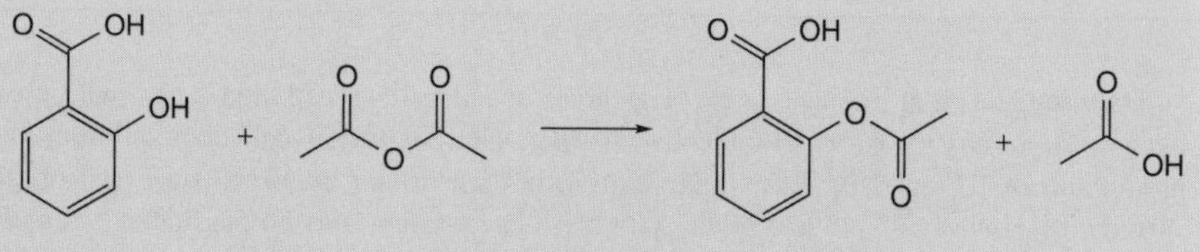

**Figure 14.2**   Aspirin.

**Table 14.5**   Main actions of aspirin

| Action | Mechanism |
|---|---|
| Anti-inflammatory | Decreased prostaglandin synthesis. |
| Analgesic | Decreased $PGE_2$ synthesis; $PGE_2$ normally sensitises nerves to histamine and bradykinin. |
| Antipyretic | Decreased $PGE_2$ synthesis; $PGE_2$ normally elevates the set-point of the hypothalamic thermoregulator, and is normally elevated during infection or inflammation. |
| Respiratory | Therapeutic doses have been shown to uncouple oxidative phosphorylation in cartilaginous and hepatic mitochondria; high doses may actually cause fever due to the heat released from uncoupled respiration. |
| Gastrointestinal | $PGE_2$ normally acts to inhibit gastric acid production, and also stimulates secretion of mucous. Reduction in $PGE_2$ (by aspirin) may lead to ulceration and gastric distress. |
| Platelets | Low doses of aspirin seem to preferentially inhibit synthesis of thromboxane $A_2$, which normally promotes platelet aggregation. |
| Kidney | Prostaglandins ($PGE_2$) normally maintain the blood flow to the kidneys, particularly in the presence of vasoconstrictors. Aspirin reduces prostaglandin levels and increases water and sodium retention. |

Often used as an analgesic to relieve minor aches and pains, as an antipyretic to reduce fever, and as an antiinflammatory medication, aspirin also has an antiplatelet or 'anticoagulate' effect through inhibition of thromboxane synthesis. Under normal circumstances thromboxane binds platelets together to repair damaged blood vessels; for this reason aspirin is used in long-term low doses to prevent heart attacks, stroke and blood-clot formation in susceptible individuals. Low doses of aspirin may be given immediately after a heart attack to reduce the risk of another heart attack or of the death of cardiac tissue.

The main undesirable side effects of aspirin are gastrointestinal ulcers, stomach bleeding and tinnitus (the perception of sound within the human ear in the absence of corresponding external sound), especially at higher doses. Aspirin is no longer used in children and adolescents due to the risk of Reye's syndrome (a potentially fatal disease that causes numerous detrimental effects to many organs, especially the brain and liver. It is associated with aspirin consumption by children with viral diseases such as chickenpox). Paracetamol or ibuprofen are now used instead. In the United Kingdom, the only indications for aspirin use in children and adolescents under 16 are Kawasaki disease and prevention of blood-clot formation. Kawasaki disease is an inflammation of the middle-sized arteries. It affects many organs, including the skin, mucous membranes, lymph nodes and blood-vessel walls, but the most serious effect is on the heart, where it can cause severe aneurysmal dilations.

Today, aspirin is one of the most widely used medications in the world, with an estimated 40 000 metric tons being consumed each year.

Aspirin was the first discovered member of the class of drugs known as non-steroidal anti-inflammatory drugs, not all of which are salicylates, although most owe their effects to inhibition of the enzyme cyclooxygenase (COX). Aspirin owes its ability to suppress the production of prostaglandins and thromboxanes to its irreversible inactivation of the COX enzyme (Figure 14.3). Aspirin acts as an acetylating agent; an acetyl group is covalently attached to a serine residue in the active site of the COX enzyme. This makes aspirin different from other NSAIDs (such as ibuprofen), which are reversible inhibitors.

There are at least two different types of COX, COX-1 and COX-2. Aspirin irreversibly inhibits COX-1 (which is constitutive); this leads to an inhibition of prostanoid synthesis, and reduced levels of $PGE_2$ in particular lead to an increase in gastric acid production (Chapter 4), as well as a reduction in mucous secretion. Aspirin also modifies the enzymatic activity of COX-2 (which is inducible); normally COX-2 produces prostanoids, most of which are pro-inflammatory, but aspirin-modified COX-2 produces lipoxins, which are anti-inflammatory.

Lipoxins are a series of anti-inflammatory, nonclassic eicosanoids ('nonclassic' in the sense that they are synthesised by oxygenation of 20-carbon fatty acids other than the classic eicosanoids) whose appearance in inflammation normally signals its resolution. At present two lipoxins have been identified: lipoxin $A_4$ ($LXA_4$) and lipoxin $B_4$ ($LXB_4$).

Prostaglandins whose synthesis involves COX-1 are largely responsible for maintenance and protection of the gastrointestinal tract (inhibiting acid production and stimulating mucous production), while prostaglandins whose synthesis involves COX-2 are responsible for inflammation and pain.

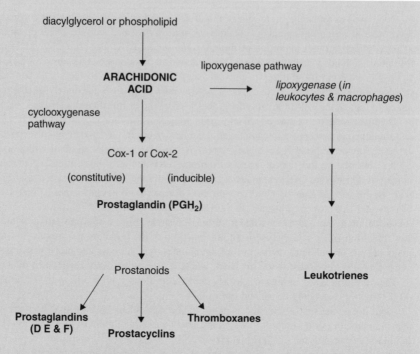

**Figure 14.3**   The cyclooxygenase pathway.

The intention has been to develop new NSAID drugs, so-called COX-2-selective inhibitors, which inhibit only COX-2, with the aim of reducing the incidence of gastrointestinal side effects. Selectivity for COX-2 reduces the risk of peptic ulceration, and is the main feature of celecoxib, rofecoxib and other members of this drug class. However, several of the new COX-2-selective inhibitors have been withdrawn on suspicion that there might be an increase in the risk for heart attack, thrombosis and stroke through a relative increase in thromboxane levels. It is proposed that endothelial cells lining the microvasculature in the body express COX-2, and by selectively inhibiting COX-2, prostaglandins (specifically $PGI_2$ and prostacyclin) are downregulated with respect to thromboxane levels, because COX-1 in platelets is unaffected. Therefore, the protective anticoagulative effect of $PGI_2$ is decreased, increasing the risk of thrombus and associated heart attacks and other circulatory problems. Since platelets have no DNA, they are unable to synthesize new COX once aspirin has irreversibly inhibited the enzyme: an important difference from reversible inhibitors.

The mechanism of action of aspirin may be more involved than is indicated above. Aspirin also induces the formation of NO-radicals in the body, which in mice have been shown to have an independent mechanism of reducing inflammation; salicylic acid and its derivatives can modulate signalling through NF-$\kappa$B, a transcription factor complex that plays a central role in many processes, including inflammation.

NF-$\kappa$B (nuclear factor kappa-light-chain-enhancer of activated B-cells) is a protein complex that acts as a transcription factor. It is found in almost all animal cell types and is involved in cellular responses to stimuli such as stress, cytokines, free radicals, UV radiation, oxidised lipoproteins and bacterial or viral antigens. It plays a key role in regulating the immune response to infection. Consistent with this role, incorrect regulation of NF-$\kappa$B has been linked to cancer, inflammatory and autoimmune disease, septic shock, viral infection and improper immune development.

# CHAPTER 15

# The immune response

## 15.1 Leukocytes

White blood cells, or leukocytes, are cells of the immune system (Table 15.1). Five different and diverse types of leukocytes exist, but they are all produced and derived from multipotent cells in the bone marrow known as haematopoietic stem cells. Leukocytes are found throughout the body, including the blood and lymphatic system.

The number of leukocytes in the blood is often an indicator of disease. There are normally between 4 and $11 \times 10^9$ white blood cells per litre of blood, making up approximately 1% of blood in a healthy adult. In conditions such as leukaemia (cancer of the blood or bone marrow), the number of leukocytes may be much higher than normal, while in types of leukopaenia the number is much lower (associated for example with chemo/radiotherapy, aplastic anaemia, human immunodeficiency virus (HIV) and malaria).

The immune system may be described as innate, passive or acquired (adaptive). Acquired immunity is further divided into humoral and cell-mediated types.

## 15.2 Innate immunity

The innate immunity system includes those non-specific physical/chemical barriers that prevent entry of antigens to the body.

1. Arguably the most important barrier is the skin; skin cannot be penetrated by most organisms unless it is damaged.

2. Mechanically, pathogens are expelled from the lungs by ciliary action as cilia move in an upward motion; coughing and sneezing abruptly eject living and nonliving things from the respiratory system. The flushing action of tears, saliva and urine forces out pathogens, as does the sloughing off of skin.

3. Mucus in the respiratory and gastrointestinal tract traps many microorganisms.

*Essential Biochemistry for Medicine*   Dr Mitchell Fry
© 2010 John Wiley & Sons, Ltd

**Table 15.1** Types of leukocyte

| Type | % of leukocyte count | Main target | Lifetime |
|---|---|---|---|
| Neutrophil | 54–62 | Bacteria/fungi | 6 hours to a few days |
| Eosinophil | 1–6 | Parasites/allergic reactions | 8–12 days |
| Basophil | <1 | Allergic reactions | – |
| Lymphocyte – further divided into: | 25–33 | – | Weeks to years |
| B-cells | – | Various pathogens | – |
| T-cells: | | | |
| CD4+ helper | – | Extracellular bacteria broken down into peptides presented by MHC class II molecule | – |
| CD8+ cytotoxic | – | Virus-infected and tumour cells | – |
| $\gamma\delta$ T-cells | – | Gamma delta T-cells. Complex and poorly understood function, abundant in the gut mucosa | – |
| Natural killer cells | – | Virus infected cells and tumour cells | – |
| Monocyte | 2–8 | Monocytes migrate from the bloodstream to other tissues and differentiate into tissue-resident macrophages and dendritic cells | Hours to days |
| Macrophage | – | Phagocytosis (engulfment and digestion) of cellular debris and pathogens, and stimulation of lymphocytes and other immune cells that respond to the pathogen | Days (if activated) to years (as immature) |
| Dendritic cells | – | Main function is as an antigen-presenting cell that activates T- lymphocytes | Similar to macrophages |

4. The acid pH (<7.0) of skin secretions inhibits bacterial growth; hair follicles secrete sebum that contains lactic acid and fatty acids, both of which inhibit the growth of some pathogenic bacteria and fungi. Saliva, tears, nasal secretions and perspiration all contain lysozyme, an enzyme that destroys the cell wall of Gram-positive bacteria. Vaginal secretions are also slightly acidic (after the onset of menses). Spermine and zinc in semen destroy some pathogens. Lactoperoxidase, in mothers' milk, has both antimicrobial and antioxidant activities.

5. The hydrochloric acid and protein-digesting enzymes of the stomach kill many pathogens.

## 15.2.1   The complement system

The complement system is not adaptable and does not change over the course of an individual's lifetime; as such it belongs to the innate immune system. However, it can be recruited and brought into action by the adaptive immune system. The complement system is a biochemical cascade that helps clear pathogens from an organism.

The basic functions of the complement system are to:

- lyse bacteria or cells containing viruses

- opsonise cells or antigens to promote their phagocytosis

- bind to specific complement receptors on the cells of the immune system, triggering specific cell functions

- promote immune clearance, the removal of immune complexes and their deposition in the spleen and liver.

An opsonin is any molecule that acts as a binding enhancer for the process of phagocytosis.

Over 20 different proteins constitute the complement system, many circulating in the blood as inactive zymogens; they are synthesised mainly in the liver and account for about 5% of the globin fraction of blood serum. Proteases in the system cleave specific proteins to release cytokines and initiate an amplifying cascade of further cleavages, resulting in the activation of the cell-killing membrane-attack complex.

There are three pathways to complement activation:

- the classical pathway

- the alternative pathway

- the lectin (mannose-binding) pathway.

The classical complement pathway requires activation by circulating antibodies, IgM or IgG (specific immune response), while the alternative and lectin pathways can be activated in the absence of antibody (non-specific immune response).

In all three pathways, a central role is played by complement protein 3 (C3). C3 is first cleaved and activated, then causes a cascade of further activation events. C3b promotes opsonisation, C5a is a chemotactic protein, C3a and C5a trigger degranulation of mast cells and C5b initiates the membrane-attack pathway.

The membrane-attack complex (MAC) is the cytolytic end product of the complement cascade; it forms a transmembrane channel, leading to osmotic lysis of the target cell (Figure 15.1).

## 15.2.2 Complement deficiency

Complement deficiency is a condition of absent or suboptimal functioning of any one of the complement system proteins, due to:

- Disorders of the proteins which act to inhibit the complement system (such as C1-inhibitor), which can lead to an overactive response, causing conditions such as hereditary angioedaema and haemolytic-uraemic syndrome.

- Disorders of the proteins which act to activate the complement system (such as C3), which can lead to an underactive response, causing greater susceptibility to infections.

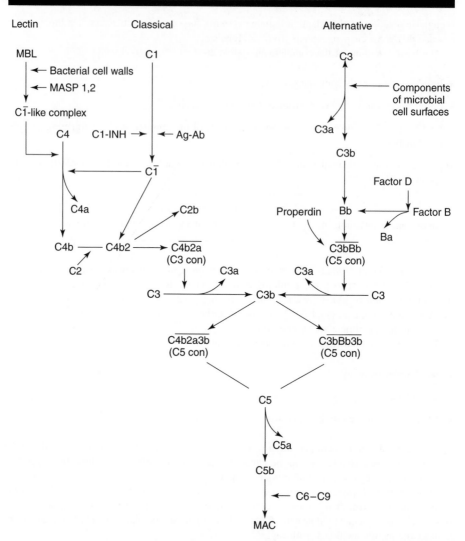

**Figure 15.1** Complement activation pathways. The classical, lectin and alternative pathways converge into a final common pathway when C3 convertase (C3 con) cleaves C3 into C3a and C3b. Ab = antibody, Ag = antigen, C1-INH = C1 inhibitor, MAC = membrane attack complex, MASP = MBL-associated serine protease, MBL = mannose-binding lectin, P = properdin. Overbar indicates activation.

As there are redundancies in the immune system, many complement disorders are never diagnosed.

Hereditary angioedaema (also known as 'Quincke oedaema') is characterised by local swelling in subcutaneous tissues. It does not respond to antihistamines, corticosteroids or adrenaline (epinephrine). It is caused by the unregulated inhibition of C1, factor XII and kallikrein (by C1 inhibitor), giving rise to vasoactive substances.

### 15.2.3 Natural killer cells

Natural killer cells are a type of cytotoxic lymphocyte that constitute a major component of the innate immune system (they should not be confused with cytoxic or killer T-cells; see below). They play a major role in the rejection of tumours and cells infected by viruses. The cells 'kill' by releasing small cytoplasmic granules of proteins called perforin and granzyme, which cause the target cell to die by apoptosis or necrosis. Natural killer cells do not express T-cell antigen receptors. They were so named because of the initial notion that they did not require activation in order to kill cells that were missing 'self' markers of major histocompatibility complex (MHC) class I (infected cells tend to down-regulate their MHC class I, i.e. 'missing self'). Natural killer cells, along with macrophages and several other cell types, express the Fc receptor that binds the Fc portion of antibodies. This allows natural killer cells to target cells against which a humoral response has been mobilised and to kill cells via antibody-dependent cellular cytotoxicity (ADCC). ADCC is a mechanism of cell-mediated immunity whereby an effector cell of the immune system actively kills a target cell that has been bound by specific antibodies. It is one of the mechanisms through which antibodies, as part of the humoral immune response, can act to limit and contain an infection. Classical ADCC is mediated by natural killer cells; monocytes and eosinophils can also mediate ADCC. For example, eosinophils can kill certain parasitic worms (helminths) through ADCC. ADCC is part of the adaptive immune response, due to its dependence on a prior antibody response.

> There is an important immunological distinction between 'apoptosis' and cell 'lysis'. For example, lysis of a virus-infected cell will only release virions to spread infection, whereas apoptosis leads to the death of the cell and the contained pathogen.

## 15.3 Passive immunity

Passive immunity involves antibodies produced outside the body. Infants have passive immunity because they acquire antibodies through the placenta from the mother; these antibodies disappear at between 6 and 12 months of age. Passive immunisation involves transfusion of antiserum, containing antibodies formed in another person or animal; it provides immediate protection against an antigen, but does not provide long-lasting protection. Gamma globulin and equine (horse) tetanus antitoxin are examples of passive immunisation.

## 15.4 Acquired immunity

Acquired immunity is immunity that develops with exposure to antigens; the immune system builds a defence that is specific to that antigen. B-cell lymphocytes, produced in the stem cells of the bone marrow, synthesise and release antibody; they oversee the humoral immune response. T-cell lymphocytes, produced in the bone marrow but sensitised in the thymus, are the basis of the cell-mediated immune response.

### 15.4.1 The humoral immune response

The human body makes millions of different types of B-cell each day, which circulate in the blood and lymphatic system, performing the role of immune surveillance. They do not

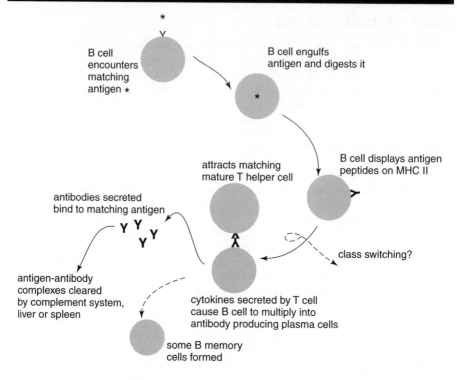

**Figure 15.2** Clonal expansion of B-cells.

produce antibodies until they become fully activated. Each B-cell has a unique membrane-bound immunoglobulin that will bind to a specific antigen. Once a B-cell encounters its cognate antigen and receives an additional signal from a T-helper cell, it can further differentiate (via clonal expansion; Figure 15.2) into plasma B-cells or memory B-cells. Prior to that expansion, it may undergo an intermediate differentiation step, leading to class switching; for example, synthesis of IgM antibody might be switched to IgG antibody.

## 15.4.2    The clonality of B-cells

B-cells exist as clones; they derive from a particular cell (see Table 15.2). Therefore the antibodies produced by a particular clone can recognise and/or bind the same components (epitope) of a given antigen. The great diversity in immune response comes about because there are up to $10^9$ clones, which recognise that many different antigens; the phenomenon of immunogenic memory relies on this clonality of B-cells.

Activated B-cells, after an initial lag, produce highly specific (monoclonal) antibodies at a rate of up to 2000 molecules per second, over four to five days.

## 15.4.3    Antibodies

Antibodies (immunoglobulins) constitute the gamma globulin part of the blood proteins. They are soluble proteins secreted by the plasma offspring (clones) of activated B-cells.

**Table 15.2** Main types of B-cell

| | |
|---|---|
| Plasma B-cells (plasma cells) | Large B-cells that have been exposed to antigen and are secreting antibodies. Short-lived cells that undergo apoptosis once the inciting agent has been eliminated. |
| Memory B-cells | Formed from activated B-cells that are specific to the antigen encountered in the primary immune response. Long-lived cells that can respond quickly following a second exposure to the same antigen. |
| B-1 cells | Thought to mediate B-cell–B-cell interaction. They express IgM in greater quantities than IgG and their receptors show polyspecificity, meaning that they have low affinities for many different antigens, but have a preference for other immunoglobulins, self-antigens and common bacterial polysaccharides. Present in low numbers in the lymph nodes and spleen, but found predominantly in the peritoneal and pleural cavities. |
| Marginal-zone (MZ) B-cells | Non-circulating mature B-cells that segregate anatomically into the marginal zone of the spleen. This region contains multiple subtypes of macrophages, dendritic cells and the MZ B-cells. MZ B-cells can be rapidly recruited into the early adaptive immune responses in a T-cell-independent manner. |
| Follicular B-cells | A readily distinguished subtype, organized into the primary follicles of B-cell zones focused around follicular dendritic cells in the white pulp of the spleen and the cortical areas of peripheral lymph nodes. |

Antibodies inactivate antigens by:

• initiating complement fixation (MAC)

• neutralisation (binding to specific sites preventing attachment)

• agglutination (clumping)

• precipitation (forcing insolubility and settling out of solution).

Antibodies come in a range of isotypes or classes (Table 15.3). The basic structural units (Figure 15.3) contain two large heavy chains and two small light chains, which together can form monomers, dimers or pentamers. A small region at the tip of the protein is extremely variable, allowing millions of antibodies with slightly different tip structures to exist; this region is known as the hypervariable region. Each of these variants can bind to a different target, known as an antigen. The unique part of the antigen recognised by an antibody is called an epitope. This huge diversity of antibodies allows the immune system to recognise an equally wide diversity of antigens.

The blood constituents of gamma globulin are: IgG-76%, IgA-15%, IgM-8%, IgD-1% and IgE-0.002% (responsible for autoimmune responses, such as allergies, and diseases like arthritis, multiple sclerosis and systemic lupus erythematosus (SLE)). IgG is the only antibody that can cross the placental barrier to the foetus, and it is responsible for the 3–6 month passive immunity of newborns that is conferred by the mother.

**Table 15.3**   Human antibody isotypes

| Antibody type | Description |
| --- | --- |
| IgADimer | Found in mucosal areas, such as the gut, respiratory tract and urogenital tract, and prevents colonisation by pathogens. Also found in saliva, tears and breast milk. |
| IgDMonomer | Functions mainly as an antigen receptor on B-cells that have not been exposed to antigens (naive). Its function is less defined than other isotypes. |
| IgEMonomer | Binds to allergens and triggers histamine release from mast cells and basophils; also protects against parasitic worms. |
| IgGMonomer | Provides the majority of antibody-based immunity against invading pathogens. The only antibody capable of crossing the placenta to give passive immunity to the foetus. |
| IgMPentamer | Expressed on the surface of B-cells and in a secreted form with very high avidity. Eliminates pathogens in the early stages of B-cell-mediated (humoral) immunity before there is sufficient IgG. |

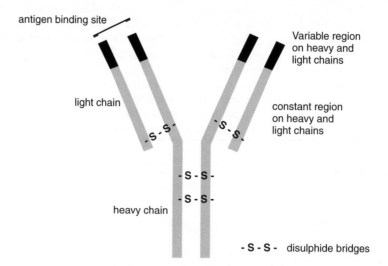

**Figure 15.3**   General antibody structure. The upper part, $F_{ab}$ (antigen binding) portion of the antibody molecule attaches to specific regions on the protein antigen, called epitopes. Thus the antibody recognises the epitope and not the entire antigen. The $F_c$ (crystallisable) region is responsible for effector functions; that is, the end to which immune cells can attach.

IgM is the dominant antibody produced in primary immune responses, while IgG dominates in secondary immune responses. IgM is physically much larger than the other immunoglobulins.

## 15.4.4   The Cell-mediated immune response

Macrophages, and other antigen-presenting cells, engulf antigens, process them internally and then display parts of them on their surface (MHC class II; see below). This sensitises the T-cells to recognise these antigens.

A critical difference between B-cells and T-cells resides in how each lymphocyte type recognises its antigen. B-cells recognise their cognate antigen in its native form; they recognise free (soluble) antigen in the blood or lymph through their membrane-bound immunoglobulin. In contrast, T-cells recognise their cognate antigen in a processed form as a peptide fragment, presented by an antigen-presenting cell MHC molecule to the T-cell receptor.

T-cells are primed in the thymus, where they undergo two selection processes (central tolerance): the first, positive selection process selects only those T-cells with the correct set of receptors that can recognise the MHC molecules responsible for self-recognition; then a negative selection process begins, whereby only T-cells that can recognise MHC molecules complexed with foreign peptides are allowed to pass out of the thymus.

- **Helper T-cells, $T_H$(CD4+),** serve as managers, directing the immune response. They secrete chemicals called lymphokines which stimulate cytotoxic T-cells and B-cells to grow and divide, attract neutrophils and enhance the ability of macrophages to engulf and destroy microbes. Lymphokines are cytokines, secreted by helper T-cells in response to stimulation by antigens, which act on other cells of the immune system.

- **Regulatory T-cells** (sometimes known as suppressor T-cells) are a specialised subpopulation of T-cells that act to suppress activation of the immune system and thereby maintain immune-system homeostasis and tolerance to self-antigens.

- **Memory T-cells** are programmed to recognise and respond to a pathogen once it has invaded and been repelled.

- **Cytotoxic or killer T-cells, $T_C$ (CD8+),** release lymphotoxins which initiate cell death. They belong to a sub-group of T-lymphocytes that are capable of inducing the death of infected somatic or tumour cells and cells that are infected with viruses (or other pathogens), or are otherwise damaged or dysfunctional. Most cytotoxic T-cells express T-cell receptors that can recognise a specific antigenic peptide bound to class I MHC molecules, present on all nucleated cells, and a glycoprotein called CD8, which is attracted to non-variable portions of the class I MHC molecule. The affinity between CD8 and the MHC molecule keeps the $T_C$-cell and the target cell bound closely together during antigen-specific activation.

The strategy of T- and B-cell interaction with antigens is summarised in Figure 15.4.

The cluster of differentiation (or cluster of designation), often abbreviated as CD, is a protocol used for the identification and investigation of cell-surface molecules present on leukocytes. CD molecules can act in numerous ways, often as receptors or ligands (the molecules that activate a receptor). A signal cascade is usually initiated, altering the behaviour of the cell. Some CD proteins do not play a role in cell signalling, but have other functions, such as cell adhesion. There are over 250 known CD proteins. The CD system is commonly used for cell markers; this allows cells to be defined based on what molecules are present on their surface and to associate them with certain immune functions or properties. Cell populations are usually defined using a + or a − symbol to indicate whether a certain cell fraction expresses or lacks a CD molecule. For example, a 'CD34+, CD31−' cell is one that expresses CD34 but not CD31. Two commonly used CD molecules are CD4 and

CD8, which are generally used as markers for helper and cytotoxic T-cells, respectively. These molecules are defined in combination with CD3+, as some other leukocytes also express these CD molecules (some macrophages express low levels of CD4; dendritic cells express high levels of CD8). HIV binds CD4 and a chemokine receptor on the surface of a T-helper cell to gain entry. The number of CD4 and CD8 T-cells in blood is often used to monitor the progression of HIV infection.

Some types of cell and their CD markers:

- stem cells: CD34+,CD31−

- leukocytes: CD45+

- granulocyte: CD45+, CD15+

- monocyte: CD45+, CD14+

- T-lymphocyte: CD45+, CD3+

- T-helper cell: CD45+,CD3+,CD4+

- cytotoxic T-cell: CD45+,CD3+, D8+

- B-lymphocyte: CD45+, CD19+ or CD45+, CD20+

- natural killer cell: CD16+, CD56+, CD3−.

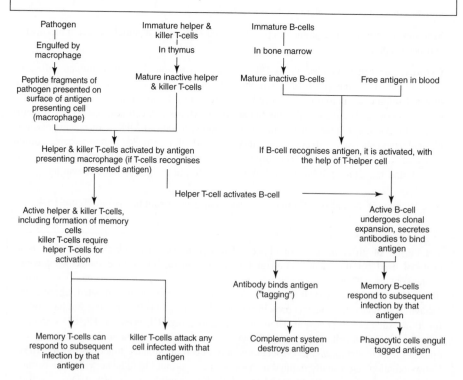

**Figure 15.4** The strategy of T- and B-cell interaction.

## 15.4.5 Antigen-presenting cells

Antigen-presenting cells (APCs) may be professional or non-professional. All nucleated cells in the body express peptides on their surface, which are derived from cytosolic proteins (MHC class I); this is the basis of self-recognition. Only those cells that are able to express 'foreign-derived' peptides on their surface (MHC class II) are able to activate naive T-cells to form CD8+ cytotoxic T-cells. These are the 'professional APCs'. Professional APCs are efficient at internalising antigen, either by phagocytosis or by receptor-mediated endocytosis, and then displaying a fragment of the antigen, bound to a class II MHC molecule, on their surface membrane. T-cells recognise and interact with the antigen–class II MHC molecule complex on the membrane of the APC. An additional co-stimulatory signal is then produced by the APC, leading to activation of the T-cell. There are three main types of professional APC:

- Dendritic cells, which have the broadest range of antigen presentation, and are probably the most important APCs. Activated dendritic cells are especially potent $T_H$-cell activators because they express co-stimulatory molecules, such as B7 (the 'second signal' required for T-cell stimulation and proliferation).

- Macrophages.

- B-cells, which express antibody, can very efficiently present the antigen to which their antibody is directed, but are an inefficient APC for most other antigens.

Non-professional APCs do not constitutively express the MHC proteins required for interaction with naive T-cells; these are expressed only upon stimulation of the non-professional APC by certain cytokines, such as IFN-$\gamma$ (interferon-gamma). Non-professional APCs include fibroblasts (skin), thymic epithelial cells, thyroid epithelial cells, glial cells (brain), pancreatic $\beta$-cells and vascular endothelial cells.

## 15.4.6 Interaction of APCs with T-cells

Once an APC has phagocytosed pathogenic material, they usually migrate, via the lymph network, to arrive at the draining lymph nodes (collectively the lymphatic system). The lymph nodes become a collection point to which APCs can interact with T-cells. During this migration the APCs undergo a process of maturation; they lose most of their ability to further engulf pathogens, and develop an increased ability to communicate with T-cells.

## 15.4.7 Major histocompatibility complex (MHC)

The MHC is a large genomic region (gene family); it is the most gene-dense region of the mammalian genome. It plays an important role in the immune, autoimmunity and reproductive systems. The proteins encoded by the MHC are expressed on the surface of cells, and display both self-antigens (peptide fragments from the cell itself) and non-self-antigens. (e.g. fragments of an invading bacterium).

The MHC region is divided into three subgroups, class I, class II and class III. Class III has a function very different from that of class I and class II (Table 15.4), but since it has a locus between the other two (on chromosome 6 in humans), the three classes are frequently discussed together; MHC class III encodes for immune components such as complement (C2, C4, factor B), as well as cytokines (e.g. tumour necrosis factor (TNF)).

**Table 15.4**    Major histocompatibility complexes class I and II

| | |
|---|---|
| MHC class I <br>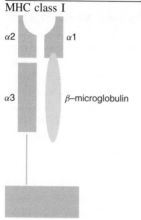 | MHC class I molecules are found on almost every nucleated cell of the body. Because MHC class I molecules present peptides derived from cytosolic proteins, the pathway of MHC class I presentation is often called the cytosolic or endogenous pathway. <br><br> MHC class I molecules are heterodimers, consisting of a single transmembrane polypeptide chain (the $\alpha$-chain) and a $\beta 2$ microglobulin (which is encoded elsewhere, not in the MHC). The $\alpha$-chain has three polymorphic domains, $\alpha 1$, $\alpha 2$, $\alpha 3$. Between $\alpha 1$ and $\alpha 2$ is the peptide-binding groove, which binds peptides derived from cytosolic proteins. The groove consists of eight $\beta$-pleated sheets on the bottom and two $\alpha$-helices making up the sides. The peptide in the groove remains bound for the life of the class I molecule, and is typically eight to nine amino acids in length. |
| MHC class II <br>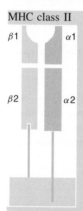 | MHC class II molecules are only expressed on professional antigen-presenting cells. The peptides presented by class II molecules are derived from extracellular proteins (not cytosolic as in class I); hence the MHC class II-dependent pathway of antigen presentation is called the endocytic or exogenous pathway. <br><br> Like MHC class I molecules, class II molecules are also heterodimers, but in this case consist of two homologous peptides, one $\alpha$- and one $\beta$-chain, both of which are encoded in the MHC. Because the antigen-binding groove of MHC class II molecules is open at both ends, the antigens presented by MHC class II molecules are longer, generally between 15 and 24 amino acid residues long. <br><br> Class II molecules interact exclusively with CD4+ ('helper') T-cells (THs). The THs then help to trigger an appropriate immune response, which may include localised inflammation and swelling due to recruitment of phagocytes, or may lead to a full-force antibody immune response due to activation of B-cells. |

Cells constantly process endogenous proteins and present self-peptide fragments, within the context of MHC class I, on their surfaces. Immune effector cells do not react to self-peptides within MHC class I, but as such are able to recognise when foreign peptides are instead presented, for example during an infection or cancer.

The MHC proteins act as 'signposts' that display fragmented pieces of an antigen on the host cell's surface; these antigens may be self or non-self. Non-self-antigens may arise in the following ways:

- If a host cell is infected by a bacterium or virus, or becomes cancerous, it may display the non-self-antigen on its surface within a class I MHC molecule. Regardless of which type of MHC molecule it is displayed on, it will initiate the cell-mediated immune response.

• Phagocytic cells, such as macrophages, neutrophils and monocytes, may engulf foreign particles and present non-self-peptides on MHC class II molecules.

## 15.4.8 Autoimmunity

Autoimmunity is caused by an adaptive immune response against self-antigen. It is the random generation of so many diverse T-cell and B-cell receptors that makes autoimmunity possible. Clonal deletion and anergy of self-specific lymphocytes greatly reduces, but does not eliminate, the possibility of low-affinity self-specific responses. Transient autoimmune responses are common but usually cause no lasting damage. Risk factors for autoimmune disease include the presence of certain MHC alleles, sex hormone levels, infection and other environmental factors. Autoimmune diseases can be caused by antibodies or T-cells, and may be organ-specific or systemic (Figure 15.5).

---

Anergy is a term in immunobiology that describes a lack of reaction by the body's defence mechanisms to foreign substances, and consists of a direct induction of peripheral lymphocyte tolerance.

---

Autoimmune diseases form a spectrum ranging from organ-specific conditions, in which one organ only is affected, to systemic diseases, in which the pathology is diffused throughout the body. The extremes of this spectrum result from quite distinct underlying mechanisms, but there are many conditions in which there are components of both organ-specific and systemic damage.

Autoimmune diseases are initiated by activation of antigen-specific T-cells. Th2-cells activate B-cells to make autoantibodies, which (by activating complement) damage tissues directly or initiate prolonged inflammation. Cytotoxic lymphocytes and macrophages activated by Th1-cells are directly cytotoxic and also promote inflammation. The events that initiate specific autoimmune diseases are not known, but include:

• **Genetic susceptibility**. This plays an important role in almost all autoimmune conditions. The most significant influence is that of the MHC. The linkage between particular MHC class II allotypes and specific diseases has in some cases been determined; in almost all cases these diseases have a strong autoimmune component.

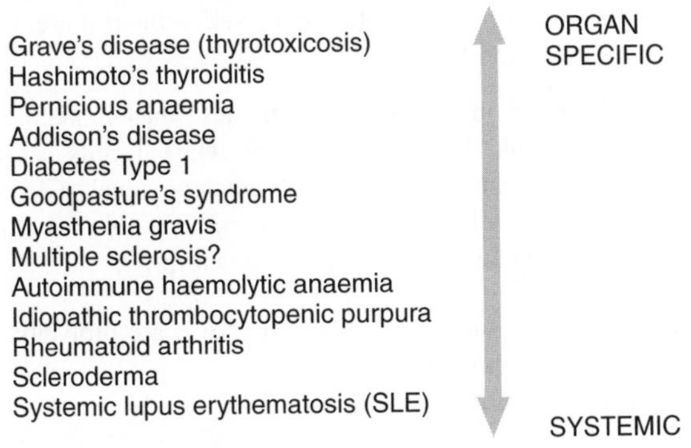

Grave's disease (thyrotoxicosis)      ORGAN
Hashimoto's thyroiditis                SPECIFIC
Pernicious anaemia
Addison's disease
Diabetes Type 1
Goodpasture's syndrome
Myasthenia gravis
Multiple sclerosis?
Autoimmune haemolytic anaemia
Idiopathic thrombocytopenic purpura
Rheumatoid arthritis
Scleroderma
Systemic lupus erythematosis (SLE)     SYSTEMIC

**Figure 15.5** Examples of organ-specific and systemic autoimmune disease.

HLA (human leukocyte antigen) DR2 is strongly positively correlated with SLE and multiple sclerosis, and negatively correlated with diabetes mellitus type 1. HLA DR3 is correlated strongly with Sjögren's syndrome, myasthenia gravis, SLE and diabetes mellitus type 1. HLA DR4 is correlated with the genesis of rheumatoid arthritis and diabetes mellitus type 1.

Fewer correlations exist with MHC class I molecules. The most notable and consistent is the association between HLA B27 and ankylosing spondylitis.

---

The best-known genes of the MHC region are that subset which encode antigen-presenting proteins on the cell surface. In humans, these genes are referred to as human leukocyte antigen (HLA) genes. In the biomedical literature, 'HLA' is often used to refer specifically to the HLA protein molecules, while 'MHC' is reserved for the region of the genome that encodes for this molecule. It is these proteins that are used in tissue matching.

---

- **Endocrine factors**. Most autoimmune diseases occur with unequal frequency in males and females. For example, Graves' and Hashimoto's are 4–5 times, and SLE 10 times, more common in females, while ankylosing spondylitis is 3–4 times more frequent in males. These differences are believed to be the result of hormonal influences. A second well-documented hormonal effect is the marked reduction in disease severity seen in many autoimmune conditions during pregnancy. Rheumatoid arthritis is perhaps the classic example of this effect. In some cases there is also a rapid exacerbation (rebound) after birth.

- **Environment**. It is clear that environmental factors also play a role in autoimmune disease. Factors include diet, but infectious organisms are the most significant environmental factor. Evidence exists for a direct link between a specific infection and an autoimmune disease, for example that of rheumatic fever following streptococcal infection. A variety of other organisms have been implicated, for example *Yersinia*, *Shigella*, *Chlamydia*.

## 15.4.9 Tolerance

Natural tolerance, the inability to invoke an immune response to an antigen, occurs principally because of:

- **Clonal deletion** of self-reactive T- and B-cells during development, which removes lymphocytes with high-avidity receptors for ubiquitous self-antigen present in the thymus and marrow. This is known as central tolerance.

- **Clonal anergy,** which occurs in the peripheral tissues when immature B-cells encounter soluble antigens that cross-link B-cell receptors, or when T-cells encounter unprocessed antigen or processed antigen in the absence of co-stimulatory signals. Clonal anergy maintains tolerance to some (but not all) self-antigens that are not available for clonal deletion in the thymus and marrow. This is known as peripheral tolerance.

Specific autoimmune diseases, such as those listed above, are probably not caused by any general failure of clonal deletion or clonal anergy.

Other aspects of tolerance include:

- **Immunological ignorance**. Most self-antigen is presented with insufficient avidity to induce either clonal deletion or lymphocyte activation. 'Immunological ignorance' refers to the common occurrence of low numbers of low-avidity self-specific T-cells existing in the presence of low levels of presented self-peptides without becoming activated to respond. Humans make at least $10^5$ proteins (average size: 300 amino acids), which can be processed to generate $3 \times 10^7$ distinct peptides for presentation to T-cells. Each APC has a maximum of $10^5$ MHC molecules per cell with which to present these peptides; T-cells must bind at least 10–100 identical peptides on an APC to become activated. Most peptides presented on an APC will therefore be below the threshold for T-cell detection.

- **Regulatory (suppressor) T-cells**. These maintain tolerance. Cells which can transfer tolerance are $CD4^+$ and $CD25^+$ T-cells. Depletion of cells, with this phenotype, from normal mice or from normal cells given to athymic mice, results in the development of autoimmune disease. This tolerance is dominant.

- **Immunologically privileged sites**. These sites, which include the brain, testis, eye and uterus (the foetus can be considered an unusually successful allograft), do not elicit immune rejection. Cells and proteins do leave these sites and circulate in the body, although they do not travel in the lymphatics. Naive lymphocytes are excluded from these sites. Immunosuppressive factors, probably transforming growth factor $\beta$ (TGF $\beta$), are produced in these sites and are released from them with the cells and proteins. Tissues in these sites also express Fas ligand; binding of Fas ligand with its receptor induces apoptosis, killing any effector ($Fas^+$) T-cells which enter.

## 15.4.10   Overcoming tolerance

Antigens in immunologically privileged sites can be targeted by the immune system. In sympathetic opthalmia, damage to one eye can on rare occasions result in an autoimmune response to eye proteins that can damage the uninjured eye. When trauma or other events cause damage to the barriers which protect such special sites, this can lead to the release of novel autoantigens and the production of autoantibodies.

Autoimmunity might arise as a result of T-cell tolerance being bypassed. This could occur in a number of ways:

- **Modification**. This can happen when a small molecule (e.g. a drug) binds to a protein and alters an MHC-binding peptide so that it becomes a neoantigen recognised by T-cells. This provides T-cell help, through linked recognition, for antibody production which need not be (and usually is not) directed against the neoantigen.

- **Inflammation**. During an inflammatory response an immunostimulatory environment is created by the release of cytokines which recruit and activate professional antigen-presenting cells and provide support for T-cell activation, rather than anergy. As a result, autoreactive T-cells which were anergic or ignorant might become activated.

- **Molecular mimicry**. This is a rather specialised version of the above in which an epitope of an invading microorganism cross-reacts with a self-protein. The T-cell help provided by the other microbial antigens permits the activation of B-cells which make a cross-reactive antibody, which either escapes tolerance or acquires sufficient self-reactivity through somatic mutation and selection driven by the cross-reactive antigen. The classic example is rheumatic fever following infection with *Streptococcus pyogenes*; antibodies to Streptococcal antigen binds host heart tissue and can damage it. The response is usually transient, since the T-cells are specific for the Streptococcal antigen and not for self. Other examples of autoimmunity following infection include reactive arthritis following infection with *Shigella*, *Salmonella*,

*Yersinia* and *Camplyobacter*, as well as chronic Lyme arthritis. Human studies are currently underway to investigate a possible link between coronary artery disease and infection with *Chlamydia pneumoniae*.

Spontaneous human autoimmunity seems to be almost entirely restricted to the autoantibody responses produced by B-lymphocytes. Loss of tolerance by T-cells has been extremely hard to demonstrate, and where there is evidence for an abnormal T-cell response it is usually not to the antigen recognised by the autoantibody. This disparity has led to the idea that human autoimmune disease is in most cases (with probable exceptions including type I diabetes) based on a loss of B-cell tolerance, which makes use of normal T-cell responses to foreign antigens in a variety of aberrant ways.

## 15.4.11  Treating autoimmune disease

Treatments for autoimmune disease have traditionally been immunosuppressive, anti-inflammatory or palliative. Non-immunological therapies, such as hormone replacement in Hashimoto's thyroiditis, treat the outcomes of the autoaggressive response. Dietary manipulation limits the severity of coeliac disease. Steroidal or NSAID treatment limits the inflammatory symptoms of many diseases.

Autoantibodies are used to diagnose many autoimmune diseases. The levels of autoantibodies are measured to determine the progress of the disease.

Infliximab (Remicade) is a drug used to treat autoimmune diseases (e.g. psoriasis, Crohn's disease, ankylosing spondylitis, rheumatoid arthritis, ulcerative colitis). It works by blocking tumour necrosis factor alpha (TNF-$\alpha$). TNF-$\alpha$ is a chemical messenger (cytokine) and a key part of the autoimmune reaction. Infliximab blocks the action of TNF-$\alpha$ by preventing it from binding to its receptor on the cell. It is an artificial antibody, originally developed in mice; because humans have immune reactions to mouse proteins, it was later developed into a human (humanised) antibody. Produced from cloned cells, it is a monoclonal antibody. As a combination of mouse and human antibody, it is called a chimeric monoclonal antibody (the $F_c$ is human-derived, the $F_{ab}$ mouse-derived). It is administered by intravenous injection, typically at six- to eight-week intervals. It cannot be administered orally because it is degraded in the gut. It neutralises the biological activity of TNF-$\alpha$ by binding with high affinity to both the soluble and the transmembrane forms of TNF-$\alpha$ and prevents the effective binding of TNF-$\alpha$ with its receptors (see 'Focus on Tumour Necrosis Factor').

# Focus on: type I hypersensitivity – anaphylaxis

'Hypersensitivity' (hypersensitivity reaction) refers to undesirable reactions produced by the normal immune system (Table 15.5). Hypersensitivity reactions require a pre-sensitised (immune) state of the host. Exposure may be by ingestion, inhalation, injection or direct contact.

**Table 15.5**  Types of hypersensitivity

| Type | Examples |
|---|---|
| I Allergy (immediate) | Anaphylaxis, atopy, asthma. Mediated by IgE. |
| II Cytotoxic (antibody-dependent) | Autoimmune haemolytic anaemia, Goodpasture's syndrome, thrombocytopaenia. Mediated by IgM or IgG and complement. |
| III Immune-complex disease | Systemic lupus erythematosus (SLE or lupus), serum sickness, arthus reaction. Mediated by IgG and complement. |
| IV Delayed-type hypersensitivity | Contact dermatitis, multiple sclerosis. Mediated by T-cells. |
| V A distinction of type II | Graves disease, myasthenia gravis. Instead of binding to cell-surface components, the antibodies recognise and bind to the cell-surface receptors, which either prevents the intended ligands binding with the receptor, or mimics the effects of the ligand, thus impairing cell signalling. |

The difference between a normal immune response and a type I hypersensitive response is that in the latter, plasma cells secrete IgE. This class of antibody binds to $F_c$ receptors on the surface of tissue mast cells and blood basophils. Mast cells and basophils coated by IgE are 'sensitised'. Later exposure to the same allergen cross-links the bound IgE on sensitised cells, resulting in degranulation and the secretion of pharmacologically active mediators such as histamine, leukotrienes and prostaglandins. The principal effects of these products are vasodilation and smooth-muscle contraction (Table 15.6).

**Table 15.6**  Mediators released by mast cells in type I hypersensitivity

| Mediators | Action |
|---|---|
| Histamine, platelet-activating factor, leukotrienes (C4, D4, E4), prostaglandin D2, neutral proteases | Vasodilation and increased permeability of vessels |
| Histamine, platelet-activating factor, leukotrienes, prostaglandin | Smooth-muscle spasm |
| Cytokines (chemokines and TNF), leukotriene B4, chemotactic factors for neutrophils and eosinophils | Leukocyte extravasation (movement of leukocytes out of the circulatory system) |

The reaction may be either local or systemic. Symptoms vary from mild irritation to sudden death from anaphylactic shock. Treatment usually involves intramuscular injection of adrenaline (epinephrine), antihistamines and corticosteroids.

'Activation' of mast cells is achieved only when IgE, bound to the high-affinity $F_{c\varepsilon}$ receptors ($F_{c\varepsilon}$R1s), is cross-linked by multivalent antigen. The $F_{c\varepsilon}$R1 is a tetrameric receptor composed of a single $\alpha$-chain, responsible for binding the IgE, a single $\beta$-chain and a disulfide-linked homodimer of $\gamma$-chains that initiates the cell signal pathway. Once the $F_{c\varepsilon}$R1s are aggregated by the cross-linking process, phosphorylation of motifs in both the $\beta$- and $\gamma$-chains initiates a cell-signalling cascade, acting on scaffold proteins of the cytoskeleton to promote degranulation (exocytosis) of the mast cell.

Anaphylactic shock, the most severe type of anaphylaxis, occurs when an allergic response triggers a quick release from mast cells of large quantities of immunological mediators (histamines, prostaglandins, leukotrienes), leading to systemic vasodilation (associated with a sudden drop in blood pressure) and bronchoconstriction (difficulty in breathing). Anaphylactic shock can lead to death in a matter of minutes if untreated.

An estimated 1–17% of the population of the United States is considered 'at risk' for having an anaphylactic reaction if exposed to one or more allergens, especially penicillin and insect stings. Most affected individuals successfully avoid such allergens and will never experience anaphylaxis. Of those who actually experience anaphylaxis, up to 1% may die as a result. Anaphylaxis results in approximately 18 deaths per year in the USA (compared to 2.4 million deaths from all other causes each year). The most common presentation includes sudden cardiovascular collapse (88% of reported cases of severe anaphylaxis).

Anaphylaxis is a severe, whole-body allergic reaction. After an initial exposure ('sensitising dose') to a substance such as bee sting toxin, the immune system becomes sensitised to that allergen. On a subsequent exposure ('shocking dose'), an allergic reaction occurs. This reaction is sudden, severe, and involves the whole body.

Anaphylaxis can occur in response to any allergen. Common causes include insect bites, food allergies (peanuts, brazil and hazelnuts are the most common) and drug allergies. Pollens and other inhaled allergens rarely cause anaphylaxis.

Symptoms of anaphylaxis are related to the action of IgE and other anaphylatoxins, which act to release histamine and other mediators from mast cells (degranulation; Figure 15.6). In addition to other effects, histamine induces vasodilation of arterioles and constriction of bronchioles in the lungs (a bronchospasm). Constriction of the airways results in wheezing and difficulty in breathing; gastrointestinal symptoms include abdominal pain, cramps, vomiting and diarrhoea. Histamine causes the blood vessels to dilate (lowering blood pressure) and fluid to leak from the bloodstream into the tissues (lowering blood volume). These effects result in shock. Fluid can leak into the alveoli of the lungs, causing pulmonary oedema.

Primary (emergency) treatment for anaphylaxis is administration of adrenaline (epinephrine). Adrenaline prevents worsening of the airway constriction, and stimulates the heart to continue beating. Adrenaline (epinephrine) acts on $\beta$-2 adrenergic receptors in the lung as a powerful bronchodilator (opening the airways), relieving allergic or histamine-induced acute asthmatic attack or anaphylaxis. Patients at risk often carry injectable adrenaline.

For further information, see Chapter 14.

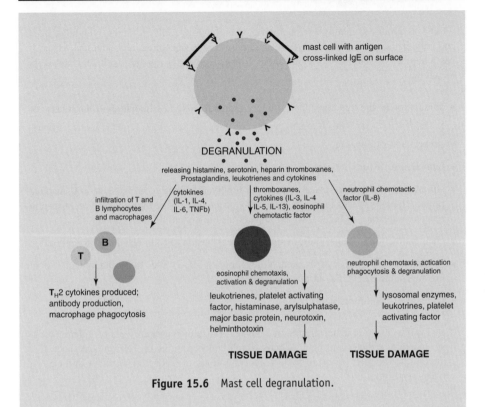

**Figure 15.6**  Mast cell degranulation.

# Focus on: tumour necrosis factor

TNF-$\alpha$ is a cytokine involved in systemic inflammation and is a member of a group of cytokines that stimulate the acute-phase reaction.

Acute-phase proteins are a class of proteins whose plasma concentrations increase (positive acute-phase proteins) or decrease (negative acute-phase proteins) in response to inflammation. This response is called the acute-phase reaction (or acute-phase response). In response to injury, local inflammatory cells secrete a number of cytokines into the bloodstream, the most notable of which are the interleukins (IL-1, IL-6, IL-8) and TNF-$\alpha$. The liver responds by producing a large number of acute-phase reactants or reducing the production of others.

TNF is produced mainly by macrophages, but also by cells including lymphoid, mast and endothelial cells, cardiac myocytes, adipose tissue, fibroblasts and neuronal tissue. It has a number of actions on various organ systems, generally together with IL-1 and IL-6, including:

- stimulation of the hypothalamic–pituitary–adrenal axis via stimulation of the release of corticotrophin-releasing hormone

- suppression of appetite

- induction of fever

- stimulation of the acute-phase response in the liver, leading to an increase in C-reactive protein. It also induces insulin resistance by promoting serine-phosphorylation of insulin receptor substrate-1 (IRS-1), which impairs insulin signalling (see Chapter 13).

- chemoattraction of neutrophils, mediating their adhesion to endothelial cells for migration

- stimulation of phagocytosis and production of IL-1 and the inflammatory lipid prostaglandin E2 ($PGE_2$) in macrophages

- increase of insulin resistance.

A local increase in concentration of TNF will cause the cardinal signs of inflammation: heat, swelling, redness and pain. Whereas high concentrations of TNF induce shock-like symptoms, prolonged exposure to low concentrations of TNF can result in cachexia, a wasting syndrome.

---

Cachexia is loss of weight, muscle atrophy, fatigue, weakness and significant loss of appetite. It is seen in patients with cancer, acquired immunodeficiency syndrome (AIDS), chronic obstructive pulmonary disease and congestive heart failure. Underlying causes are poorly understood, but there is an involvement of inflammatory cytokines, such as TNF-$\alpha$, IFN-$\gamma$, IL-6 and tumour-secreted proteolysis-inducing factor. Related syndromes are kwashiorkor and marasmus, although these are most often symptomatic of severe malnutrition.

---

The TNF gene maps to chromosome 6p21.3, is about 3 kB in size and contains four exons. TNF is primarily produced as a 212-amino acid transmembrane protein; a soluble homotrimeric cytokine is released via proteolytic cleavage by the metalloprotease TNF-$\alpha$-converting enzyme (TACE, also called ADAM17).

Two receptors, TNF-R1 and TNF-R2, bind to TNF. TNF-R1 is expressed in most tissues, and can be fully activated by both the membrane-bound and the soluble trimeric forms of TNF, whereas TNF-R2 is found only in cells of the immune system, and responds to the membrane-bound form of the TNF homotrimer.

The importance of TNF-$\alpha$ is noted in the activation of the endothelium. TNF-$\alpha$ induces endothelial cells to present E-selectin and intercellular adhesion molecule 1 (ICAM-1), both of which are cell-adhesion molecules that mediate the mechanism of leukocyte extravasation, termed diapedesis (see Chapter 14). While this process is essential for the recruitment of leukocytes to a localised area during an inflammatory response, it can be catastrophic in cases of, for example, systemic infection. Sepsis (infection in the bloodstream) is accompanied by the release of TNF-$\alpha$ by macrophages in liver, spleen and other systemic sites. The systemic release of TNF-$\alpha$ will result in vasodilation, leading to loss of blood pressure, increased vascular permeability, loss of plasma volume and eventually shock.

ICAM-1, also known as CD54 (cluster of differentiation 54), is a human gene. The protein encoded by this gene is a type of intercellular adhesion molecule, normally present in low concentrations in the membranes of leukocytes and endothelial cells; upon cytokine stimulation (particularly IL-1 and TNF-$\alpha$) the concentrations greatly increase. ICAM-1 is a ligand for LFA-1 (integrin), a receptor found on leukocytes. When activated, leukocytes bind to endothelial cells via ICAM-1/LFA-1 and then transmigrate into tissues.

E-selectin, also known as CD62E, is a cell-adhesion molecule, expressed only on endothelial cells activated by cytokines.

## Focus on: HIV – invasion of the immune system

HIV is a lentivirus (a member of the retrovirus family) that can lead to AIDS, a condition in humans in which the immune system begins to fail, leading to life-threatening opportunistic infections. Lentiviruses are transmitted as single-stranded, positive-sense, enveloped RNA-viruses. Upon entry of the target cell, the viral RNA genome is converted to double-stranded DNA by a virally encoded reverse transcriptase that is present in the virus particle. This viral DNA is then integrated into the cellular DNA by a virally encoded integrase, along with host cellular co-factors, so that the genome can be transcribed. After the virus has infected the cell, two outcomes are possible; either the virus becomes latent and the infected cell continues to function, or the virus becomes active and replicates, and a large number of virus particles are liberated, which can then infect other cells.

There are two known strains of HIV, HIV-1 and HIV-2 (Table 15.7). HIV-1 is the cause of the majority of HIV infections globally.The term 'viral tropism' refers to the type of cell HIV infects. HIV primarily infects vital cells in the human immune system, such as T-helper cells (specifically CD4+ T-cells), macrophages and dendritic cells. HIV-1

**Table 15.7** Types of HIV

| Species | Virulence | Transmission | Prevalence | Origin(?) |
|---------|-----------|--------------|------------|-----------|
| HIV-1 | High | High | Global | Common chimpanzee |
| HIV-2 | Lower | Low | West Africa | Sooty mangabey |

entry to macrophages and CD4+ T-cells is mediated through interaction of the virion envelope glycoprotein (gp120) with the CD4 molecule on the target cells, and also with chemokine co-receptors. HIV infection leads to low levels of CD4+ T-cells through three main mechanisms:

- direct viral killing of infected cells

- increased rates of apoptosis in infected cells

- killing of infected CD4+ T-cells by CD8 cytotoxic lymphocytes that recognise infected cells.

When CD4+ T-cell numbers decline below a critical level, cell-mediated immunity is compromised and the body becomes progressively more susceptible to opportunistic infections.

HIV infection in humans is now pandemic. Over 25 million people are believed to have died of the infection since its recognition in 1981. It is estimated that about 0.6% of the world's population is infected with HIV. Eventually most HIV-infected individuals develop AIDS. These individuals mostly die from opportunistic infections or malignancies. Without treatment, about 9 out of 10 people with HIV will progress to AIDS within 10–15 years.

There is currently no vaccine or cure for HIV or AIDS. A course of antiretroviral treatment administered immediately after exposure, referred to as post-exposure prophylaxis, is believed to reduce the risk of infection if begun as quickly as possible. Current treatment for HIV infection consists of highly active antiretroviral therapy (HAART). This has been highly beneficial to many HIV-infected individuals since its introduction in 1996. Current HAART options are combinations (or 'cocktails') consisting of at least three drugs belonging to at least two types, or 'classes', of antiretroviral agent. Typically, these classes are two nucleoside analogue reverse transcriptase inhibitors, plus either a protease inhibitor or a non-nucleoside reverse transcriptase inhibitor. New classes of drugs, such as entry inhibitors, provide treatment options for patients who are infected with viruses already resistant to common therapies, although they are not widely available and not typically accessible in resource-limited settings. HAART neither cures the patient nor uniformly removes all symptoms; high levels of HIV-1, often HAART-resistant, return if treatment is stopped. Moreover, it would take more than a lifetime for HIV infection to be cleared using HAART. Despite this, many HIV-infected individuals have experienced remarkable improvements in their general health and quality of life through HAART, which has led to a large reduction in HIV-associated morbidity and mortality in the developed world.

# CHAPTER 16
# Mitochondrial dysfunction

The first sub-cellular organelle to be isolated (other than the nucleus), mitochondria are the 'powerhouse' of the cell, generating ATP through aerobic oxidative phosphorylation; the TCA (Krebs) cycle (the 'hub of metabolism') and fatty acid oxidation take place entirely within mitochondria. Other pathways and cycles (urea cycle, haem biosynthesis, cardiolipin synthesis, quinone and steroid biosynthesis) include steps both outside and inside the mitochondria.

## 16.1   Mitochondrial DNA

Mitochondrial DNA (mtDNA) is present in mitochondria as a circular molecule and in most species codes for 13 or 14 proteins involved in the electron transfer chain, 2 rRNA subunits and 22 tRNA molecules (all necessary for protein synthesis) (Table 16.1). Given that 80% of mtDNA codes for functional mitochondrial proteins involved in energy production, it is not surprising that mtDNA mutations commonly lead to functional problems that manifest as muscle disorders (myopathies).

Mutation rates for mtDNA are high, about 10 times those of nuclear DNA, leading to a high variation between mitochondria, not only between different species but even within the same species. The number of affected mitochondria varies from cell to cell, depending both on the number inherited from the mother cell and on environmental factors, which may favour mutant or wildtype mtDNA.

## 16.2   Non-Mendelian inheritance

In sexually reproducing organisms, mitochondria are normally inherited exclusively from the mother. Mitochondria in mammalian sperm are usually destroyed by the egg cell after fertilisation; furthermore, most mitochondria are present at the base of the sperm's tail, which is

*Essential Biochemistry for Medicine*   Dr Mitchell Fry
© 2010 John Wiley & Sons, Ltd

**Table 16.1**  Mitochondrial gene products

| Product | Mitochondrial gene |
|---|---|
| NADH dehydrogenase (complex I) | ND1, ND2, ND3, ND4, ND4L, ND5, ND6 |
| Coenzyme Q–cytochrome c reductase/cytochrome b (complex III) | CYB |
| Cytochrome c oxidase (complex IV) | CO1, CO2, CO3 |
| ATP synthase | ATP6, ATP8 |
| rRNA | RNR1 (12S), RNR2 (16S) |
| tRNA | TA, TC, TD, TE, TF, TG, TH, TI, TK, TL1, TL2, TM, TN, TP, TQ, TR, TS1, TS2, TT, TV, TW, TY, 1X |

used for propelling the sperm cell, but which is often lost during fertilisation. Paternal sperm mitochondria are marked with ubiquitin to select them for later destruction inside the embryo. Mitochondrial inheritance is therefore non-Mendelian (Mendelian inheritance presumes that half the genetic material of a fertilised egg derives from each parent). Some *in vitro* fertilisation techniques, such as the injection of a sperm into an oocyte, may interfere with this pattern.

Since mtDNA is not highly conserved and has a rapid mutation rate, it is useful for studying the evolutionary relationships (phylogeny) of organisms. Biologists can determine and then compare mtDNA sequences among different species and use the comparisons to build an evolutionary tree for the species examined.

# 16.3   Mitochondrial cytopathies

Mitochondrial cytopathies can be very varied. Since the distribution of defective mtDNA may vary from organ to organ, a mutation that in one person would cause liver disease might in another cause a brain disorder. The severity of the defect may be great or small: some may cause 'exercise intolerance', with no serious illness or disability; other defects can have severe body-wide impacts.

Mitochondrial disease begins to become apparent once the number of affected mitochondria reaches a certain level; this phenomenon is called 'threshold expression'.

---

Since cells have multiple mitochondria, different mitochondria in the same cell can have different variations of the mtDNA genome; this condition is referred to as heteroplasmy. Heteroplasmy is the presence of a mixture of more than one type of organellar genome (mtDNA or plastid DNA) within a cell or individual. Since eukaryotic cells contain hundreds of mitochondria with hundreds of copies of mtDNA, it is possible and indeed very likely for mutations to affect only some of the copies, while the remainder are unaffected.

Table 16.2 lists some common mitochondrial cytopathies, caused by mutations within the mitochondrial genome, the nuclear genome that encodes mitochondrial proteins, or a combination of the two; they may exhibit non-Mendelian or Mendelian inheritance.

**Table 16.2**  Common mitochondrial cytopathies

| | |
|---|---|
| NARP (neuropathy, ataxia and retinitis pigmentosa) | A variety of signs and symptoms chiefly affecting the nervous system. Mutations in the MT-ATP6 gene ($F_0$-subunit 6 of the ATP synthase) cause neuropathy, ataxia and retinitis pigmentosa. |
| Kearns–Sayre syndrome (ragged red fibre myopathy or oculocraniosomatic syndrome) | Caused by a 5000 base deletion in mtDNA. It is not maternally inherited but rather occurs sporadically. Starts before the age of 20. |
| Leigh syndrome (sub-acute necrotising encephalomyelopathy) | Rare neurometabolic disorder that affects the central nervous system. An inherited disorder that usually affects infants between the age of three months and two years, but in rare cases teenagers and adults as well. Mutations in mtDNA and/or nuclear DNA lead to degradation of motor skills and eventually death. |
| MELAS syndrome (mitochondrial encephalopathy lactic acidosis with stroke-like episodes) | Mutations in the MT-ND1, MT-ND5, MT-TH, MT-TL1 and MT-TV genes. |
| MERRF (myoclonic epilepsy and ragged red fibres) | Clumps of diseased mitochondria accumulate in the subsarcolemmal region of the muscle fibre which appear as 'ragged red fibres' when the muscle is stained. |
| MNGIE (mitochondrial neurogastrointestinal encephalopathy) | Rare mitochondrial disease typically appearing between the second and fifth decades of life. It is a multisystem disorder. |
| Progressive external ophthalmoplegia | Multiple mtDNA deletions in skeletal muscle. Clinical features include adult onset of weakness of the external eye muscles (ophthalmoplegia) and exercise intolerance. Both autosomal dominant and autosomal recessive inheritance can occur, but in most cases it appears to be due to a sporadic deletion or duplication within the mtDNA. |
| Leber's hereditary optic neuropathy (Leber optic atrophy) | Results in degeneration of retinal ganglion cells and their axons, causing an acute or sub-acute loss of central vision; affects predominantly young adult males. Due to one of three point mutations in the ND4, ND1 or ND6 subunit genes of complex I. |
| Pearson's syndrome | Characterised by sideroblastic anaemia and exocrine pancreas dysfunction. Usually fatal in infancy. The few patients who survive into adulthood often develop symptoms of Kearns–Sayre syndrome. It is very rare; less than 100 cases have been reported. |

A myopathy is a neuromuscular disease resulting in muscular weakness; 'myopathy' simply means muscle disease.

## 16.4    Common symptoms of mitochondrial dysfunction

Common symptoms of mitochondrial dysfunction frequently involve the nerve, brain and muscle cells, since these are particularly dependent on a sustained energy supply. Symptoms of mitochondrial myopathies include:

- muscle weakness or exercise intolerance
- heart failure or rhythm disturbances
- dementia
- movement disorders
- stroke-like episodes
- deafness
- blindness
- droopy eyelids
- limited mobility of the eyes
- vomiting or seizures.

Mitochondrial disease is difficult to identify; symptoms may be apparent at birth or appear later in adult life. Many diseases are suspected to be caused in part by dysfunction of mitochondria, such as diabetes mellitus, forms of cancer and cardiovascular disease, lactic acidosis, specific forms of myopathy, osteoporosis, Alzheimer's disease, Parkinson's disease, stroke and many others.

## 16.5    Mitochondria and ageing

Mitochondria are frequently implicated in the ageing process. The mitochondrial respiratory chain is a source of reactive oxygen species. A number of changes occur to mitochondria during ageing: tissues from elderly patients show a decrease in enzymatic activity of the proteins of the respiratory chain; large deletions in the mitochondrial genome can lead to high levels of oxidative stress. Hypothesised links between aging and oxidative stress are not new, but there is much debate over whether mitochondrial changes are causes or merely characteristics of ageing.

With a high mtDNA mutation rate, the prevalence of mitochondrial myopathies might be expected to be high. They are however relatively rare, having an incidence of approximately 2 in 10 000 births. As part of the normal ageing process, the accumulation of mtDNA mutations might lead to abnormalities that are normally considered to be associated with ageing.

## 16.6 Diagnosis of mitochondrial myopathies

Diagnosis of mitochondrial myopathies is initially clinical, involving phenotypic (observable expression of characters and traits) evaluation, followed by laboratory evaluation. If an mtDNA mutation is detected, diagnosis is relatively straightforward; in the absence of an mtDNA mutation, diagnosis becomes difficult. Laboratory studies may include blood plasma or cerebral spinal fluid measurement for lactic acid, ketone bodies, plasma acylcarnitines and organic acids in the urine. If they are abnormal, a muscle biopsy is performed.

# CHAPTER 17

# Nerve and muscle systems

## 17.1 Nerves

The nervous system is composed of neurons and non-neuronal specialised cells called glial cells (neuroglia or simply glia). Neurons are electrically excitable cells that process and transmit information. They are the core components of the brain, the vertebrate spinal cord and the peripheral nerves. Interneurons connect neurons to other neurons within the brain and spinal cord. Neurons are 'maintained' by glial cells; glial cells provide support, nutrients and oxygen, electrical insulation in the form of myelin, and destroy pathogens and remove dead neurons. Table 17.1 documents some of the approaches to classifying the nervous system.

## 17.2 The nerve message

The plasma membrane of neurons, like all other cells, has an unequal distribution of ions and electrical charges between its two sides. Sodium–potassium ATPase pumps maintain this unequal concentration by actively transporting ions against their concentration gradients: sodium in, potassium out. The membrane is positive outside and negative inside. This charge difference is referred to as the resting potential and is measured in millivolts ($=-65\,\mathrm{mV}$).

A change in polarity of the membrane, an action potential, results in propagation of the nerve impulse along the membrane. An action potential is a temporary reversal of the electrical potential along the membrane that lasts for a few milliseconds. Sodium gates and potassium gates open in the membrane to allow their respective ions to cross. Sodium and potassium ions reverse positions by passing through membrane protein channel gates; sodium crosses first, to the outside, followed by potassium, to the inside. The changed ionic distributions are reset by the continuously running sodium–potassium pump, eventually restoring the original resting potential.

The junction between two neurons is called a synapse. Synapses are 'terminals' at which the action potential may be arrested, redirected or relayed; synapses at muscle fibres are called neuromuscular junctions (or myoneural junctions). There is a very narrow gap, the synaptic

**Table 17.1**   Classifying the nervous system

| Method of classification | |
| --- | --- |
| Anatomical | Central and peripheral. |
| Functional | The **somatic nervous system** is responsible for coordinating voluntary body movements (i.e. activities that are under conscious control). The **autonomic nervous system** is responsible for coordinating involuntary functions, such as breathing and digestion. |
| Directional | **Afferent system** by sensory neurons, which carry impulses from a somatic receptor to the CNS. **Efferent system** by motor neurons, which carry impulses from the CNS to an effector. **Relay system** by interneurons (also called 'relay neurons'), which transmit impulses between the sensory and motor neurons (in both the CNS and the PNS). |
| Electrophysiology | **Tonic or regular-spiking neurons** are typically constantly (or tonically) active, for example the interneurons in the neurostriatum. **Phasic or bursting neurons** that fire in bursts. **Fast-spiking neurons** are notable for their fast firing rates, for example some types of cortical inhibitory interneurons, cells in the globus pallidus and retinal ganglion cells. **Thin-spike neurons** have narrow action potentials, for example interneurons in the prefrontal cortex. |
| Neurotransmitter | Cholinergic neurons – acetylcholine GABAergic neurons – gamma aminobutyric acid Glutamatergic neurons – glutamate Dopaminergic neurons – dopamine Serotonin neurons – serotonin |

CNS = central nervous system, PNS = peripheral nervous system.

cleft (about 20 nm), between the neurons. Transmitting the nerve impulse across the synaptic cleft involves an electrical to chemical to electrical event (Figure 17.1):

1. Action potentials travel down the axon of the neuron to its end(s), the axon terminal(s).

2. Each axon terminal is swollen, forming a synaptic knob.

3. The synaptic knob is filled with membrane-bounded vesicles containing a neurotransmitter.

4. Arrival of an action potential at the synaptic knob opens voltage-gated $Ca^{2+}$ channels in the plasma membrane.

5. The influx of $Ca^{2+}$ triggers the exocytosis of some of the vesicles.

6. Their neurotransmitter is released into the synaptic cleft.

7. The neurotransmitter molecules bind to receptors on the postsynaptic membrane. These receptors are ligand-gated ion channels which initiate the next electrical impulse.

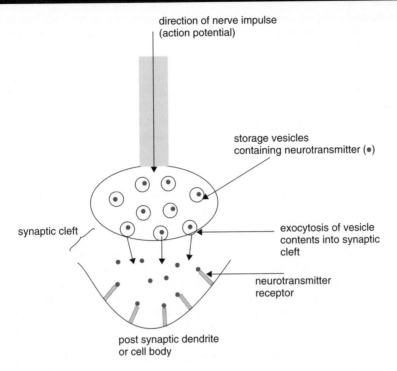

direction of nerve impulse
(action potential)

storage vesicles
containing neurotransmitter (●)

synaptic cleft

exocytosis of vesicle
contents into synaptic
cleft

neurotransmitter
receptor

post synaptic dendrite
or cell body

**Figure 17.1**  Transmission of the action potential across a synapse.

8. The neurotransmitter is broken down by a specific enzyme in the synaptic cleft; for example acetylcholinesterase breaks down the neurotransmitter acetylcholine. The breakdown products are re-absorbed by the pre-synaptic neurone via endocytosis and used to resynthesise more neurotransmitter. This stops the synapse from being permanently 'on'.

The neuroreceptors (ligand-gated ion channels) on the postsynaptic membrane determine the type of synapse:

- **Excitatory ion-channel synapses** have neuroreceptors that are sodium channels. When the channels open, sodium ions flow in, causing a local depolarisation and making an action potential more likely. Typical 'excitatory' neurotransmitters are acetylcholine, glutamate and aspartate.

- **Inhibitory ion-channel synapses** have neuroreceptors that are chloride channels. When the channels open, chloride ions flow in, causing a local hyperpolarisation and making an action potential less likely. Typical 'inhibitory' neurotransmitters are glycine and GABA.

- **Neuromuscular junctions** are the synapses formed between motor neurones and muscle cells. They always use the neurotransmitter acetylcholine and are always excitatory. Motor neurones also form specialised synapses with secretory cells.

- **Electrical synapses,** in which two cells actually touch, allow the action potential to pass directly from one membrane to the next. They are very fast, but are quite rare, found only in the heart and the eye.

- **Non-channel synapses** have membrane-bound neuroreceptors (which are not ion channels). When activated by the neurotransmitter they initiate an intracellular signalling pathway; in particular they can alter the number and sensitivity of the ion-channel receptors in the same cell. These synapses are involved in slow and long-lasting responses such as learning and memory. Typical neurotransmitters are adrenaline (epinephrine), noradrenaline, dopamine, serotonin, endorphin, angiotensin and acetylcholine.

Voltage-gated ion channels, a class of transmembrane ion channel, are especially critical in neurons, but are common in many types of cell. They include the $Na^+$, $K^+$ and $Ca^{2+}$ channels of neurons.

---

Sodium channels consist of two types of protein subunit, $\alpha$ and $\beta$. An $\alpha$-subunit forms the core of the channel; it has four repeat domains, I through IV, each containing six membrane-spanning regions, S1 through S6. The highly conserved S4 region acts as the channel's voltage sensor. The voltage sensitivity of this channel is caused by positive amino acids located at every third position. When stimulated by a change in transmembrane voltage, this region moves towards the extracellular side of the cell membrane, allowing the channel to become permeable to ions. The ions are conducted through a pore, which can be divided into two regions. The more external (more extracellular) portion of the pore is formed by the 'P-loops' (the region between S5 and S6) of the four domains. This region is the narrowest part of the pore and is responsible for its ion selectivity. The inner portion (more cytoplasmic) of the pore is formed by the combined S5 and S6 regions of the four domains. The region linking domains III and IV acts to 'plug' the channel after prolonged activation, so inactivating it.

Potassium channels are the most widely distributed type of ion channel and are found in virtually all living organisms. Potassium channels have a tetrameric structure in which four identical protein subunits associate to form a fourfold symmetric ($C_4$) complex arranged around a central ion-conducting pore (a homotetramer). Alternatively, four related but not identical protein subunits may associate to form a heterotetramer with pseudo $C_4$ symmetry. All potassium-channel subunits have a distinctive pore-loop structure that lines the top of the pore and is responsible for potassium-selective permeability.

Voltage-dependent calcium channels (VDCCs) are a group of voltage-gated ion channels found in excitable cells (e.g. muscle, glial cells, neurons, etc.). At resting membrane potential, VDCCs are normally closed; they are activated (opened) at depolarised membrane potentials. Activation of particular VDCCs allows $Ca^{2+}$ entry into the cell, which, depending on the cell type, results in muscular contraction, excitation of neurons, up-regulation of gene expression, or release of hormones or neurotransmitters. VDCCs are formed as a complex of several different subunits: $\alpha_1$, $\alpha_2\delta$, $\beta_{1-4}$ and $\gamma$. The $\alpha_1$ subunit forms the ion-conducting pore, while the associated subunits have several functions, including modulation of gating.

---

# 17.3 Diseases of the nervous system

All categories of neurological disorder are somewhat artificial because more than one aspect of brain function may be affected. Table 17.2 gives a brief summary.

**Table 17.2**  Categories of neurological disorder

| | |
|---|---|
| Seizures (epilepsy) | An epileptic seizure is a transient symptom of abnormal, excessive or synchronous neuronal activity in the brain (normal brain activity is marked by unsynchronised firing of neurons). It can manifest as an alteration in mental state, tonic or clonic movements and convulsions. The medical syndrome of recurrent, unprovoked seizures is termed epilepsy, but seizures can occur in people who do not have epilepsy. |
| Coma | Coma is the most severe form of depressed consciousness, and may result from a variety of conditions, including intoxication, metabolic abnormalities, CNS diseases, and acute neurologic injuries such as stroke and hypoxia. |
| Infection | Infectious agents can be fungal, protozoal, bacterial or viral. |
| Vascular | Brain metabolic activity is very high (a major component of which is the activity of the $Na^+$-$K^+$ ATPase); interference with the blood supply or $O_2$ delivery will have prompt consequences. Such events are called strokes; they are the third most common cause of death in developed countries. |
| Degenerative | Although degenerative diseases generally affect mental capacity, some only affect motor capacity, for example Fiedreich's ataxia and amyotrophic lateral sclerosis. |
| Demyelinating | Multiple sclerosis (together with other idiopathic inflammatory demyelinating diseases) is an autoimmune and multifactorial disorder. |
| Movement | Parkinson's disease is caused by the premature degeneration of the dopaminergic neurons of the substantia nigra. Huntington's disease is known to be an inherited disorder with an autosomal dominance pattern. |
| Dementias | Dementias cause a loss of previously normal intellectual functions (in mental retardation such functions are not attained). Major causes of dementia are Alzheimer's disease and multiple strokes or infarcts. |
| Mental retardation | Generally defined as a non-progressive intellectual deficit; frequently the result of processes that begin *in utero*. |
| Metabolic | Can include a variety of effects, but neurodegenerative disorders are primarily hereditary in nature (in-born errors of metabolism), for example lysosomal storage diseases. |

# 17.4  Specific neural disorders

- **Depression,** associated with reduced levels of the neurotransmitter serotonin (Figure 17.2). Selective serotonin reuptake inhibitors (SSRIs) are widely used to treat depression. SSRIs increase the extracellular level of serotonin by inhibiting its reuptake into the presynaptic cell, thereby increasing its availability to the postsynaptic receptor.

  Fluoxetine hydrochloride (Prozac, Fontex, etc.; Figure 17.2) is an antidepressant of the SSRI class. As a structural analogue of serotonin, it competes for its re-uptake at the

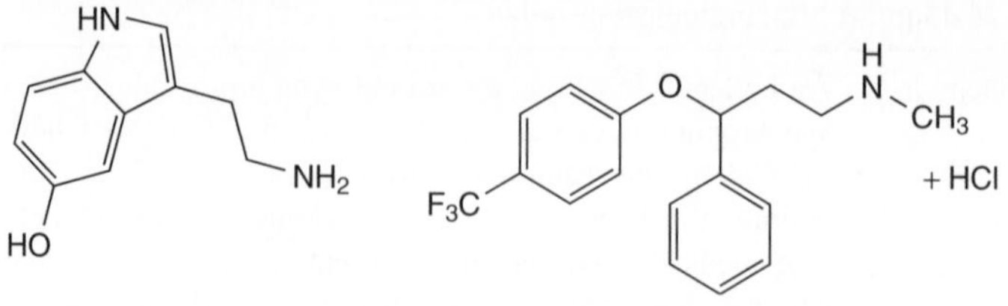

**Figure 17.2** Structures of serotonin and fluoxetine hydrochloride.

presynaptic cell. Fluoxetine is approved for the treatment of major depression (including pediatric depression), obsessive-compulsive disorder (in both adult and pediatric populations), bulimia nervosa, anorexia nervosa, panic disorder and premenstrual dysphoric disorder. Fluoxetine is metabolised in the liver by isoenzymes of the cytochrome 450 system, including CYP2D6. The role of CYP2D6 may be clinically important, as there is great genetic variability in the function of this enzyme between individuals (Chapter 19). Only one metabolite of fluoxetine, norfluoxetine (demethylated fluoxetine), is biologically active.

- **Myasthenia gravis** (literally 'serious muscle-weakness'), a neuromuscular disease leading to fluctuating muscle weakness and fatigue (see below).

- **Neurotoxins,** such as the clostridial neurotoxins responsible for tetanus and botulism. These are metallo-proteases that enter nerve cells and block neurotransmitter release via zinc-dependent cleavage of protein components of the neuroexocytosis apparatus.

  Tetanus is characterised by a prolonged contraction of skeletal muscle fibres; the neurotoxin responsible is from *Clostridium tetani*. The toxin initially binds to peripheral nerve terminals and is then transported within the axon and across synaptic junctions until it reaches the central nervous system (CNS). Here it attaches to gangliosides at the presynaptic inhibitory motor nerve endings and is taken up into the axon by endocytosis. The effect of the toxin is to block the release of inhibitory neurotransmitters (glycine and gamma-amino butyric acid), which are required to check the nervous impulse, leading to the generalised muscular spasms characteristic of tetanus.

  Botulinum toxin is produced by the bacterium *Clostridium botulinum*; it is the most toxic protein known. The toxin consists of a heavy chain and a light chain; the light chain has protease activity, which degrades a cellular protein (SNAP-25) required for the normal release of neurotransmitters from the axon endings. The consequent lack of release of acetylcholine results in muscle paralysis.

- **Prions** are infectious agents that are proteins. All known prion agents appear to propagate by transmitting a misfolded protein state; the protein itself does not self-replicate and the process is dependent on the presence of the normal polypeptide in the host organism. The prion is hypothesised to cause normal molecules in the host to misfold. The misfolded form of the prion protein has been implicated in a number of diseases in a variety of mammals, including bovine spongiform encephalopathy (BSE, 'mad cow disease') in cattle and Creutzfeldt−Jakob disease (CJD) in humans. All known prions induce the formation of an amyloid fold, in which the protein polymerises into an aggregate consisting of tightly packed $\beta$-sheets. Amyloid is characterised by a cross-$\beta$-sheet-quaternary structure; the $\beta$-strands of the stacked $\beta$-sheets come from different protein monomers and align perpendicular to the axis of the fibril. The protein that prions are made of (referred to as PrP$^c$) is normally found throughout the body. However, prions found in infectious material (PrP$^{Sc}$) have a different structure that is resistant to degradation by proteases. The human form of PrP$^c$ is 209 amino acids in length, 35–36 kDa, mainly $\alpha$-helical in structure, and contains one disulphide bond. Several topological forms exist, including a membrane-bound form that is anchored by a glycolipid and two transmembrane forms. PrP$^C$ is known to bind copper with high affinity. There is evidence that it might function in cell–cell adhesion of neural cells and/or be involved in cell–cell signalling in the brain. All prion diseases affect the structure of the brain or other neural tissue, and all are currently untreatable and always fatal. The reason for amyloid plaque (aggregates) association with such diseases is unclear. In some cases the deposits may physically disrupt tissue architecture. An emerging consensus implicates pre-fibrillar intermediates, rather than mature amyloid fibres, in causing cell death. Studies have also shown that amyloid deposition is associated with mitochondrial dysfunction, which can initiate a signalling pathway leading to apoptosis.

# Focus on: nicotine addiction

Nicotinic acetylcholine receptors (nAChRs) are ligand-gated ion channels, and one of the best-studied of the ionotropic receptors. They are found in both the CNS and the peripheral nervous system, as well as in the neuromuscular junctions of somatic muscles.

Nicotinic receptors are made up of five subunits, arranged symmetrically around a central pore. They share similarities with GABA$_A$ receptors, glycine receptors and type 3 serotonin receptors. They are broadly classified into two subtypes based on their primary sites of expression, namely muscle type and neuronal type; to date 17 nAChR subunits have been identified. Of these $\alpha 2$–7 and $\beta 2$–4 have been cloned in humans; the remaining genes have been identified in chick and rat genomes.

Opening of the nAChR channel pore requires the binding of its endogenous ligand, acetylcholine. Agonists of acetylcholine include nicotine, epibatidine and choline (an agonist mimics the response of the normal ligand, an antagonist opposes the response), Figure 17.3.

In neuronal nAChRs, the acetylcholine binding site is formed from amino acid residues of both $\alpha$- and $\beta$-subunits, in the extracellular domain near the N-terminus. Agonist binding causes a conformational change resulting in channel opening (a pore of about 0.65 nm is formed). Binding of an agonist stabilises the open and desensitised states. Opening of the channel allows positively charged ions to move across it; in particular, Na$^+$ enters and K$^+$ exits, with a net flow of positively charged ions inward.

Agonist binding modifies the state of the neurons through two main mechanisms. First, movement of cations through the channel causes a depolarisation of the plasma membrane (which results in an excitatory postsynaptic potential in neurons), as well as the activation

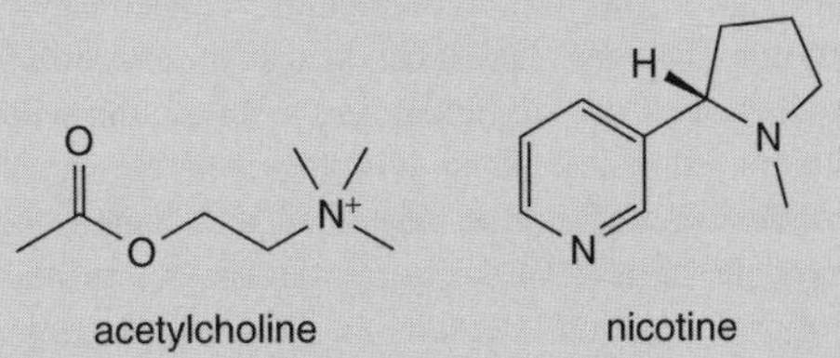

acetylcholine                    nicotine

Figure 17.3 Structures of acetylcholine and nicotine.

of other voltage-gated ion channels. Second, entry of $Ca^{2+}$ through the channel may directly or indirectly initiate intracellular cascades, leading, for example, to gene regulation or release of neurotransmitters. Given that each nicotinic receptor is a pentamer, there is an immense potential of types, with highly variable kinetic, electrophysiological and pharmacological properties. Some subunit combinations are notable, specifically $(\alpha 1)_2\beta 1\delta\varepsilon$ (muscle type), $(\alpha 3)_2(\beta 4)_3$ (ganglion type), $(\alpha 4)_2(\beta 2)_3$ (CNS type) and $(\alpha 7)_5$ (another CNS type). It is the $\alpha 4$ $\beta 2$ CNS variant to which nicotine binds most strongly.

The current consensus is that nicotine is the psychoactive drug primarily responsible for the addictive nature of tobacco use. Addiction is a complex behavioural phenomenon with causes and effects that range from molecular mechanisms to social interactions. Ultimately the process of drug addiction begins with molecular interactions which alter the activity and metabolism of the neurons that are sensitive to that drug. Over time this alters the properties of individual neurons and circuits, leading to complex behaviours such as dependence, tolerance, sensitisation and craving. A common feature of many addictive drugs, including nicotine, is that they increase dopamine levels in the nucleus accumbens (NAcc). The principal dopaminergic projections to the NAcc arise from neurons in the midbrain ventral tegmental area (VTA). The VTA is strongly implicated in the response to natural rewards, such as food or sex, as well as the reinforcing effects of various drugs of abuse. All addictive drugs elicit release of the neurotransmitter dopamine, which is manufactured in neurons of the VTA, and rodents will self-administer nicotine (or cocaine or morphine) directly into this region. Evidence that NAcc dopamine levels are important in reward has come from VTA lesion studies of the NAcc and study of dopamine-receptor antagonists, both of which result in reduced self-administration of many addictive drugs, including nicotine.

Important functional properties of these receptors that contribute to their physiological effects include activation, desensitisation and up-regulation following nicotine exposure. There is considerable diversity in the sensitivity of different receptor subtypes to nicotine. *In vitro* experiments demonstrate that chronic exposure to nicotine can 'desensitise' these receptors, making them less sensitive to the natural neurotransmitter acetylcholine. It is thought that the up-regulation of binding sites might reflect an increased number of receptors that compensate for nicotine-induced desensitisation. Up-regulation of nAChR function and ligand binding, following pre-exposure to nicotine, varies with cell type and receptor subtype. Physiologically relevant nicotine concentrations have been shown to up-regulate $\alpha 4$-$\beta 2$-containing receptors. Up-regulation of other receptor subtypes can occur with higher nicotine concentrations in some cells. Nicotinic receptor up-regulation has previously been reported to involve an increase in the number of receptors, but apparently not with an associated change in mRNA; this may reflect increased assembly or change in receptor state, rather than receptor number.

The positive pleasurable effects of nicotine are instantaneous and short-lasting, while the negative effects are delayed and long-lasting. In rats, chronic nicotine use recruits a major brain-stress system, the extrahypothalamic corticotropin releasing factor (CRF) system, which contributes to continued tobacco use by exacerbating anxiety and craving upon withdrawal. It was found that administering a compound which blocked the receptors involved in this stress system alleviated withdrawal symptoms. This may underlie the transition from nicotine use to nicotine dependence. Changes in the CRF system were measured in the amygdala, an area of the brain that plays a primary role in the processing and memory of emotional reactions. Results suggest long-lasting neuroadaptations of the CRF system, possibly through gene regulation, which may help explain why many cigarette smokers relapse even after a long abstinence from smoking.

# Focus on: anaesthetics

Anaesthesia has traditionally meant the condition of having sensation (including pain) temporarily blocked. The simple but very diverse molecular structures of general anaesthetics, together with their diverse side effects, made it difficult to comprehend that they acted in a specific manner. However, general anaesthetics may be more selective than previously thought, binding to only a small number of targets in the CNS. At surgical concentrations their principal effects appear to be on ligand-gated (rather than voltage-gated) ion channels. Although the role of second messengers remains uncertain, it is clear that anaesthetics act directly on proteins, rather than on lipids as was previously thought.

Much of our current understanding of how anaesthetics work has been obtained using genetic approaches, in particular knock-out or knock-in mice. Anaesthetic drugs can be grouped into volatile and intravenous anaesthetics, according to their route of administration. Common volatile anaesthetics induce immobility via molecular targets in the spinal cord, including GABA receptors, glycine receptors, glutamate receptors and TREK-1 potassium channels. In contrast, intravenous anaesthetics cause immobility almost exclusively via GABA receptors. Hypnosis is predominantly mediated by three $\beta$-subunit-containing GABA receptors in the brain, whereas two $\beta$-subunit-containing receptors, which make up more than 50% of all GABA receptors in the CNS, mediate sedation. At clinically relevant concentrations, ketamine and nitrous oxide block $N$-methyl $D$-aspartate (NMDA) receptors (an ionotropic glutamate receptor, see below); unlike all other anaesthetics in clinical use they produce analgesia. Not only the desired actions of anaesthetics, but also their undesired side effects, are linked to certain receptors. Respiratory depression involves three $\beta$-subunit-containing GABA receptors, whereas hypothermia is largely mediated by two $\beta$-subunit GABA receptors.

The GABA receptor structure is typical of most ligand-gated (ionotropic) receptors. It is made up of five protein subunits arranged to form a pore, or channel, that remains closed until its specific ligand (in this case, GABA) binds to the recognition site. The GABA receptor requires both $\alpha$- and $\beta$-subunit components in order to bind GABA, and GABA receptors are typically made up of two $\alpha$- and two $\beta$-subunits among the five subunits, though the particular subunit composition often varies widely between brain regions and species. In fact, 17 different combinations of subunits have been identified. Having a family of receptors that all share the same basic structure, but differ in specific subunit compositions, allows for a greater level of functional diversity. Each of the receptor varieties varies in binding affinity, channel activity and the degree to which GABA binding is affected by different endogenous modulators. As a result, inhibitory signals can be more finely controlled. Such receptor heterogeneity also allows for more control at the genomic level, allowing post-synaptic cells to respond to changing developmental needs or variable activity at the synapse. GABA receptors are chloride channels; when activated by GABA they allow the flow of chloride ions across the membrane of the cell. Whether this chloride flow is excitatory/depolarising (makes the voltage across the cells membrane less negative), shunting (has no effect on the cells membrane) or inhibitory/hyperpolarising (makes the cells membrane more negative) depends on the direction of the flow of chloride. When net chloride flows out of the cell, GABA is excitatory or depolarising; when net chloride flows into the cell, GABA is inhibitory or hyperpolarising. When the net flow of chloride is close to zero, the action of GABA is shunting. Shunting inhibition has no direct effect on the membrane potential of the cell; however, it minimises the effect of any coincident synaptic input essentially by reducing the electrical resistance of the cell's membrane.

A third class of GABA receptor (GABA$_B$) is the G-protein-coupled receptor, which opens ion channels via intermediaries (G-proteins).

Neurons that produce GABA as their output are called GABAergic neurons and have chiefly inhibitory action at receptors in the adult vertebrate. Drugs that act as agonists of GABA receptors (GABAergic drugs), or increase the available amount of GABA, typically have relaxing, anti-anxiety and anti-convulsive effects.

The glycine receptor is one of the most widely distributed inhibitory receptors in the CNS and has important roles in a variety of physiological processes, especially in mediating inhibitory neurotransmission in the spinal cord and brain stem. There are presently four known isoforms of the $\alpha$-subunit ($\alpha_{1-4}$) that are essential to the binding of ligands, and a single $\beta$-subunit. The adult form of the glycine receptor is the heteromeric $\alpha_1\beta$ receptor, which has a stoichiometry of three $\alpha_1$-subunits and two $\beta$-subunits, or four $\alpha_1$-subunits and one $\beta$-subunit. Glycine activates its receptor channel to increase Cl$^-$ ion conductance; like GABA, the increase in Cl$^-$ ion conductance results in a hyperpolarisation of the neuronal membrane and an antagonism of other depolarising stimuli.

Glutamate receptors exist as a number of specific subtypes, both ionotropic (directly open ion channels) and metabotropic (G-protein-coupled receptors).

Glutamate is the most prominent neurotransmitter in the body, present in over 50% of nervous tissue.

The NMDA receptor is a specific type of ionotropic glutamate receptor. NMDA is the name of a selective agonist that binds to NMDA receptors but not to other glutamate receptors. Activation of NMDA receptors results in the opening of an ion channel that is nonselective to cations. A unique property of the NMDA receptor is its voltage-dependent activation, a result of ion-channel block by extracellular Mg$^{2+}$ ions. This allows voltage-dependent flow of Na$^+$ and small amounts of Ca$^{2+}$ ions into the cell, and of K$^+$ out of the cell. Calcium flux through NMDA receptors is thought to play a critical role in synaptic plasticity, a cellular mechanism for learning and memory. The NMDA receptor is distinct in two ways: first, it is both ligand-gated and voltage-dependent, and second, it requires co-activation by two ligands, glutamate and glycine.

## 17.5    Muscle types

There are three types of muscle:

- **Voluntary skeletal muscle** is under conscious control. Each fibre is a large, multi-nucleate cell, formed by fusing hundreds of myoblasts end to end. They show a striated pattern, reflecting the regular arrangement of sarcomeres (see Section 17.8) within each cell.

- **Cardiac muscle** is similar to skeletal muscle, but is not under conscious control. Its mono-nucleate cells are much smaller, but still show a striated pattern.

- **Smooth muscle** has no regular striations and the contractions are much slower. Smooth muscle is found in the blood vessels, gut, skin, eye pupils and urinary and reproductive tracts.

Voluntary muscles contain a variety of fibre types which are specialised for particular tasks:

- **Type 1** or **slow oxidative fibres** have a slow contraction speed and a low myosin ATPase activity. These cells are specialised for steady, continuous activity and are highly resistant to fatigue. Their motor neurones are often active, with a low firing frequency. These cells are thin (high surface-to-volume ratio), with a good capillary supply for efficient gas exchange. They are rich in mitochondria and myoglobin, which gives them a red colour. They are built for aerobic metabolism and prefer to use fat as a source of energy. These are the marathon runner's muscle fibres.

- **Type 2A** or **fast oxidative-glycolytic fibres** have a fast contraction speed and a high myosin ATPase activity. They are progressively recruited when additional effort is required, but are still very resistant to fatigue. Their motor neurones show bursts of intermittent activity. These cells are thin (high surface-to-volume ratio) with a good capillary supply for efficient gas exchange. They are rich in mitochondria and myoglobin, which gives them a red colour. They are built for aerobic metabolism and can use either glucose or fats as a source of energy. These are general-purpose muscle fibres which give the edge in athletic performance, but they are more expensive to operate than type 1.

- **Type 2B** or **fast glycolytic fibres** have a fast contraction speed and a high myosin ATPase activity. They are only recruited for brief maximal efforts and are easily fatigued. Their motor neurones transmit occasional bursts of very high-frequency impulses. These cells are large (poor surface-to-volume ratio) and their limited capillary supply slows the delivery of oxygen and removal of waste products. They have few mitochondria and little myoglobin, resulting in a white colour (e.g. chicken breast). They generate ATP by the anaerobic fermentation of glucose to lactic acid. These are the sprinter's muscle fibres, of no use for sustained performance.

---

Myosins are a large family of motor proteins; they are powered by the hydrolysis of ATP and convert chemical energy into mechanical work. They are responsible for actin-based motility. Most myosin molecules are composed of a head, a neck and a tail domain.

- **The head domain** binds actin, and uses ATP hydrolysis to generate force and 'walk' along the actin filament.

- **The neck domain** acts as a linker and lever for transducing force generated by the catalytic head domain. It also has regulatory functions.

- **The tail domain** mediates interaction with other molecules and/or other myosin subunits and may play a role in regulating motor activity.

---

# 17.6 The neuromuscular junction

The neuromuscular junction is the synapse (junction) of an axon terminal of a motorneuron, which terminates in a depression of the sarcolemma (the cell membrane of a muscle cell), the motor end plate. It is here that the initiation of action potentials across the muscle surface ultimately leads to muscle contraction. In vertebrates, the neurotransmitter is acetylcholine.

The sarcolemma consists of a plasma membrane and an outer coat made up of a thin layer of polysaccharide material that contains numerous thin collagen fibrils. At each end of the muscle fibre this outer coat of the sarcolemma fuses with a tendon fibre, and the tendon fibres in turn collect into bundles to form the muscle tendons that then insert into bones. The membrane is designed to receive and conduct stimuli, is extensible and encloses the contractile substance of a muscle fibre. The sarcolemma is attached to the cytoskeleton on its cytoplasmic surface. It invaginates into the cytoplasm, forming membranous tubules called transverse tubules; sarcoplasmic reticulum (enlarged smooth endoplasmic reticulum) lies either side of the transverse tubules. The transverse tubules and sarcoplasmic reticulum transmit altered membrane permeability down the tubules and into the muscle.

- An action potential results in opening of VDCCs at the axon terminal. Entrance of $Ca^{2+}$ triggers a biochemical cascade to cause neurotransmitter-containing vesicles to fuse with the cell membrane and release acetylcholine into the synaptic cleft.

- Acetylcholine receptors on the motor end plate bind acetylcholine, which opens ligand-gated ion channels. This allows movement of $Na^+$ into and $K^+$ out of the myocyte, producing a local depolarisation of the motor end plate (the end-plate potential).

- This depolarisation spreads across the surface of the muscle fibre, into the transverse tubules, eliciting the release of calcium from the sarcoplasmic reticulum that initiates muscle contraction.

- The action of acetylcholine is terminated when the enzyme acetylcholinesterase degrades the neurotransmitter.

## 17.7   Neuromuscular disease

Neuromuscular disease is a very broad term that encompasses many diseases and ailments which either directly, via intrinsic muscle pathology, or indirectly, via nerve pathology, impair the functioning of the muscles. Diseases of the motor end plate include myasthenia gravis and its related condition Lambert–Eaton myasthenic syndrome. Tetanus and botulism are bacterial infections in which bacterial toxins cause increased or decreased muscle tone, respectively.

Myasthenia gravis is an autoimmune reaction against acetylcholine receptors; the end-plate potential fails to activate the muscle fibre, resulting in muscle weakness and fatigue. A protein known as muscle specific kinase (MuSK), which appears to be important in the formation of the neuromuscular junction, is also the target of autoimmune antibodies. Antibodies directed against this protein are found in those patients with myasthenia gravis who do not demonstrate antibodies to the acetylcholine receptor (sero-negative). Lambert–Eaton myasthenic syndrome is a rare autoimmune disorder which is usually associated with presynaptic antibodies to the VDCCs.

Botulinum toxin is both a medication and a neurotoxin, produced by the bacterium *Clostridium botulinum*. It is the most toxic protein known. It can be used to treat muscle spasms, and is sold commercially under various names (Botox, Dysport, Myobloc, etc.). Botox Cosmetic and Vistabel are available for cosmetic treatment. The toxin protein consists of a heavy and a light chain. The heavy chain is important in targeting the toxin to axon terminals. The toxin enters neurons by endocytosis. The light chain of the toxin has protease activity; SNAP-25 protein, which is required for release of neurotransmitter at the axon endings, is proteolytically degraded, resulting in muscle paralysis.

## 17.8 Sarcomeres and focal adhesions

A sarcomere is the basic unit of a muscle's cross-striated myofibril (Figure 17.4). Sarcomeres are the 'motor units' of skeletal and cardiac muscle. They are multi-protein complexes composed of three different filament systems:

- The thick filament system, which comprises myosin protein, connected from the M-line to the Z-disc by titin (connectin), and myosin-binding protein C, which binds at one end to the thick filament and at the other to actin.

- The thin filaments, which are assembled by actin monomers bound to nebulin; they also involve tropomyosin (a dimer which coils itself around the F-actin core of the thin filament).

- Nebulin and titin, which give stability and structure to the sarcomere.

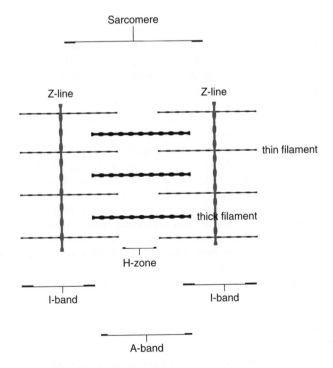

**Figure 17.4** Skeletal and cardiac sarcomere structure.

The sarcomeres are responsible for the striated appearance of skeletal and cardiac muscle. The myofibrils of smooth-muscle cells are not arranged into sarcomeres.

The relationship between the proteins and the regions of the sarcomere are as follows:

- Actin filaments are the major component of the I-band and extend into the A-band.

- Myosin filaments extend throughout the A-band and are thought to overlap in the M-band.

- The giant protein titin extends from the Z-line of the sarcomere, where it binds to the thin filament system, to the M-band, where it is thought to interact with the thick filaments. Titin (along with its splice isoforms) is the biggest single highly elasticated protein found in nature. It provides binding sites for numerous proteins and is thought to play an important role in the assembly of the sarcomere.

- Several proteins important for the stability of the sarcomeric structure are found in the Z-line as well as in the M-band of the sarcomere.

- Actin filaments and titin molecules are cross-linked in the Z-disc via the Z-line protein $\alpha$-actinin.

- The M-band proteins myomesin, as well as M-protein, cross-link the thick-filament system (myosins) and the M-band part of titin (the elastic filaments).

- The interaction between actin and myosin filaments in the A-band of the sarcomere is responsible for the muscle contraction (sliding-filament model).

Focal adhesions (in muscle often referred to as costameres) are regions that are associated with the sarcolemma of skeletal muscle fibres and comprise proteins of the dystrophin–glycoprotein complex and vinculin–talin–integrin system. Focal adhesions play both a mechanical and a signalling role, transmitting force from the contractile apparatus to the extracellular matrix in order to stabilise skeletal-muscle fibres during contraction and relaxation. Several focal adhesion constituent proteins have been shown to be defective in muscular dystrophies and cardiomyopathies.

---

Focal adhesions are large macromolecular assemblies through which both mechanical force and regulatory signals are transmitted. They can be considered as sub-cellular macromolecules that mediate the regulatory effects (e.g. cell anchorage) of extracellular matrix adhesion on cell behaviour. Focal adhesions serve as the mechanical linkages to the extracellular matrix, and as a biochemical signalling hub to concentrate and direct numerous signalling proteins at sites of integrin binding and clustering. Integrins are cell-surface receptors that interact with the extracellular matrix and mediate various intracellular signals. They define cellular shape and mobility, and regulate the cell cycle. Vinculin is a membrane-cytoskeletal protein in focal adhesions that is involved in linkage of integrin adhesion molecules to the actin cytoskeleton.

---

# 17.9    Dystrophin

Dystrophin is a rod-shaped cytoplasmic protein that is a vital part of the focal adhesion, or the dystrophin-associated protein complex. Other muscle proteins, such as $\alpha$-dystrobrevin, syncolin,

**Table 17.3** Skeletal myopathies

| Inheritance | Chromosome | Gene | Protein product | Skeletal myopathy |
|---|---|---|---|---|
| X-linked | Xp21 | Dystrophin | Dystrophin | Duchenne/Becker muscular dystrophy |
| X-linked | Xq28 | G4.5 | Tafazzin | Barth syndrome |
| Autosomal dominant | 15q14 | Actin | Actin | Nemaline myopathy |
| Autosomal dominant | 2q35 | Desmin | Desmin | Desmin myopathy |
| Autosomal dominant | 5q33 | $\delta$-Sarcoglycan | $\delta$-Sarcoglycan | Limb girdle muscular dystrophy |
| Autosomal dominant | 1q32 | Troponin T | Troponin T | – |
| Autosomal dominant | 14q11 | $\beta$-Myosin heavy chain | $\beta$-Myosin heavy chain | – |
| Autosomal dominant | 15q2 | $\alpha$-Tropomyosin | $\alpha$-Tropomyosin | – |

synemin, sarcoglycan, dystroglycan and sarcospan, co-localise with dystrophin at the costamere. The dystrophin gene (Xp21) is one of largest known, covering some 2.4 megabases with 79 exons. Although dystrophin is not required for the assembly of focal adhesions, its absence in humans and mice leads to a disorganised focal adhesion lattice and disruption of sarcolemmal integrity.

Deficiency of dystrophin is the main cause of muscular dystrophy; mutation in the gene causes Duchenne muscular dystrophy, a severe recessive X-linked form of muscular dystrophy characterised by rapid progression of muscle degeneration, which eventually leads to loss of ambulation and death. This affliction affects 1 in 3500 males, making it the most prevalent of muscular dystrophies. In general, only males are afflicted, though females can be carriers.

Normal tissue contains small amounts of dystrophin (about 0.002% of total muscle protein), but its absence leads to both muscular dystrophy and fibrosis, a condition of muscle hardening. A different mutation of the same gene causes defective dystrophin, leading to Becker's muscular dystrophy.

Table 17.3 documents a number of skeletal myopathies.

# 17.10  Intrinsic cardiomyopathy

Intrinsic cardiomyopathy can have a number of causes, including drug and alcohol toxicity, certain infections (e.g. hepatitis C) and various genetic and idiopathic (i.e. unknown) causes. Intrinsic cardiomyopathies are generally classified into a number of types, but dilated cardiomyopathy is the most common form, and one of the leading indications for heart transplantation; approximately 40% of cases are familial, with mutations of genes encoding cytoskeletal, contractile or other proteins present in myocardial cells. The disease is genetically heterogeneous, but the most common form of its transmission is an autosomal dominant pattern.

# 17.11    Metabolic diseases of muscle

Muscles require a significant amount of energy in the form of ATP. When energy levels become too low, muscle weakness and exercise intolerance with muscle pain or cramps may occur. Metabolic muscle diseases that have their onset in infancy tend to be the most severe, and some forms are fatal. Those that begin in childhood or adulthood tend to be less severe, and changes in diet and lifestyle can help most people with the milder forms adjust. The more common metabolic diseases are summarised below.

- **Glycogen storage disease type II (Pompe's disease or acid maltase deficiency)** is a neuromuscular, autosomal recessive metabolic disorder, in the family of lysosomal storage diseases caused by a deficiency in the enzyme acid $\alpha$-glucosidase, which is needed to break down glycogen to provide energy. It is the only glycogen storage disease with a defect in lysosomal metabolism, and was the first glycogen storage disease to be identified. The build-up of glycogen causes progressive muscle weakness (myopathy) throughout the body and affects various body tissues, particularly in the heart, skeletal muscles, liver and nervous system.

- **Glycogen storage disease type V (McArdle's disease)** is caused by a deficiency of myophosphorylase. It is the most common of the various types of glycogen storage disease, but is still considered rare (about 1 in 100 000).

- **Glycogen storage disease type III (Cori's or Forbes'disease)** is inherited in an autosomal recessive pattern, and occurs in about 1 of every 100 000 live births. Forbes' disease is one of several glycogen storage disorders that are inherited as autosomal recessive traits. Symptoms are caused by a lack of the enzyme amylo-1,6 glucosidase (debrancher enzyme). This enzyme deficiency causes excess amounts of an abnormal glycogen to be deposited in the liver, muscles and in some cases heart. There are two forms of this disorder: type IIIA affects about 85% of patients with Forbes' disease and involves both the liver and the muscles; type IIIB affects only the liver. The disease typically presents during infancy with hypoglycaemia and failure to thrive.

- **Primary carnitine deficiency** is caused by a deficiency in the plasma-membrane carnitine transporter. Intracellular carnitine deficiency impairs the entry of long-chain fatty acids into the mitochondrial matrix. Consequently, long-chain fatty acids are not available for $\beta$-oxidation and energy production, and the production of ketone bodies (which are used by the brain) is also impaired. Regulation of intramitochondrial free CoA is also affected, with accumulation of acyl-CoA esters in the mitochondria. This in turn affects the pathways of intermediary metabolism that require CoA, for example the TCA cycle, pyruvate oxidation, amino acid metabolism, and mitochondrial and peroxisomal $\beta$-oxidation. Cardiac muscle is affected by progressive cardiomyopathy (the most common form of presentation), the CNS is affected by encephalopathy caused by hypoketotic hypoglycaemia, and skeletal muscle is affected by myopathy.

- **Carnitine palmitoyltransferase II** deficiency is a metabolic disorder characterised by an enzymatic defect that prevents long-chain fatty acids from being transported into the mitochondria for utilisation as an energy source. The adult myopathic form is the most common inherited disorder of lipid metabolism affecting the skeletal muscles; it is also the most frequent cause of hereditary myoglobinuria. Symptoms of this disease are commonly provoked by prolonged exercise or periods without food.

- **Lactate dehydrogenase** deficiency interferes with the processing of carbohydrates for energy production and the conversion of pyruvate to lactate, and vice versa. Onset is early adulthood.

Symptoms include exercise intolerance and episodes of myoglobinuria (rust-coloured urine, indicating breakdown of muscle tissue); a skin rash is common. Inheritance is autosomal recessive.

- **Myoadenylate deaminase** deficiency is a recessive disorder that affects approximately 1–2% of populations of European descent, but appears considerably rarer in Asian populations. Myoadenylate deaminase, also called adenosine monophosphate (AMP) deaminase, is an enzyme that converts AMP to inosine monophosphate (IMP). Its deficiency results in excess AMP, which is lost by excretion with disturbances in energy generation. Symptoms of severe fatigue and muscle pain can result.

- **Phosphofructokinase deficiency (Tarui's disease)** is an inborn error of glycogen metabolism characterised by a phosphofructokinase deficiency in the muscles, and associated with abnormal deposition of glycogen in muscle tissues, occasionally with myoglobinuria. The symptoms are similar to those seen in McArdle's disease. Classic Tarui's disease typically presents in childhood with exercise intolerance and anaemia. The fatal infantile variant presents in the first year of life. All reported patients have died by age four years. A late-onset variant manifests itself during later adulthood with progressive limb weakness without myoglobinuria or cramps. It is an autosomal recessive inheritance. Males are slightly more often affected than females.

- **Phosphoglycerate kinase deficiency,** the seventh enzyme step of the glycolytic pathway, is an inherited X-linked recessive disorder, meaning it mostly affects males, although females are carriers. Onset is infancy to early adulthood. Symptoms may include anaemia, enlargement of the spleen, mental retardation and epilepsy (seizures); more rarely, weakness, exercise intolerance, muscle cramps and episodes of myoglobinuria occur.

- **Phosphoglycerate mutase deficiency,** the eighth enzyme step of glycolysis, is inherited autosomal recessive. Onset occurs in childhood to early adulthood. Symptoms include exercise intolerance, cramps, muscle pain and sometimes myoglobinuria.

---

Myoglobinuria is the presence of myoglobin in the urine, often associated with rhabdomyolysis. Rhabdomyolysis is the rapid breakdown of skeletal muscle tissue. The destruction of the muscle leads to the release of the breakdown products of damaged muscle cells into the blood stream; some of these, such as myoglobulin, are harmful to the kidney and may lead to acute kidney failure.

# CHAPTER 18

# The cytoskeleton

The cytoskeleton is a cellular 'scaffold' contained within the cytoplasm. It is present in all cells, including prokaryotes. It is a dynamic structure that maintains cell shape, protects the cell and mediates cellular motion and intracellular transport, as well as cell division.

Eukaryotic cells contain three main kinds of cytoskeletal filament:

- actin filaments/microfilaments
- intermediate filaments
- microtubules.

## 18.1   Actin filaments/microfilaments

Actin filaments are about 6 nm in diameter and composed of two intertwined actin chains. They are mostly concentrated just beneath the cell membrane, and are responsible for resisting tension and maintaining cellular shape, and participating in cell–cell and cell–matrix junctions; in these latter roles they are essential to intracellular signal transduction processes. They are also important for cytokinesis (division of the cytoplasm in mitosis) and cytoplasmic streaming in most cells.

## 18.2   Intermediate filaments

Intermediate filaments are about 10 nm in diameter, and are more stable (strongly bound) than actin filaments. Like actin filaments they function in the maintenance of cell shape by bearing tension. Intermediate filaments organise the internal tridimensional structure of the cell, anchoring organelles and serving as structural components of the nuclear lamina (a dense fibrillar network inside the nucleus) and sarcomeres. They also participate in some cell–cell and cell–matrix junctions.

*Essential Biochemistry for Medicine*   Dr Mitchell Fry
© 2010 John Wiley & Sons, Ltd

The following are the different types of intermediate filament:

- vimentins, the common structural support of many cells

- keratin, found in skin, hair and nails

- neurofilaments of neural cells

- lamin, providing structural support to the nuclear envelope.

## 18.3   Microtubules

Microtubules are hollow cylinders about 23 nm in diameter, most commonly comprising 13 protofilaments, which in turn are polymers of $\alpha$- and $\beta$-tubulin. They have a very dynamic behaviour, binding GTP for polymerisation. They are commonly organised by the centrosome. In nine triplet sets (star-shaped), they form the centrioles, and in nine doublets oriented about two additional microtubules (wheel-shaped) they form cilia and flagella. The latter formation is commonly referred to as a '9+2' arrangement, wherein each doublet is connected to another by the protein dynein. Microtubules play key roles in:

- intracellular transport, associated with dyneins and kinesins, in the transport of organelles and vesicles

- the axoneme of cilia and flagella

- the mitotic spindle.

The three-dimensional intracellular network, formed by the filamentous polymers that comprise the cytoskeleton, affects the way cells sense their extracellular environment and respond to stimuli. The cytoskeleton is viscoelastic, so it provides a continuous mechanical coupling throughout the cell which changes as the cytoskeleton remodels. Mechanical effects, based on network formation, can influence ion-channel activity at the plasma membrane of cells and may conduct mechanical stresses from the cell membrane to internal organelles. As a result, both rapid responses, such as changes in intracellular $Ca^{2+}$, and slower responses, such as gene transcription or the onset of apoptosis, can be elicited or modulated by mechanical perturbations. In addition to mechanical features, the cytoskeleton also provides a large negatively charged surface on which many signalling molecules, including protein and lipid kinases, phospholipases and GTPases, can localise in response to activation of specific transmembrane receptors. The resulting spatial localisation and concomitant change in enzymatic activity can alter the magnitude and limit the range of intracellular signalling events.

Diseases, in which cytoskeletal components play a crucial role in pathogenesis, are probably numerous, but classification of such disorders on the basis of phenotypic changes that occur in microfilaments, intermediate filaments and microtubules is presently not possible. Some disorders that have known or suspected cytoskeletal involvement are given below.

## 18.4   Spectrin

Spectrin is a cytoskeletal protein that lines the intracellular side of the plasma membrane of many cell types, in a pentagonal or hexagonal arrangement, forming a scaffold and playing an important role in maintenance of plasma-membrane integrity and cytoskeletal structure.

The hexagonal arrangements are formed by tetramers of spectrin associating with short actin filaments at either end of the tetramer. These short actin filaments act as junctional complexes, allowing the formation of the hexagonal mesh.

In certain types of brain injury, such as diffuse axonal injury, spectrin is irreversibly cleaved by the proteolytic enzyme calpain. This destroys the cytosketelon, causing the membrane to form blebs, irregular bulges in the plasma membrane of a cell caused by localised decoupling of the cytoskeleton from the plasma membrane, ultimately leading to degradation and usually death of the cell.

---

Diffuse axonal injury is one of the most common and devastating types of traumatic brain injury; it refers to the extensive lesions in white-matter tracts and is one of the major causes of unconsciousness and persistent vegetative state after head trauma.

Though the processes involved in secondary brain injury are still poorly understood, it is now accepted that stretching of axons during injury causes physical disruption to and proteolytic degradation of the cytoskeleton. It also results in opening of sodium channels in the axolemma, which causes voltage-gated calcium channels to open and $Ca^{2+}$ to flow into the cell. The intracellular presence of $Ca^{2+}$ initiates several different pathways, including activation of phospholipases and proteolytic enzymes, damage to mitochondria and the cytoskeleton and activation of secondary messengers, which together can lead to separation of the axon and death of the cell.

---

Spectrin forms the meshwork that provides red blood cells their shape. Its importance in the erythrocyte is demonstrated through spectrin mutations leading to hereditary elliptocytosis and hereditary spherocytosis. Hereditary elliptocytosis is an inherited blood disorder in which an abnormally large number of erythrocytes are elliptical rather than the typical biconcave disc shape.

## 18.5 Alzheimer's disease

Alzheimer's disease is the most common form of dementia occurring in mid-to-late life. Late onset of the disease is influenced by the genetic risk factor apolipoprotein E. However, most of the early-onset, familial forms of Alzheimer's are caused by mutations associated with amyloid precursor protein and the presenilins (PSs). Inherited mutations located on chromosomes 14 and 1 are associated with PS1 and PS2. PSs are membrane-bound proteins that participate in the Notch-like cleavage of the amyloid precursor protein, which eventually accumulate as extracellular plaques. An interacting partner of PS1 was recently found to be $\delta$-catenin, an adheren junction protein involved with cell motility. Cell motility is associated with a massive restructuring of the actin cytoskeleton. Therefore, it is possible that defects in PSs may stimulate structural alterations in the cytoskeleton of neurons preceding the formation of fibril-containing dystrophic plaques. A fundamental alteration of the cytoskeleton as an underlying cause for Alzheimer's may in part explain why accumulation of amyloid precursor protein and plaque formation cannot be definitively confirmed as causative events in the disease.

Neuropathology of Alzheimer's is also defined by accumulation of another form of insoluble protein, the neurofibrillary tangles (NFTs). NFTs are fibrillar structures largely composed of tau, a microtubule-binding protein that stabilises the microtubule tracts necessary for vesicular trafficking, endo- and exocytosis and axonal polarity. No tau mutations have yet been identified in Alzheimer's families. Although the relationship between tau, NFT and amyloid precursor protein remains to be elucidated, it is possible that defects may influence aggregation of tau protein, leading to impairment or misdirection of recycling endosomes that contain the amyloid precursor protein.

Tau forms up to six different isoforms by alternative splicing, and all six isoforms have been found in NFTs. Tau in NFT is typically hyperphosphorylated and in this state tubulin assembly is impaired. Hyperphosphorylated forms of tau have lower binding affinities to microtubules and may destabilise them.

---

Pick's disease is a rare neurodegenerative disease which causes progressive destruction of nerve cells in the brain and causes tau proteins to accumulate into 'Pick bodies', which are a defining characteristic of the disease.

---

## 18.6 Amyotrophic lateral sclerosis

Amyotrophic lateral sclerosis, sometimes referred to as Lou Gehrig's disease, is a progressive, usually fatal, neurodegenerative disease caused by the degeneration of motor neurons. The disease has been linked to mutations in the copper-zinc superoxide dismutase, known to underlie 2% of familial cases. Superoxide dismutase mutations may be directly linked to defects in both cytoskeleton components and vesicular transport motors. Aggregates, containing both neurofilament and kinesin, are hallmarks of amyotropic lateral sclerosis. Kinesin and dynein facilitate transport of organelles along microtubules in an ante-retrograde and a retrograde direction, respectively. In amyotropic lateral sclerosis there is not only selective loss of kinesin motors, but a measurable slowing of axonal transport in motor neurons.

## 18.7 Synapsins

Synapsins are a family of proteins which have long been implicated in the regulation of neurotransmitter release at synapses. They are thought to be involved in regulating the number of synaptic vesicles available for release via exocytosis. Synapsins are suggested to bind synaptic vesicles to components of the cytoskeleton, preventing them from migrating to the presynaptic membrane and releasing transmitter. During an action potential, synapsins are phosphorylated by $Ca^{2+}$/calmodulin-dependent protein kinase II, releasing the synaptic vesicles and allowing them to move to the membrane and release their neurotransmitter. Mutations in this gene may be associated with X-linked disorders with primary neuronal degeneration, such as Rett syndrome.

Neurons cannot synthesise proteins along the axon and are particularly dependent on vesicular transport to provide them. Many neurodegenerative disorders show examples of defects in the cytoskeletal tracts, which sustain neuronal shape and trafficking, or defects in the motors, which provide energy for vesicle/organelle movement, including mitochondria.

# CHAPTER 19
# Genes and medicine

## 19.1  Chromosomes

Chromosomes are composed of protein and DNA (chromatin) and are distinct dense bodies
found in the nucleus of cells. The DNA in an individual chromosome is one long molecule,
highly coiled and condensed. Genetic information in the DNA is defined by the linear sequences
of bases (A, T, C and G), the genetic code. There may be 50–250 million bases in an individual chromosome. The DNA sequence for a single trait is called a gene. During cell division
(mitosis), the chromosomes become highly condensed and are visible as dark distinct bodies
within the nuclei of cells. The number of chromosomes in human cells is 46; 22 autosomal
pairs (same in both sexes) and 2 sex chromosomes, 2 X chromosomes in females and 1 X and
1 Y in males.

Chromatin is the complex combination of DNA, RNA and protein that makes up chromosomes inside the nuclei of eukaryotic cells; it is divided between heterochromatin
(condensed) and euchromatin (extended) forms. The functions of chromatin are to package
DNA into a smaller volume to fit into the cell, to support the DNA to allow mitosis and
meiosis, and to serve as a mechanism to control expression and DNA replication. Changes
in chromatin structure are affected by chemical modifications of histone proteins, such
as methylation (DNA and proteins) and acetylation (proteins), and by non-histone DNA-
binding proteins. Chromatin is easily visualised by staining, hence its name, which literally
means *coloured, lightened material*.

## 19.2  Chromosome banding

Chromosome banding is evident when chromosomes are treated with chemical dyes, such as
Giemsa; chromosomes appear as a series of alternate dark (G-band or G-positive band) and

*Essential Biochemistry for Medicine*   Dr Mitchell Fry
© 2010 John Wiley & Sons, Ltd

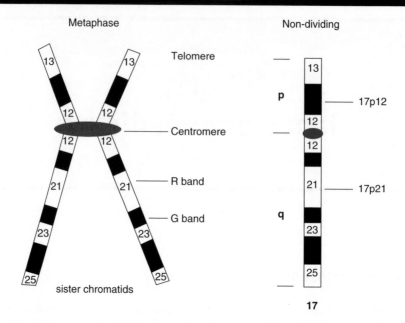

**Figure 19.1** During the metaphase of cell division, a chromosome becomes two sister chromatids attached at the centromere. Chromosome banding exemplified using human chromosome 17.

light (R-band or G-negative band) bands. The banding pattern is depicted as an ideogram (Figure 19.1). Each chromosome consists of two arms separated by the centromere. The long arm and short arm are labelled q (for queue) and p (for petit), respectively. At the lowest resolution, only a few major bands can be distinguished, which are labelled q1, q2, q3...; p1, p2, p3..., counting from the centromere. Higher resolution reveals sub-bands, labelled q11, q12, q13... Sub-sub-bands identified by even higher resolution are labelled q11.1, q11.2, q11.3... Traditionally, the short arm (p) is displayed on top of the long arm (q).

## 19.3 Karyotypes

A karyotype is the representation of the entire metaphase chromosomes in a cell, arranged in order of size (Figure 19.2).

The band width and order of bands is characteristic of a particular chromosome, and identifiable by a trained cytogeneticist.

## 19.4 The spectral karyotype: fluorescence *in situ* hybridisation (FISH)

So-called 'chromosome painting' refers to the hybridisation of fluorescently labelled chromosome-specific probes (Figure 19.3). Chromosome painting allows the visualisation of individual chromosomes in metaphase or interphase cells, and the identification of both

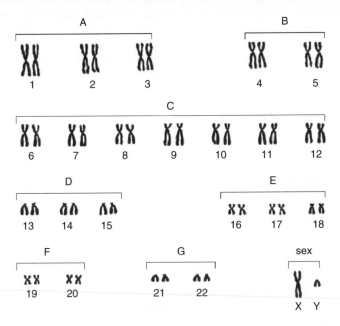

**Figure 19.2** A human somatic cell contains two sets of homologous chromosomes, which may be divided into two types: autosomes and sex chromosomes. Autosomes are further divided into seven groups: A–G. During the metaphase of cell division, each chromosome has been duplicated. Therefore, this karyotype consists of 92 chromosomes.

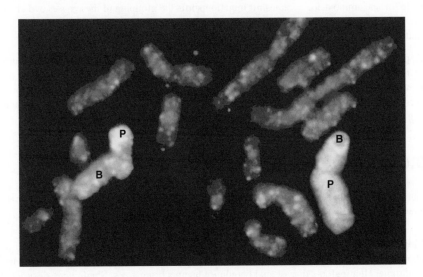

**Figure 19.3** Human chromosomes 'painted' by fluorescent dyes to detect abnormal exchange of genetic material frequently present in cancer (note exchange between 'pink – P' and 'light blue – B' chromosomes). Source: US Department of Energy Genome Programs (http://genomics.energy.gov).

numerical and structural chromosomal aberrations in human pathology. The use of fluorescent dyes that bind to specific regions of chromosomes can impart unique spectral characteristics. Slight variations in colour, undetectable by the human eye, can be quantified by digital imaging; multicolour (24-colour) fluorescence *in situ* hybridisation (mFISH) can provide specific information to improve diagnosis and prognosis of genetic diseases.

## 19.5   Gene mutations

Genetic mutations of the DNA can include:

- **Missense mutation**. A change in one DNA base pair, resulting in the substitution of one amino acid for another in the protein made by a gene.

- **Nonsense mutation**. A change in one DNA base pair, but here the altered DNA sequence prematurely signals the cell to stop building a protein, resulting in a shortened protein that may function improperly or not at all.

- **Insertion**. Changes the number of DNA bases in a gene by adding a piece of DNA.

- **Deletion**. Changes the number of DNA bases by removing a piece of DNA. Small deletions may remove one or a few base pairs, larger deletions can remove an entire gene or several neighbouring genes.

- **Duplication**. A piece of DNA that is abnormally copied one or more times.

- **Frameshift mutation**. This type of mutation occurs when the addition or loss of DNA bases changes a gene's reading frame. A reading frame consists of groups of three bases that each code for one amino acid; a frameshift mutation shifts the grouping of these bases and changes the code for amino acids. Insertions, deletions and duplications can all result in frameshift mutations.

- **Repeat expansion**. Nucleotide repeats are short DNA sequences that are repeated a number of times in a row. For example, a trinucleotide repeat is made up of 3 bp sequences, and a tetranucleotide repeat is made up of 4 bp sequences. A repeat expansion is a mutation that increases the number of times that the short DNA sequence is repeated. This type of mutation can cause the resulting protein to function improperly.

## 19.6   Genetic disorders

Genetic disorders may be single-gene, multifactorial, chromosomal or mitochondrial. Both environmental and genetic factors have roles in the development of any disease.

- **Single-gene** (also called Mendelian) disorders number more than 6000, with an incidence of about 1 in 200 births. Some examples include cystic fibrosis, sickle cell anaemia, Marfan syndrome, Huntington's disease and hereditary haemochromatosis. Single-gene disorders are inherited in recognisable patterns: autosomal dominant, autosomal recessive and X-linked.

- **Multifactorial** (also called complex or polygenic) disorders are caused by a combination of environmental factors and mutations in multiple genes. For example, different genes that influence breast cancer susceptibility have been found on chromosomes 6, 11, 13, 14, 15, 17 and 22; this complexity makes breast cancer far more difficult to analyse. Some of the most common chronic disorders are multifactorial, for example heart disease, high blood pressure, Alzheimer's disease, arthritis, diabetes, cancer and obesity. Multifactorial inheritance is associated with heritable traits such as fingerprint patterns, height, eye colour and skin colour.

• **Chromosomal abnormalities**, such as missing or extra copies, gross breaks and rejoinings (translocations) can result in disorders. Down's syndrome, or trisomy 21, is a common disorder that is a result of having three copies of chromosome 21.

• **Mitochondria-associated** disorders are a relatively rare type of genetic disorder caused by mutations in the non-chromosomal DNA of mitochondria (see Chapter 16).

## 19.7 Gene testing

Gene testing may involve direct examination of the DNA molecule, biochemical tests for the gene products or chromosomal analysis. Genetic tests are used for several reasons, including:

• carrier screening, which involves identifying unaffected individuals who carry one copy of a gene for a disease that requires two copies in order for the disease to be expressed

• preimplantation genetic diagnosis (screening embryos for disease)

• prenatal diagnostic testing

• newborn screening

• presymptomatic testing for predicting adult-onset disorders such as Huntington's disease

• presymptomatic testing for estimating the risk of developing adult-onset cancers and Alzheimer's disease

• confirmational diagnosis of a symptomatic individual

• forensic/identity testing.

DNA probes (Section 19.18) whose sequences are complementary to the mutated sequences are used to examine an individual's genome, or to make comparisons of the sequence of DNA bases in one of their genes with a normal version of that gene. There are currently more than 1000 genetic tests, with an increasing number becoming available commercially. Some examples include:

• alpha-1-antitrypsin deficiency (emphysaema and liver disease)

• amyotrophic lateral sclerosis (Lou Gehrig's disease; progressive motor function loss)

• Alzheimer's disease (APOE; late-onset variety of senile dementia)

• Gaucher's disease (enlarged liver and spleen, bone degeneration)

• inherited breast and ovarian cancer (BRCA 1 and 2; early-onset tumours)

• congenital adrenal hyperplasia (hormone deficiency; ambiguous genitalia)

• cystic fibrosis (disease of lung and pancreas)

- Duchenne muscular dystrophy/Becker muscular dystrophy (severe to mild muscle wasting, deterioration, weakness)

- dystonia (muscle rigidity, repetitive twisting movements)

- factor V-Leiden (blood-clotting disorder)

- fragile X syndrome (leading cause of inherited mental retardation)

- galactosaemia (metabolic disorder affects ability to metabolise galactose)

- haemophilia A and B (bleeding disorders)

- hereditary haemochromatosis (excess iron storage disorder)

- Huntington's disease (progressive, lethal, degenerative neurological disease)

- Marfan syndrome (connective tissue disorder)

- mucopolysaccharidosis (deficiency of enzymes in breaking down glycosaminoglycans)

- myotonic dystrophy (progressive muscle weakness; adult muscular dystrophy)

- phenylketonuria (missing enzyme, phenylalanine hydroxylase, correctable by diet)

- sickle cell disease (blood-cell disorder)

- spinal muscular atrophy (severe, usually lethal, progressive muscle-wasting disorder)

- Tay–Sachs disease (fatal neurological disease of early childhood)

- thalassaemias (anaemias).

### 19.7.1　Advantages and disadvantages of gene testing

Gene tests can be used to clarify a diagnosis, allow families to avoid having children with devastating diseases, or identify people at high risk for conditions that may be preventable. Commercialised gene tests for adult-onset disorders, such as Alzheimer's disease and some cancers, are the subject of much debate. Such tests are targeted to healthy, presymptomatic individuals who are identified as being at high risk because of a strong family medical history for the disorder. The tests give only a probability for developing the disorder. A serious limitation of such susceptibility tests is the difficulty in interpreting a positive result, because some people who carry a disease-associated mutation may never develop the disease. Since genetic information is shared, there are implications for family members as well. Uncertainties surrounding test interpretation, the current lack of available medical options for these diseases, the tests' potential for provoking anxiety, and risks for discrimination and social stigmatisation are important aspects for consideration.

## 19.8　The human genome project

The Human Genome Project has provided some interesting facts and figures about the human genome:

- The human genome contains 3164.7 million chemical nucleotide bases (A, T, C and G).

- The average gene consists of 3000 bases, but sizes vary greatly, with the largest known human gene being dystrophin, at 2.4 million bases.

- Chromosome 1 has the most genes (2968), the Y chromosome the fewest (231).

- The total number of genes is estimated at 30 000, much lower than previous estimates.

- Almost all (99.9%) nucleotide bases are exactly the same in all people.

- The functions are unknown for over 50% of discovered genes.

- Less than 2% of the genome codes for proteins.

- Repeated sequences that do not code for proteins ('junk DNA') make up at least 50% of the human genome.

- Genes appear to be concentrated in random areas along the genome, with large expanses of non-coding DNA in between.

- Stretches of up to 30 000 C and G bases, repeating over and over, often occur adjacent to gene-rich areas, forming a barrier between the genes and the junk DNA. These CpG islands are believed to help regulate gene activity.

### 19.8.1   How do we compare to other organisms?

- Compared to the human's seemingly random distribution of gene-rich areas, other organisms' genomes are more uniform, with genes evenly spaced.

- The product of our genes (proteins) is greater because of mRNA transcript 'alternative splicing' and chemical modifications to the proteins, yielding different protein products from the same gene.

- While we share most of the same protein families, the number of gene family members has expanded in humans, particularly for those proteins involved in development and immunity.

- The human genome has a greater portion of repeat sequences.

- Some 1.4 million locations at which single-base DNA differences (SNPs) occur have been identified in humans. This information promises to revolutionise the processes of finding chromosomal locations for disease-associated sequences and tracing human history.

- The ratio of germline (sperm or egg cell) mutations is 2 : 1 in males vs. females.

## 19.9   Gene therapy

Gene therapy is an evolving technique for correcting defective genes. One of several approaches may be taken:

- A normal gene may be inserted into a nonspecific location within the genome to replace a non-functional gene; this approach is most common.

- An abnormal gene can be swapped for a normal gene through homologous recombination.

- The abnormal gene can be repaired through selective reverse mutation, returning the gene to its normal function.

- The regulation (the degree to which a gene is turned on or off) of a particular gene may be altered.

**Table 19.1** Viral vectors

| | |
|---|---|
| Retroviruses | A class of viruses that can create double-stranded DNA copies of their RNA genomes. These copies can be integrated into the chromosomes of host cells. Human immunodeficiency virus (HIV) is a retrovirus. |
| Adenoviruses | A class of viruses with double-stranded DNA genomes that cause respiratory, intestinal and eye infections in humans. The virus that causes the common cold is an adenovirus. |
| Adeno-associated viruses | A class of small, single-stranded DNA viruses that can insert their genetic material at a specific site on chromosome 19. |
| Herpes simplex viruses | A class of double-stranded DNA viruses that infect a particular cell type: neurons. Herpes simplex virus type 1 is a common human pathogen that causes cold sores. |

Frequently a carrier molecule, a vector, must be used to deliver the therapeutic gene to the patient's target cells. The most common vector used is a virus that has been genetically altered to carry normal human DNA (Table 19.1). Since viruses have evolved a way of encapsulating and delivering their genes to human cells in a pathogenic manner, the rationale is to take advantage of this capability and manipulate the virus genome to remove disease-causing genes and insert therapeutic ones.

Other non-viral options for gene delivery include:

- Direct introduction of therapeutic DNA into target cells.

- Use of liposomes to carry the therapeutic DNA through the target cell's membrane.

- Delivery of therapeutic DNA by chemically linking the DNA to a molecule that will bind to target cell receptors.

Experimentation is currently underway to introduce a 47th (artificial human) chromosome into target cells. It would be a large vector capable of carrying substantial amounts of genetic code.

## 19.9.1  Current status

The current status of gene therapy is experimental; there are no approved human gene therapy products. The success of the Human Genome Project has provided information for the development of new ways to treat, cure or prevent the thousands of diseases that afflict humankind, but there are many challenges to overcome, including:

- The short-lived nature of gene therapy: therapeutic DNA introduced into target cells must remain functional, but the rapidly dividing nature of cells prevents gene therapy from achieving any long-term benefits. Consequently patients need to undergo multiple rounds of gene therapy.

- The immune response: the 'immune memory' makes repeated gene therapy more difficult.

- Viral vectors, the carrier of choice in most gene therapy studies. These present a variety of potential problems to the patient, including toxicity, immune and inflammatory responses, gene control and targeting issues.

- Multigene disorders. These account for many of the most commonly occurring disorders, such as heart disease, high blood pressure, Alzheimer's disease, arthritis and diabetes; such disorders will be especially difficult to treat effectively using gene therapy.

### 19.9.2  Some ethical questions

- What is normal and what is a disability or disorder, and who decides?

- Are disabilities diseases? Do they need to be cured or prevented?

- Does searching for a cure demean the lives of individuals presently affected by disabilities?

- Is somatic gene therapy (which is carried out in the adult cells of persons known to have a disease) more or less ethical than germline gene therapy (which is carried out in egg and sperm cells and prevents the trait from being passed on to further generations)? In cases of somatic gene therapy, the procedure may have to be repeated in future generations.

- Preliminary attempts at gene therapy are expensive. Who will have access to these therapies? Who will pay for their use?

## 19.10   The next step: functional genomics

As the quantity of genome data grows, the challenge will be to use these data to explore how DNA and proteins work with each other and the environment to create complex, dynamic living systems (functional genomics). Explorations will encompass studies in transcriptomics, proteomics, structural genomics, new experimental methodologies and comparative genomics.

- **Transcriptomics** involves the analysis of messenger RNAs, following when, where and under what conditions genes are expressed. This will provide a clearer understanding of what is actually happening in the cell.

- **Structural genomics** initiatives are being launched worldwide to generate the three-dimensional structures of one or more proteins from each protein family, thus offering clues to function and biological targets for drug design.

- **Knockout studies** are being used to inactivate genes in living organisms and monitor any changes that could reveal their functions.

- **Comparative genomics**, the analysis of DNA sequence patterns of humans and other model organisms, has become one of the most powerful strategies for identifying human genes and interpreting their function.

> The genome is an organism's complete set of DNA, and genomics is the study of this genome; the products of a cell's genome, the proteins, are its proteome, and proteomics is the study of the function and structure of these proteins. Unlike the relatively unchanging genome, the dynamic proteome changes from minute to minute in response to tens of thousands of intra- and extracellular environmental signals.

# 19.11   Pharmacogenomics

Pharmacogenomics is the study of how an individual's genetic makeup will affect their response to drugs. It holds the promise that drugs may one day be tailor-made for an individual.

A 1998 study of hospitalised patients in the USA showed adverse drug reactions (ADRs) to account for more than 2.2 million serious cases, with over 100 000 deaths. Currently there is no simple way to determine whether an individual will respond well, badly or not at all to a medication; pharmaceutical companies are therefore limited to developing drugs using a 'one size fits all' system. Heterogeneity in patient response to chemotherapy is consistently observed across patient populations. Pharmacogenomics may be especially important for oncology, where severe systemic toxicity and unpredictable efficacy are hallmarks of cancer therapies. Genetic polymorphisms in drug-metabolising enzymes are responsible for much of the inter-individual differences in the efficacy and toxicity of many chemotherapeutic agents.

Anticipated benefits of pharmacogenomics include:

- The facilitation of new drugs with specific targeting, based on the proteins, enzymes and RNA molecules associated with genes and diseases.

- Safer drugs and avoidance of ADRs, and more accurate drug dosage.

- Advanced screening for disease, to allow for lifestyle and environmental changes at an early age, and to ascertain the most appropriate timeline for therapy.

- Better vaccines, made of genetic material, which will avoid the current risks, be inexpensive, stable and easy to store, and capable of being engineered to carry several strains of a pathogen at once.

- Facilitation of drug discovery and approval, since trials will be targeted to specific genetic population groups; cost and risk should also be reduced.

- Decreasing overall costs of health care, in line with a reduction in adverse drug effects, number of failed drug trials, time for approval, length of time patients are on medication and time to find an effective therapy.

## Focus on: cytochrome P450 and pharmacogenomics

The Cytochrome P450 isoenzymes (CYPs) are a superfamily of haemoprotein enzymes, responsible for catalysing the metabolism of a large number of endogenous and exogenous compounds (see Chapter 6); they are found predominantly in the liver, but also in the intestine, lungs, kidneys and brain. DNA variations in genes that code for these enzymes can influence their ability to metabolise certain drugs. Clinical trials researchers can use genetic tests for variations in CYP genes to screen and monitor patients; additionally pharmaceutical companies may screen chemicals to see how well they are broken down by variant forms of these enzymes. They are a major determinant of the pharmacokinetic behaviour of numerous drugs.

The name 'cytochrome P450' is derived from the fact that these proteins have a haem group and an absorption spectrum characterised by a maximum absorption wavelength of

450 nm in the reduced state in the presence of carbon monoxide. An Arabic numeral denotes the CYP family (e.g. CYP1, CYP2), followed by letters A, B, C and so on to indicate the subfamily (e.g. CYP3A, CYP3C), and another Arabic numeral to represent the individual gene/isoenzyme (e.g.CYP3A4, CYP3A5). Of the 74 gene families so far described, 14 exist in all mammals; these 14 families comprise some 26 mammalian subfamilies (Table 19.2).

**Table 19.2**  Some common cytochrome P450 isoenzymes

| | |
|---|---|
| 3A | Accounts for about 30% of CYP proteins in the liver; also present in the small intestinal epithelium, making it a major contributor to presystemic elimination of orally administered drugs. There is considerable inter-individual variability in hepatic and intestinal CYP3A activity (about 5–10 fold). Since 40–50% of drugs used in humans involve 3A-mediated oxidation to some extent, the members of this subfamily are involved in many clinically important drug interactions. 3A4 is the major isoenzyme in the liver, 3A5 is present in the kidneys. |
| 2D6 | Accounts for <5% of total CYP proteins, but is of interest because of its large number of substrates (30–50 known drugs) and its genetic polymorphism. Many psychotropic, antiarrhythmic and $\beta$-adrenergic receptor-blocker drugs are substrates, as well as inhibitors of 2D6. |
| IA2 | The only isoenzyme affected by tobacco, cigarette smoking may lead to a threefold increase in 1A2 activity. Caffeine is metabolised in part by 1A2, which explains why smokers require higher doses of caffeine than non-smokers. Alcohol inhibits metabolism of caffeine and has been reported to mask the 1A2-inducing potential of smoking. Exposure to polyaromatic hydrocarbons found in charbroiled food can also induce this isoenzyme; IA2 can cause the metabolic activation of procarcinogens to carcinogens, for example aromatic and heterocyclic amines. |
| 2C9 | S-warfarin and phenytoin are both involved in a large number of drug interactions and are metabolised mainly by 2C9. St John's wart, a herbal antidepressant, has been reported to decrease levels of warfarin by induction of 2C9. |
| 2C19 | Polymorphic and involved in the metabolism of a number of clinically important drugs, for example omeprazole, diazepam, antidepressants and antimalarials. |
| 2E1 | Metabolism of low molecular-weight toxins, fluorinated ether volatile anaesthetics and procarcinogens. 2E1 is inducible by ethanol and is responsible in part for metabolism of acetaminophen to produce a highly reactive and hepatotoxic metabolite; thus alcohol-dependant patients are at increased risk of acetaminophen hepatotoxicity (see Chapter 7). |

Genetic variation in a population is termed 'polymorphism'; all genes coding CYP enzymes in families 1–3 are polymorphic. The polymorphic forms are responsible for the development of a significant number of ADRs; some 56% of drugs reported in ADR studies are metabolised by polymorphic phase 1 enzymes, of which 86% are CYPs. CYP2C9, CYP2C19 and CYP2D6 are the main polymorphic forms responsible for almost 40% of

P450 mediated drug metabolism. Major allelic variants of P450 genes of clinical importance have been identified with four phenotypes:

- poor metabolism, lacking functional enzymes

- intermediate metabolism, heterozygous for one deficient allele

- extensive metabolism, with two normal alleles

- ultrarapid metabolism, with multiple gene copies.

    Ultrarapid metabolism of an active drug could mean that therapeutic levels are not reached, while poor metabolism will increase the likelihood of adverse events at normal doses because of resulting increased levels of the drug. Conversely, for prodrugs that must first be activated by CYP enzymes, ultrarapid metabolism may increase adverse effects, while poor metabolism may result in little response.

# 19.12    Genetic engineering: recombinant DNA technology

Genetic engineering, also known as recombinant DNA technology, involves altering the genes in a living organism to produce a genetically modified organism with a new genotype. Various kinds of genetic modification are possible:

- Inserting a foreign gene from one species into another, forming a transgenic organism.

- Altering an existing gene so that its product is changed.

- Affecting the rate of gene expression.

## 19.12.1    The tools and terminology of the genetic engineer

- **Restriction enzymes** cut DNA at specific sites. These are properly called restriction endonucleases because they cut the bonds in the middle of the polynucleotide chain. Most restriction enzymes make a staggered cut in the two strands, forming 'sticky ends'. The cut ends are 'sticky' because they have short stretches of single-stranded DNA. The sticky ends will stick (or anneal) to another piece of DNA by complementary base pairing, but only if they have both been cut with the same restriction enzyme. Restriction enzymes are highly specific, and will only cut DNA at specific base sequences, 4–8 bp long. Restriction enzymes are produced naturally by bacteria as a defence against viruses (they 'restrict' viral growth), but they are very useful in genetic engineering for cutting DNA at precise places ('molecular scissors'), producing so-called restriction fragments. Thousands of different restriction enzymes are known, with over a hundred different recognition sequences. Restriction enzymes are named after the bacteria species they came from, for example *Eco*R1 is from *E. coli* strain R.

- **DNA ligase** repairs broken DNA by joining two nucleotides in a DNA strand. It is commonly used in genetic engineering to do the reverse of a restriction enzyme; that is, to join together complementary restriction fragments. While sticky ends allow two complementary restriction fragments to anneal, via weak hydrogen bonding, DNA ligase completes the DNA backbone through covalent bond formation. Restriction enzymes and DNA ligase can be used together to join lengths of DNA from different sources.

- **Vectors**, in genetic engineering, are lengths of DNA that are used to carry the gene into a host cell. Vectors must be big enough to hold the gene, be circular (or more accurately a closed loop) so they are less likely to be broken down, contain control sequences such as a transcription promoter so that the gene will be replicated or expressed, and contain marker genes so that cells containing them can be identified.

- **Plasmids** are the most common kind of vector; they are short circular pieces of DNA found naturally in bacterial cells, typically containing three to five genes. Plasmids are copied when the cell divides, so the plasmid genes are passed on to all daughter cells. They are used naturally for exchange of genes between bacterial cells, so bacterial cells will take up a plasmid. Being small, they are easy to handle *in vitro*, and foreign genes can quite easily be incorporated into them using restriction enzymes and DNA ligase. One of the most common plasmids in use is the R-plasmid (or pBR322) (Figure 19.4)

## 19.12.2  Transferring genes

Vectors must be incorporated into the host cell so that they can be replicated or expressed; since they are large molecules which do not readily cross cell membranes, different strategies have evolved to achieve this:

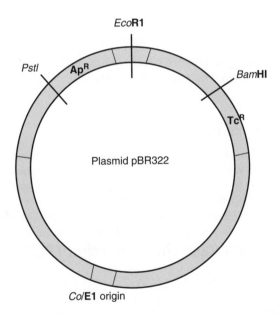

**Figure 19.4** The R-plasmid features of plasmid Pbr322: the gene conferring resistance to ampicillin (ApR) is inserted into the Pstl site, and that conferring resistance to tetracycline into the BamHI site. Replication is controlled by the ColE1 origin.

- **Heat shock**. Cells are incubated with the vector in a solution containing calcium ions at 0 °C. The temperature is suddenly raised to about 40 °C; heat shock causes some of the cells to take up the vector. This method works well for bacterial and animal cells.

- **Electroporation**. Cells are subjected to a high-voltage pulse, which temporarily disrupts the membrane and allows the vector to enter the cell. This is the most efficient method of delivering genes to bacterial cells.

- **Viruses**. The vector is first incorporated into a virus, which is then used to infect cells. Since viruses rely on getting their DNA into host cells for their survival, they are an obvious choice for gene delivery. The virus must first be genetically engineered to make it safe, so that it cannot reproduce itself or make toxins. Bacteriophages (phages) are viruses that infect bacteria; they are an effective way of delivering large genes into bacteria in culture. Adenoviruses are human viruses that cause respiratory diseases, including the common cold; their genetic material is double-stranded DNA and so they are ideal for delivering genes to patients in gene therapy. Their DNA is not incorporated into the host's chromosomes, so is not replicated, but their genes are expressed. Retroviruses are a group of human viruses that include HIV. Their genetic material is double-stranded RNA; upon infection the RNA is copied to DNA and the DNA incorporated into the host's chromosome, meaning that foreign genes are replicated into every daughter cell.

- **Micro-injection**. Foreign DNA can be injected directly into the nucleus using a fine micro-pipette; this is commonly used in fertilised animal egg cells.

- **Liposomes**. These are small membrane vesicles that can be used to encase the vector; they fuse with the cell membrane (and sometimes the nuclear membrane too), delivering the DNA into the cell. This approach is particularly useful for delivering genes to cells *in vivo* (such as in gene therapy).

- **Plant tumours**. These have been used successfully to transform plant cells.

- **Gene gun**. Microscopic gold particles, coated with the foreign DNA, are fired at the cell using a compressed air gun. This was designed to overcome the problem of the strong cell wall in plant tissue.

### 19.12.3  Genetic markers

Most of the techniques listed above result in less than 1% of the cells transforming, so a marker is needed to distinguish these cells from all the others. A common marker, used in plasmids, is a gene for resistance to an antibiotic, such as tetracycline. If the bacterial cells are grown on a medium containing tetracycline, the normal untransformed cells (99%) will die, leaving the 1% of transformed cells; these can then be grown and cloned on another plate.

## 19.13  The polymerase chain reaction (PCR)

The polymerase chain reaction (PCR) can clone (or amplify) DNA samples as small as a single molecule. If a length of DNA is mixed with the four nucleotides (A, T, C and G), and the enzyme DNA polymerase, then the DNA will be replicated many times. The principle of PCR is as follows:

1. Starting with a sample of the DNA to be amplified, add the four nucleotides and the enzyme DNA polymerase.

2. Normally (*in vivo*) the DNA double helix would be separated by the enzyme helicase, but in PCR (*in vitro*) the strands are separated by heating to 95 °C for two minutes. This breaks the hydrogen bonds between the two DNA strands.

3. Initiation of DNA polymerisation always requires short lengths of DNA (about 20 bp long) called primers. *In vivo* the primers are made during replication by DNA polymerase, but *in vitro* they must be synthesised separately and added at this stage. This means that a short length of the sequence of the DNA must already be known. The DNA must be cooled to 40 °C to allow the primers to anneal to their complementary sequences on the separated DNA strands.

4. The DNA polymerase enzyme can now extend the primers and complete the replication of the rest of the DNA. The enzyme used in PCR is derived from the thermophilic bacterium *Thermus aquaticus*, which grows naturally in hot springs at a temperature of 90 °C, so is not denatured by the high temperatures in step 2. Its optimum temperature is about 72 °C, so the mixture is heated to this temperature for a few minutes to allow replication to take place as quickly as possible.

5. Each original DNA molecule has now been replicated to form two molecules. The cycle is repeated from step 2, each time doubling the number of DNA molecules. This is why it is called a chain reaction, since the number of molecules increases exponentially. Typically PCR is run for 20–30 cycles.

PCR can be completely automated; a minute sample of DNA can be amplified millions of times with little effort. PCR is routinely used in forensic medicine to amplify DNA taken from samples of blood, hair or semen. A potential problem for PCR is obtaining a pure sample of DNA to start with; any contaminant DNA will also be amplified.

# 19.14   Complementary DNA (cDNA)

Complementary DNA (cDNA) is DNA that has been made from mRNA. This makes use of the enzyme reverse transcriptase, which does the reverse of transcription; that is, it synthesises DNA from an RNA template (it is produced naturally by a group of viruses called the retroviruses, which include HIV). Although there are some 70 000 genes in the human genome, a given cell only expresses a few, so only makes a few different kinds of mRNA molecule. Using cDNA makes these genes much easier to find. For example, the $\beta$-cells of the pancreas make insulin, and so make a lot of mRNA molecules coding for insulin. This mRNA can be isolated from these cells and used to make cDNA of the insulin gene.

# 19.15   DNA probes

DNA probes (hybridisation probes) are used to identify and label DNA fragments that contain a specific sequence. A probe is simply a short length of DNA (20–100 nucleotides long) with a label attached. There are two common types used:

- A radioactive-labelled probe (synthesised using the isotope $^{32}$P), which can be visualised using a photographic film (an autoradiograph).

- A fluorescent-labelled probe, which will emit visible light when illuminated with ultraviolet (UV) light. Probes can be made to fluoresce with different colours.

Probes are always single-stranded and can be made of DNA or RNA. If a probe is added to a mixture of different pieces of DNA (e.g. restriction fragments) it will anneal (hybridise or base pair) with any lengths of DNA containing the complementary sequence. These fragments become labelled and can be identified. DNA probes have many uses in genetic engineering, for example:

- Identification of restriction fragments containing a particular gene.

- Identification of the short DNA sequences used in DNA fingerprinting.

- Identification of genes from one species which are similar to those of another species. Most genes are remarkably similar in sequence from one species to another, so for example a gene probe for a mouse gene will probably anneal with the same gene from a human. This has aided the identification of human genes.

- Identification of genetic defects. DNA probes have been prepared which match the sequences of many human genetic disease genes, such as muscular dystrophy and cystic fibrosis. Such probes can be attached to a grid, forming a DNA microarray (or DNA chip). A sample of human DNA is added to the array and any sequences that match any of the various probes will stick to the array and be labelled. This allows rapid testing for a large number of genetic defects at a time.

---

A DNA microarray is a multiplex (performs multiple assays concurrently) technology used in molecular biology. It consists of an arrayed series of thousands of microscopic spots of DNA oligonucleotide (a short nucleic acid polymer, typically with 20 or fewer bases), called features, each containing picomoles of a specific DNA sequence. This might be a short section of a gene, or another DNA element, used as a probe to hybridise a cDNA or cRNA sample (called the target). Probe–target hybridisation is usually detected and quantified by detection of a fluorophore-, silver- or chemiluminescence-labelled target. In standard microarrays, the probes are covalently attached to a solid surface; the solid surface can be a glass or silicon chip, in which case the microarray is commonly known as a gene chip. DNA microarrays are commonly used to measure changes in expression levels in order to detect single nucleotide polymorphisms in genotyping or in resequencing mutant genomes.

---

# 19.16 DNA sequencing

DNA sequencing is the reading of the base sequence of a length of DNA. Dideoxynucleotide sequencing (commonly called Sanger sequencing) utilises $2',3'$-dideoxynucleotide triphosphates (ddNTPs), in which an H atom replaces the OH group at the $3'$ carbon (Figure 19.5). When such nucleotides are incorporated into a growing DNA chain, they will terminate chain elongation, since they are unable to form a phosphodiester bond with the next deoxynucleotide.

With all four normal nucleotides present, chain elongation will proceed until, by chance, DNA polymerase inserts a dideoxy base (shown in colour in Figure 19.5), thereby halting the

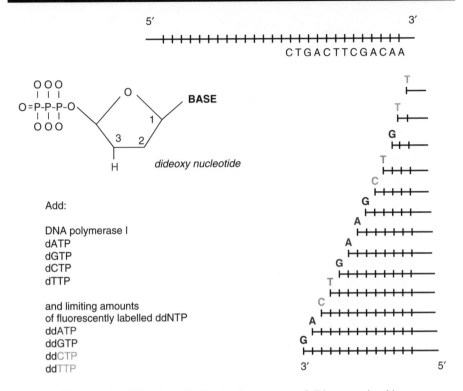

**Figure 19.5**  DNA polymerisation in the presence of dideoxy nucleotides.

process. The ratio of normal to dideoxy nucleotide is deliberately high in order for a range of DNA strand lengths to be produced.

Following incubation, fragments are separated by electrophoresis (shortest fragments move the furthest). Each of the four dideoxynucleotides fluoresce a different colour when illuminated by laser, and scanning provides a printout and automated determination of the sequence (Figure 19.6).

---

Electrophoresis is a form of chromatography used to separate different pieces of DNA according to their length. DNA samples are placed into wells at the top of a gel (usually made of agarose). An electric current is passed across the gel; since nucleotides contain a negatively charged phosphate group, the DNA is attracted to the anode (the positive electrode). The molecules have to diffuse through the gel, therefore smaller lengths of DNA move faster than larger lengths. The completed gel may be visualised with chemicals that specifically label DNA, for example ethidium bromide, or else the initial DNA samples can be radiolabelled, for example with $^{32}$P, and developed with photographic film, or labelled with a fluorescent molecule.

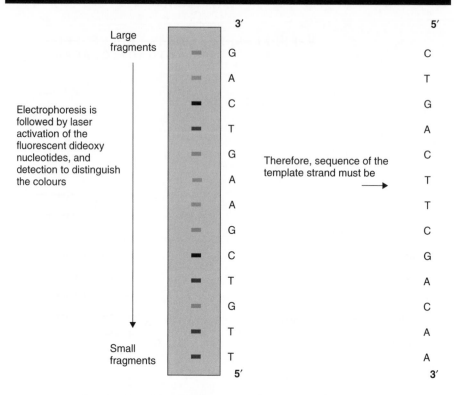

**Figure 19.6**     Representation of an acrylamide sequencing gel.

## 19.17   Genetic engineering applications

Genetic engineering applications might include:

- Commercial gene products, using genetically modified organisms (usually microbes) to produce chemicals, usually for medical or industrial applications.

- New phenotypes, using gene technology, to alter the characteristics of organisms (usually farm animals or crops).

- Gene therapy, using gene technology on humans to treat a disease.

## 19.18   Commercial gene products

Commercial gene products are the most successful form of genetic engineering, including products of medical, agricultural or commercial value. The strategy is to transfer a gene (often human) to a host organism (usually a microbe) so that it will make its product quickly, cheaply and ethically. It is also possible to make 'designer proteins' by altering gene sequences, but while this is a useful research tool, there are no commercial applications as yet. Table 19.3 shows some of the advantages and disadvantages of using different organisms for the production of genetically engineered gene products.

**Table 19.3** Organisms used to produce genetically engineered gene products

| Host organism | Advantages | Disadvantages |
|---|---|---|
| Prokaryotes (i.e. bacteria) | No nucleus so DNA easy to modify; have plasmids, small genome; genetics are well understood; asexual so can be cloned; small and fast growing; easy to grow commercially; few ethical problems. | Can't splice introns; no post-translational modification; small gene size. |
| Eukaryotes | Can do post-translational modifications; can accept large genes. | Do not have plasmids (except yeast); often diploid so two copies of genes may need to be inserted; control of expression not well understood. |
| Fungi (yeast, moulds) | Asexual so can be cloned; haploid, so only one copy needed; can be grown in vats. | Can't always make animal gene products. |
| Plants | Photosynthetic so don't need much feeding; can be cloned from single cells; products can be secreted from roots or in sap. | Cell walls difficult to penetrate by vector; slow growing; multi-cellular. |
| Animals (pharming) | Most likely to be able to make human proteins; products can be secreted in milk or urine. | Multi-cellular; slow growing. |

- **Human insulin**. Insulin-dependent diabetes was first successfully treated by injection of insulin extracted from the pancreases of slaughtered cows and pigs. However, the species difference did lead to immune rejection and side effects. The human insulin gene was isolated, cloned and sequenced in the 1970s. In humans, pancreatic cells first make pro-insulin, which is then converted to functional insulin by post-translational modification. Since bacterial cells do no perform post-translational modification, it was some time until a synthetic cDNA gene was made and inserted into the bacterium *E. coli* to make pro-insulin; the post-translational conversion to insulin was carried out chemically. This was the first genetically engineered product approved for medical use. In the 1990s the procedure was further improved by using the yeast *Saccharomyces cerevisiae* instead of *E. coli*. Since yeast is a eukaryote, it is capable of post-translational modification, and so the production of human insulin was further simplified.

- **Bovine somatotrophin (BST)**. The gene has been cloned in bacteria to produce large quantities of BST, a growth hormone produced by cattle. Injection with BST can result in up to a 10% increase in mass in beef cattle and a 25% increase in milk production in dairy cows.

- **Rennin**. This is an enzyme used in the production of cheese. It is produced in the stomach of juvenile mammals (including humans) and aids the digestion of the milk protein casein. The cheese industry used to obtain rennin from the stomach of young calves when they were slaughtered for veal, but there are moral and practical objections to this source. Now an artificial cDNA gene for rennin has been made from mRNA extracted from calf stomach cells, and this gene has been inserted into a variety of microbes. Most rennin is now sourced this way; such cheese products are sometimes labelled as 'vegetarian cheese'.

- **AAT(a-1-antitrypsin)**. This is a human protein made in the liver and found in the blood; it is an inhibitor of protease enzymes such as trypsin and elastase. A rare mutation of the AAT gene (a single base substitution) causes AAT to be inactive and the protease enzymes to be uninhibited. A consequence of this, seen in the lungs, is the digestion of tissue by elastase, leading to the lung disease emphysaema. This condition can be treated by inhaling an aerosol spray containing AAT. Bacteria cannot be used to produce AAT since it is a glycoprotein; instead it is now produced in genetically modified sheep. The AAT gene was coupled to a promoter for the milk protein $\beta$-lactoglubulin, which is only activated in mammary gland cells; AAT is therefore synthesised and secreted into sheep's milk, from which it can be harvested.

## 19.19   Gene therapy

Gene therapy is perhaps the most significant and most controversial kind of genetic engineering; it is also the least well developed. The goal of gene therapy is to genetically alter humans in order to treat a disease; this means altering the genotype of a tissue or even a whole individual. Following a number of setbacks, some promising progress is now being seen:

- There has been reported success in using gene therapy for a type of inherited blindness, Leber congenital amaurosis.

- A follow-up study of severe combined immunodeficiency (SCID) children concluded that 8 of 10 treated seemed to have been cured.

- Cystic fibrosis is a good candidate for gene therapy. The gene for CFTR was identified in 1989 and a cDNA clone was made soon after. The idea is to deliver copies of the good gene to the epithelial cells of the lung, where they can be incorporated into the nuclear DNA and make functional CFTR chloride channels. If about 10% of the cells can be corrected, this will 'cure' the disease. Two methods of delivery are being tried, liposomes and adenoviruses, both delivered with an aerosol inhaler.

As the technology to deliver genes into cells becomes safer and more efficient, advances in gene therapy are likely to be seen.

## 19.20   Controlling gene expression

Control of human gene expression occurs principally at the level of transcription. Transcription is just one step in the conversion of genetic information into a final processed gene product, which includes:

- initiation of transcription

- processing the transcript

- transport of the transcript to the cytoplasm

- translation of the transcript

- post-transcription processing.

Transcriptional initiation is the most important mode for control of eukaryotic gene expression. Specific factors that exert control include the strength of promoter elements within the DNA sequences of a given gene, the presence or absence of enhancer sequences (which enhance the activity of RNA polymerase at a given promoter by binding specific transcription factors) and the interaction between multiple activator proteins and inhibitor proteins.

Transcription of the different classes of RNA in eukaryotes is carried out by three different polymerases, namely:

- RNA pol I synthesises rRNAs, except for the 5S species.

- RNA pol II synthesises mRNAs and some small nuclear RNAs (snRNAs) involved in RNA splicing.

- RNA pol III synthesises 5S rRNA and tRNAs.

The most complex controls observed in eukaryotic genes are those that regulate the transcribed mRNA genes. Almost all eukaryotic mRNA genes contain a basic structure consisting of coding exons and non-coding introns, promoter regions and any number of different transcriptional regulatory domains (Figure 19.7). Promoter regions, such as TATA boxes and CCAAT boxes, are so named because of their sequence motifs; the TATA box resides 20–30 bases upstream of the transcriptional start site, the CCAAT box about 50–130 bases upstream.

The TATA box has the core DNA sequence 5'-TATAAA-3' or a variant, which is usually followed by three or more adenine bases. The sequence is believed to have remained consistent throughout much of the evolutionary process. It is normally bound by the TATA binding protein in the process of transcription, which unwinds the DNA and bends it through 80°. The AT-rich sequence facilitates easy unwinding (due to two hydrogen bonds between bases, as opposed to three between GC pairs). The TATA box is usually found as the binding site of RNA polymerase II. The transcription factor TFIID binds to the TATA box, followed by TFIIA binding to the upstream part of TFIID. TFIIB can then bind to the downstream part of TFIID. The polymerase then recognises this multi-protein complex and binds to it, along with various other transcription factors. Transcription is initiated and the polymerase moves along the DNA strand, leaving TFIID and TFIIA bound to the TATA box. These can then facilitate the binding of additional RNA polymerase II molecules.

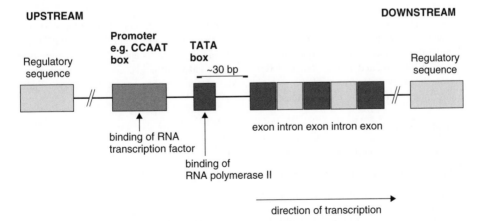

**Figure 19.7** Basic elements of gene expression.

The CCAAT box (sometimes referred to as a CAAT or CAT box) is a distinct pattern of nucleotides with a GGNCAATCT sequence that occurs 50–130 bases upstream of the initial transcription site. The CCAAT box signals the binding site for the RNA transcription factor and is typically accompanied by a conserved sequence. Genes that have this element seem to require it for transcription in sufficient quantities.

Additionally, mRNA genes contain many different regulatory sequences that bind a variety of transcription factors. Regulatory sequences are predominantly located upstream (5′) of the transcription initiation site, although some can occur downstream (3′) or even within the genes themselves. The number and type of regulatory elements varies with each mRNA gene. Different combinations of transcription factors can exert differential regulatory effects upon transcriptional initiation. Various cell types express characteristic combinations of transcription factors; this is the major mechanism for cell-type specificity in the regulation of mRNA gene expression.

## 19.21   Transcription factors

Transcription factors bind to DNA through regions referred to as DNA-binding regions, or motifs; they are highly conserved between species. There are four major structural motifs:

- **Helix-turn-helix motif**. Mediates DNA binding by fitting into the major groove of the DNA helix.

- **Helix-loop-helix motif**. Promotes both DNA binding and protein-dimer formation.

- **Zinc finger**. A DNA-binding motif consisting of specific spacings of cysteine and histidine residues that allow the protein to bind zinc atoms. The metal atom coordinates the sequences around the cysteine and histidine residues into a finger-like domain. The finger domains can interdigitate into the major groove of the DNA helix. The spacing of the zinc finger domain in this class of transcription factor coincides with a half-turn of the double helix. The classic example is the RNA pol III transcription factor, TFIIIA. Proteins of the steroid/thyroid hormone family of transcription factors also contain zinc fingers.

- **Leucine zipper**. This forms a dimer that 'grips' the DNA helix like a peg, by inserting into the major groove.

An example of gene expression can be illustrated by consideration of the action of steroid hormones, and in the control of sterol biosynthesis. Steroid hormones enter the cell by diffusion through the plasma membrane and bind to their steroid hormone receptor. These receptors are part of a large related family that includes those for glucocorticoids, oestrogens, androgens, thyroid hormone, calcitriol and the retinoids. All steroid hormone receptors are zinc finger transcription factors. The receptor must:

- bind to the hormone

- bind to a second copy of itself to form a homodimer

- be in the nucleus, moving from the cytosol if necessary

- bind to its response element on the DNA

- activate other transcription factors to start transcription.

## 19.22   Response element

The response element is a DNA sequence that is bound by the steroid-receptor complex; it is part of the promoter region of a gene. Binding by the receptor complex activates, or represses, the gene controlled by that promoter. For example, the glucocorticoid response element (GRE) is 5′-GGTACAnnnTGTTCT-3′ ('n' can be any nucleotide); this GRE is shared with mineral-corticoids, progesterone and androgens, but not oestrogen.

Sterol regulatory element binding proteins (SREBPs) are transcription factors that regulate sterol biosynthesis, such as cholesterol. SREBPs belong to the basic-helix-loop-helix leucine zipper class of transcription factors. Unactivated SREBPs are attached to the nuclear envelope and endoplasmic reticulum membranes. In cells with low levels of sterols, SREBPs are cleaved (activated) to a water-soluble N-terminal domain that is translocated to the nucleus. Activated SREBPs then bind to specific sterol regulatory element DNA sequences, up-regulating the synthesis of enzymes involved in sterol biosynthesis. Sterols in turn inhibit the cleavage of SREBPs, so downregulating enzyme synthesis through a negative-feedback loop.

---

SREBP are released from a membrane-bound form by proteolytic cleavage. In the case of cholesterol control, two separate site-specific proteolytic cleavages are necessary for release: site-1 protease (S1P) and site-2 protease (S2P). In addition, active SREBP requires the cholesterol-sensing protein and the SREBP cleavage-activating protein (SCAP) to form a complex with SREBP through interaction of their respective carboxy-terminal domains. When SCAP binds cholesterol in the endoplasmic reticulum, it cannot transport SREBP to the Golgi in order for it to be cleaved by S1P. When cellular demand for cholesterol rises, cholesterol dissociates from SCAP and the SREBP–SCAP complex exits the endoplasmic reticulum and travels to the Golgi apparatus, where S1P cleaves SREBP at site 1, cutting it into two halves; the newly generated amino-terminal half of SREBP is then cleaved at site 2 by S2P, a metalloprotease. This releases the cytoplasmic portion of SREBP, which then travels to the nucleus, where it activates transcription of target genes.

---

## 19.23   Genes and cancer: the cell cycle

Most eukaryotic cells will proceed through an ordered series of events in which they duplicate their contents and then divides into two cells. This cycle of duplication and division is called the cell cycle. In order to maintain the fidelity of the developing organism, the process of cell division must be highly ordered and tightly regulated. Any loss of control can lead to abnormal development and cancer. The eukaryotic cell cycle is composed of four phases, as depicted in Figure 19.8.

Of the four phases, the two critical ones are DNA replication, which occurs during $G_1$ to S phase, and the physical process of cell division, which occurs during $G_2$ to M (for mitosis) phase; these are critical 'checkpoints'.

In gap phases $G_1$ and $G_2$, the cell is preparing for DNA replication and cell division respectively. M phase is composed of two discrete steps: mitosis, which constitutes the pairing and separation of the duplicated chromosomes, and cytokinesis, the physical process whereby the cell splits into two daughter cells. Not all cells continue to divide during the life span of an

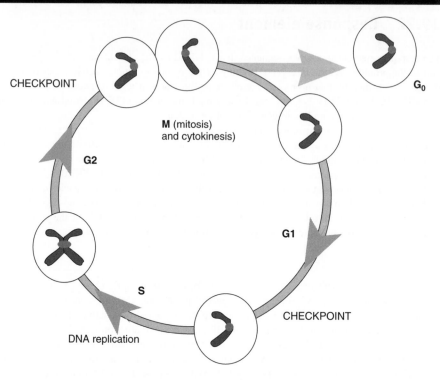

**Figure 19.8** The eukaryotic cell cycle.

organism; many undergo what is referred to as 'terminal differentiation' and become quiescent; cells in this phase are said to reside in another gap phase called $G_0$. Under certain conditions, such as that resulting from an external signal stimulating cell growth, cells can exit the quiescent state and re-enter the cell cycle.

## 19.23.1    Checkpoints and cell-cycle regulation

Progression through the different phases of a cell cycle is controlled at 'checkpoints'. There are a number of checkpoints, but the two most critical are the one that occurs near the end of $G_1$, prior to S-phase entry, and the one near the end of $G_2$, prior to mitosis.

Cell-cycle control mechanisms exert their influences at specific times during each transit through a cell cycle; this is the responsibility of a family of protein kinases called cyclin-dependent kinases (CDKs). CDKs are also involved in the regulation of transcription and mRNA processing. CDKs phosphorylate proteins on serine and threonine amino acid residues; they are serine/threonine kinases. A CDK is activated by association with a cyclin, forming a CDK complex. Cyclins are so named because their concentration varies in a cyclical fashion during the cell cycle; they are produced or degraded as needed in order to drive the cell through the different stages of the cell cycle.

CDKs are considered a potential target for anti-cancer medication; the aim is to selectively interrupt the cell-cycle regulation in cancer cells by interfering with CDK action, causing the cell to die. Currently some CDK inhibitors are undergoing clinical trials.

## 19.23.2  Initiation of cell division and differentiation

Many cells reside in a resting or quiescent state but can be stimulated by external signals to re-enter the cell cycle. These external growth-promoting signals are the result of growth factors binding to their receptors. Most growth factors induce the expression of genes that are referred to as early- and delayed-response genes. The activation of early-response genes occurs in response to growth factor receptor-mediated signal transduction, resulting in phosphorylation and activation of transcription factor proteins that are already present in the cell. Many of the induced early-response genes are themselves transcription factors which in turn activate the expression of delayed-response genes.

Growth factors and growth factor receptors play an important physiological role in the normal process of growth and differentiation. In a simplistic model, the binding of the growth factor to its receptor leads to receptor dimerisation and cross-phosphorylation, activating the receptors. The activated receptors phosphorylate a series of cytoplasmic proteins, which in turn sets off a cascade of events, leading to the activation of transcription factors in the nucleus, which then leads to increased mRNA synthesis. The translation of the mRNA results in increased protein synthesis, finally leading to either growth or differentiation.

Although a cell may respond to a vast number of growth factors and possess a variety of types of receptor, there are only a few known intracellular second messenger systems through which all these signals can be channelled, through the cytoplasm and into the nucleus. These are the cyclic AMP and the cyclic GMP systems, the control of free intracellular calcium levels usually mediated by the action of inositol 1,4,5-triphosphate, the pathways involving receptor protein tyrosine kinases and the tumour growth factor *b* (TGF*b*), which utilises receptor serine/threonine kinases. See Chapter 13 for more on intracellular signalling.

## 19.23.3  Genes controlling the cell cycle

Genes that are involved in the control of normal cell proliferation can be classified as proliferative or antiproliferative:

- Proliferative genes are proto-oncogenes; mutated proto-oncogenes may become oncogenes.

- Antiproliferative or tumour-suppressor genes act to suppress cell proliferation.

### 19.23.3.1  Proliferative genes

Most, if not all cancer cells contain genetic damage that appears to be the responsible event leading to tumourigenesis. The genetic damage present in a parental tumourigenic cell is maintained such that it is a heritable trait of all cells of subsequent generations.

The distinction between the terms 'proto-oncogene' and 'oncogene' relates to the activity of the protein product of the gene. A proto-oncogene is a gene whose protein product has the capacity to induce cellular transformation, given it sustains some genetic insult. An oncogene is a gene that has sustained some genetic damage and, therefore, produces a protein capable of cellular transformation. The process of activation of proto-oncogenes to oncogenes can include retroviral transduction or retroviral integration (see below), point mutations, insertion mutations, gene amplification, chromosomal translocation and/or protein–protein interactions.

Proto-oncogenes can be classified into many different groups, based upon their normal function within cells or based upon sequence homology to other known proteins. Proto-oncogenes have been identified at all levels of the various signal transduction cascades that control cell growth, proliferation and differentiation.

*19.23.3.2    Antiproliferative or tumour-suppressor genes*

Antiproliferative genese, or more precisely the proteins for which they code, either have a repressive effect on the regulation of the cell cycle or promote apoptosis, or sometimes both. Tumour suppressors are so called because cancer ensues as a result of a loss of their normal function; that is, these proteins suppress the ability of cancer to develop. The two most important check points in the eukaryotic cell cycle are the $G_1$ to S transition and the entry into mitosis. The former prevents DNA replication prior to repair of damaged DNA and the latter prevents damage that may have occurred to the DNA during replication from being propagated into daughter cells during mitosis. The proteins encoded by the retinoblastoma susceptibility gene (pRB) and by the p53 protein are both tumour suppressors. The function of pRB is to act as a brake, preventing cells from exiting $G_1$, and that of p53 is to inhibit progression from S phase to M phase. Under normal circumstances, p53 levels remain very low. In response to DNA damage, for example as a result of UV irradiation or $\gamma$ irradiation, cells activate several kinases, including checkpoint kinase 2 (CHK2); the target of these kinases is p53. When p53 is phosphorylated it is released to carry out its transcriptional activation functions. One target of p53 is the cyclin inhibitor $p21^{Cip1}$ gene; activation of $p21^{Cip1}$ effectively leads to stoppage of the cell cycle, either prior to S-phase entry or during S phase. The aim is to allow the cell time to repair its damaged DNA prior to replication or mitosis.

The p53 tumour suppressor protein is encoded by the *TP53* gene. Homozygous loss of p53 is found in 70% of colon cancers, 30–50% of breast cancers and 50% of lung cancers. Mutated p53 is also involved in the pathophysiology of leukaemia, lymphoma, sarcoma and neurogenic tumours. Abnormalities of the p53 gene can be inherited in Li–Fraumeni syndrome, which increases the risk of developing various types of cancers (see Focus on: p53).

# 19.24    Viruses and cancer

Tumour cells also can arise by non-genetic means through the actions of specific tumour viruses. Tumour viruses are of two distinct types, those with DNA genomes (e.g. papilloma and adenoviruses) and those with RNA genomes (termed retroviruses). RNA tumour viruses are common in chickens, mice and cats but rare in humans. The only currently known human retroviruses are the human T-cell leukaemia viruses and the related retrovirus (see Chapter 15).

Cellular transformation by DNA tumour viruses in most cases has been shown to be the result of protein–protein interaction. Proteins encoded by the DNA tumour viruses, termed tumour antigens (T antigens), can interact with cellular proteins. This interaction effectively sequesters the cellular proteins away from their normal functional locations within the cell. The predominant types of protein sequestered by viral T antigens have been shown to be of the tumour-suppressor type. It is the loss of their normal suppressor functions that results in cellular transformation.

In the retroviruses, the viral RNA genome is first converted into DNA. During this process part of the host genome may be incorporated into the viral genome (transduction). Should that host genome include a proliferative gene, the transduced gene will confer a growth advantage to the infected cell. Alternatively, the integration of a retrovirus genome into the host genome (a random process) may place the powerful viral promoter region close to a host gene that encodes a growth-regulating protein. If the protein is expressed at an abnormally elevated level it can result in cellular transformation. This is termed retroviral integration-induced transformation. It has been shown that HIV induces certain forms of cancer in infected individuals by this integration induced-transformation process.

# 19.25  Apoptosis

Apoptosis, or programmed cell death, is a normal physiological event. For example, the differentiation of fingers and toes in a developing human embryo occurs because cells between the fingers apoptose; the result is that the digits are separate. Apoptosis also functions to remove damaged cells. Apoptosis plays a major role in preventing cancer. If a cell is unable to undergo apoptosis, it continues to divide and develop into a tumour. For example, infection by papillomavirus causes a viral gene to interfere with the cell's p53 protein; this interference in the apoptotic capability of the cell plays a role in the development of cervical cancer.

The process of apoptosis is controlled by a diverse range of cell signals, which may be either extracellular or intracellular. Extracellular signals may include toxins, hormones, growth factors, nitric oxide or cytokines; these must either cross the plasma membrane or transduce to effect a response. These signals may positively or negatively induce apoptosis. A cell initiates intracellular apoptotic signalling in response to a stress. The binding of nuclear receptors by glucocorticoids, heat, radiation, nutrient deprivation, viral infection, hypoxia or increased intracellular calcium concentrations can trigger the release of intracellular apoptotic signals by a damaged cell.

Two theories for the direct initiation of apoptotic mechanisms in mammals have been suggested:

- the TNF-induced (tumour necrosis factor) model

- the Fas-Fas ligand-mediated model.

Both involve receptors of the tumour necrosis factor receptor (TNFR) family coupled to extrinsic signals.

TNF is a cytokine produced mainly by activated macrophages, and is the major extrinsic mediator of apoptosis. Most cells in the human body have two receptors for TNF: TNF-R1 and TNF-R2. The binding of TNF to TNF-R1 has been shown to initiate the pathway that leads to caspase activation via the intermediate membrane proteins TNF receptor-associated death domain (TRADD) and Fas-associated death domain (FADD). The link between TNF and apoptosis shows why an abnormal production of TNF plays a fundamental role in several human diseases, especially autoimmune diseases (see Chapter 15).

# 19.26  Caspases

Many pathways and signals lead to apoptosis, but there is only one mechanism that actually causes the death of a cell, namely the activation of proteolytic caspases. Caspases (cysteine-aspartic acid proteases) are a family of cysteine proteases, first synthesised as inactive procaspases. Eleven caspases have so far been identified in humans, taking one of two forms: the initiator (apical) caspases and the effector (executioner) caspases. Initiator caspases (e.g. CASP2, CASP8, CASP9 and CASP10) cleave inactive pro-forms of effector caspases, thereby activating them. Effector caspases (e.g. CASP3, CASP6 and CASP7) in turn cleave other protein substrates within the cell, to trigger the apoptotic process. The initiation of this cascade reaction is regulated by caspase inhibitors. Caspases are regulated at a post-translational level, ensuring that they can be rapidly activated.

# Focus on: p53

p53 is a transcription factor which in humans is encoded by the *TP53* gene, located on the short arm of chromosome 17 (17p13.1). It regulates the cell cycle and thus functions as a tumour suppressor involved in preventing cancer; p53 has been described as 'the guardian of the genome'. Its name derives from its apparent molecular mass, 53 KDa, on SDS-PAGE.

The 393 amino acid-long human p53 has seven domains:

1. N-terminal transcription-activation domain (TAD), also known as activation domain 1 (AD1), which activates transcription factors: residues 1–42.

2. Activation domain 2 (AD2), important for apoptotic activity: residues 43–63.

3. Proline-rich domain, important for the apoptotic activity of p53: residues 80–94.

4. Central DNA-binding core domain (DBD), which contains one zinc atom and several arginine amino acids: residues 100–300.

5. Nuclear localisation signalling domain, residues 316–325.

6. Homo-oligomerisation domain (OD): residues 307–355. Tetramerisation is essential for the activity of p53 *in vivo*.

7. C-terminal, which is involved in downregulation of DNA binding of the central domain: residues 356–393.

Mutations that deactivate p53 in cancer usually occur in the DBD. Most of these mutations prevent binding of the protein to its target DNA sequences, and thus prevent transcriptional activation of these genes. As such, mutations in the DBD are recessive loss-of-function mutations.

p53 has many anti-cancer mechanisms (Figure 19.9):

• It can activate DNA repair proteins when DNA has sustained damage.

• It can induce growth arrest by holding the cell cycle at the $G_1/S$ checkpoint on DNA damage recognition. (If it holds the cell here for long enough, the DNA repair proteins will have time to fix the damage and the cell will be allowed to continue the cell cycle.)

• It can initiate apoptosis if the DNA damage proves to be irreparable.

In a normal cell, p53 is inactivated by its negative regulator, mdm2 (mdm2 protein functions both as an E3 ubiquitin ligase, which recognises the N-terminal TAD of the p53 tumour suppressor, and as an inhibitor of p53 transcriptional activation). Upon DNA damage or other stress, various pathways will lead to the dissociation of the p53 and mdm2 complex. Once activated, p53 will either induce a cell-cycle arrest or initiate apoptosis. Activated p53 binds DNA and activates expression of several genes, including one encoding for p21 protein, a potent CDK inhibitor. The p21 protein binds to and inhibits the activity of cyclin-CDK2 or -CDK4 complexes, and thus functions as a regulator of cell-cycle progression at $G_1$. When p21 is complexed with CDK2, the cell is unable to pass through to the next stage of cell division. Mutant p53 can no longer bind DNA in an effective way, and as a consequence the p21 protein is not made available to act as the 'stop signal' for cell division; thus cells divide uncontrollably and cancer will ensue.

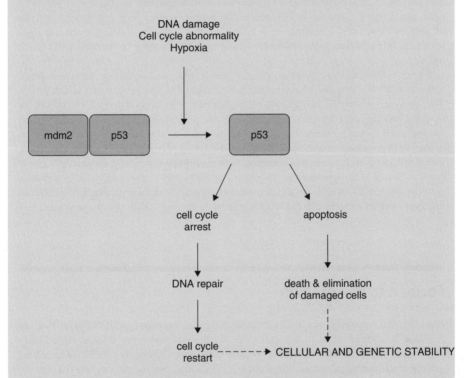

**Figure 19.9** p53 roles in the cell.

p53 can be activated by a variety of stress types, including DNA damage (induced by UV, IR or chemical agents), oxidative stress, osmotic shock, ribonucleotide depletion and deregulated oncogene expression. Activation is marked by two major events: first, the half-life of the p53 protein is increased, leading to a rapid accumulation of p53 in stressed cells; second, a conformational change forces p53 to take on an active role as a transcription regulator in these cells. The critical event leading to the activation of p53 is the phosphorylation of its N-terminal domain. The N-terminal transcriptional activation domain contains a large number of phosphorylation sites and can be considered the primary target for protein kinases that transduce stress signals.

In unstressed cells, p53 levels are kept low through a continuous degradation of p53. The protein mdm2 binds to p53, preventing its action and transporting it from the nucleus to the cytosol. Also, mdm2 acts as a ubiquitin ligase and covalently attaches ubiquitin to p53, thus marking it for degradation by the proteosome. Phosphorylation of the N-terminal end of p53 disrupts mdm2 binding.

If the *TP53* gene is damaged, tumour suppression is severely reduced. Individuals who inherit only one functional copy of the *TP53* gene will most likely develop tumours in early adulthood, a disease known as Li–Fraumeni syndrome. The *TP53* gene can also be damaged in cells by mutagens, increasing the likelihood that the cell will begin uncontrolled division. More than 50% of human tumours contain a mutation or deletion of the *TP53* gene. Increasing the amount of p53, which may initially seem a good way to treat tumours or prevent them from spreading, is in actuality not a usable method of treatment, since it can cause premature aging. However, restoring endogenous p53 function holds much promise.

Certain pathogens can also affect the p53 protein. One such pathogen, the human papillomavirus (HPV), encodes a protein, E6, which binds the p53 protein and inactivates it. This, in synergy with the inactivation of another cell-cycle regulator, p105RB, allows for repeated cell division, which is manifested in the clinical disease of warts. Infection by oncogenic HPV types, especially HPV16, can also lead to progression from a benign wart to low- or high-grade cervical dysplasia, reversible forms of precancerous lesions. Persistent infection causes irreversible changes, leading to carcinoma *in situ* and eventually invasive cervical cancer. This results from the effects of HPV genes, particularly those encoding proteins E6 and E7, which are the two viral oncoproteins that are preferentially retained and expressed in cervical cancers, by integration of the viral DNA into the host genome.

# Focus on: *Ras* protein

*Ras* (reticular activating system) is a family of genes encoding small GTPases that are involved in cellular signal transduction; the *ras* protein belongs to a large super family of proteins known as 'low-molecular weight G-proteins'. These G-proteins, which bind guanine nucleotides (guanosine triphosphate (GTP) and guanosine diphosphate (GDP)), are referred to as 'low-molecular weight' in order to distinguish them from another, distinct class of guanine nucleotide-binding proteins, the heterotrimeric G-proteins (see Chapter 13). They are single-subunit proteins, related in structure to the $G_\alpha$ subunit of heterotrimeric G-proteins (large GTPases).

Activation of *ras* signalling can cause cell growth and differentiation. Since *ras* communicates signals from outside the cell to the nucleus, mutations in *RAS* genes can permanently activate it and cause inappropriate transmission inside the cell, even in the absence of extracellular signals. Deregulated *ras* signalling can ultimately lead to oncogenesis and cancer. Activating mutations in *ras* are found in 20–25% of all human tumours, and in up to 90% of specific tumour types.

There are more than a hundred proteins in the *ras* super family. Each subfamily shares the common core G domain, which provides essential GTPase and nucleotide exchange activity.

*Ras* activates several pathways, of which the mitogen-activated protein (MAP) kinase cascade has been most well studied. This cascade transmits signals downstream and results in the transcription of genes involved in cell growth and division.

*Ras* G-proteins function as signalling switches, with 'on' and 'off' states. In the off state they are bound to the nucleotide GDP, while in the on state they are bound to GTP. In the GTP-bound conformation, *ras* has high affinity for numerous effectors, which allow it to carry out its functions.

*Ras* is attached to the cell membrane by prenylation; prenylation is the addition of hydrophobic prenyl groups (3-methyl-2-buten-1-yl) to the protein to facilitate its attachment to the cell membrane (forming a 'lipid anchor').

The intrinsic GTPase activity of *ras* is stimulated by a family of proteins collectively known as GAPs (GTPase-activating proteins). The tumour suppressor gene NF1 encodes a *ras*-GAP; its mutation is evident in neurofibromatosis. *Ras* oncogenes can be activated by point mutations, such that their GTPase reaction can no longer be stimulated by GAP; this increases the half-life of active *ras*-GTP mutants. Constitutively active *ras* (*ras*$^D$) contains mutations that prevent GTP hydrolysis, thus locking *ras* in a permanent on state.

Mutations in the *ras* family of proto-oncogenes (comprising H-*ras*, N-*ras* and K-*ras*) are very common. The *ras* inhibitor, trans-farnesylthiosalicylic acid (FTS, salirasib), exhibits profound anti-oncogenic effects in many cancer cell lines.

Inappropriate activation of the gene has been shown to play a key role in signal transduction, proliferation and malignant transformation. Mutations in a number of different genes, as well as *ras* itself, can have this effect. Oncogenes such as p210BCR-ABL and the growth receptor erbB are upstream of *ras*, so if they are constitutively activated their signals will transduce through *ras*.

# CHAPTER 20

# Antibacterial drug resistance

## 20.1 Horizontal gene transfer

Antibiotic resistance in bacteria is likely a result of horizontal gene transfer, and also of unlinked point mutations in the pathogen's genome, at a rate of about 1 in $10^8$ per chromosomal replication.

> Horizontal gene transfer (or lateral gene transfer) is any process in which an organism incorporates genetic material from another organism without being its offspring. Vertical transfer occurs when an organism receives genetic material from its ancestor. Amongst single-celled organisms, horizontal gene transfer may be the dominant form of genetic transfer.

The bacterial protein LexA has been identified as playing a key role in the acquisition of bacterial mutations. Repressor LexA represses SOS response genes that code for those DNA-polymerases required for repairing DNA damage. The SOS response is a post-replication DNA-repair system that allows DNA replication to bypass lesions or errors in the DNA.

During normal growth, the SOS genes are negatively regulated by LexA repressor protein dimers; LexA binds to a 20-bp consensus sequence (the SOS box) in the operator region for those genes.

Antibiotics, such as ciprofloxacin, that inhibit normal DNA replication in bacteria can be counteracted by the polymerase repair molecules from the SOS response. However, certain polymerases in the SOS pathway are error-prone in their copying of DNA, which leads to mutations. While such mutations are often lethal to the cell, they can also improve the bacteria's survival.

*Essential Biochemistry for Medicine*   Dr Mitchell Fry
© 2010 John Wiley & Sons, Ltd

The main methods by which microorganisms exhibit resistance to antimicrobials are:

- Drug inactivation or modification, for example the enzymatic deactivation of penicillin G in some penicillin-resistant bacteria, through the production of $\beta$-lactamases.

- Alteration of target site, for example alteration of penicillin-binding proteins (PBPs); PBPs are a group of proteins that are characterised by their affinity for and binding of penicillin. They are a normal constituent of many bacteria, and all $\beta$-lactam antibiotics bind to PBPs to exert their effect.

- Alteration of a metabolic pathway, for example some sulphonamide-resistant bacteria 'switch' to using pre-formed folic acid, rather than synthesise the precursor para-aminobenzoic acid (PABA), which is the reaction inhibited by sulphonamides.

- Reduced drug accumulation, by decreasing the permeability and uptake and/or increasing active efflux (pumping out) of the drug across the cell membrane.

## 20.2   Penicillin resistance

The $\beta$-lactam antibiotics are a broad class of antibiotics which include penicillin derivatives, cephalosporins, monobactams, carbapenems and $\beta$-lactamase inhibitors; that is, any antibiotic agent that contains a $\beta$-lactam nucleus in its molecular structure (Figure 20.1). They are the most widely-used group of antibiotics.

If the bacterium produces the enzyme $\beta$-lactamase (penicillinase), the $\beta$-lactam ring of the antibiotic will be enzymatically 'opened' and rendered ineffective. Genes encoding these enzymes may be inherently present on the bacterial chromosome, or may be acquired via plasmid transfer (horizontal gene transfer); $\beta$-lactamase gene expression may also be induced by exposure to $\beta$-lactams. The production of a $\beta$-lactamase by a bacterium does not necessarily rule out all treatment options with $\beta$-lactam antibiotics. In some instances $\beta$-lactam antibiotics may be co-administered with a $\beta$-lactamase inhibitor.

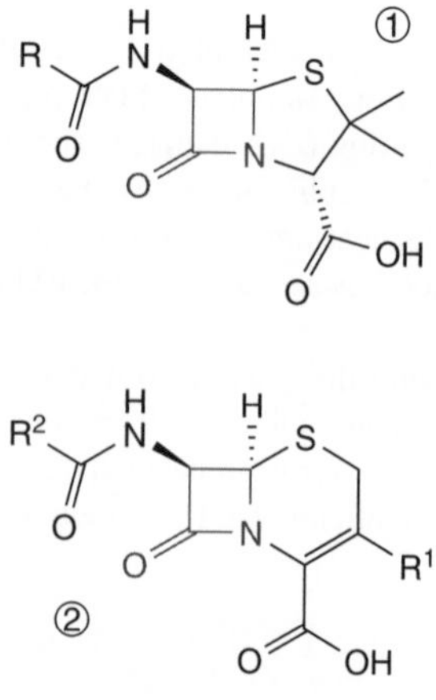

**Figure 20.1**   $\beta$-lactam structures: (1) penicillin and (2) cephalosporin; $\beta$-lactam ring shown in red.

Clavulanic acid (Figure 20.2) is a $\beta$-lactamase inhibitor used to overcome resistance in bacteria that secrete $\beta$-lactamase. It is usually combined with amoxicillin (Figure 20.2).

Clavulanic acid                                    Amoxicillin

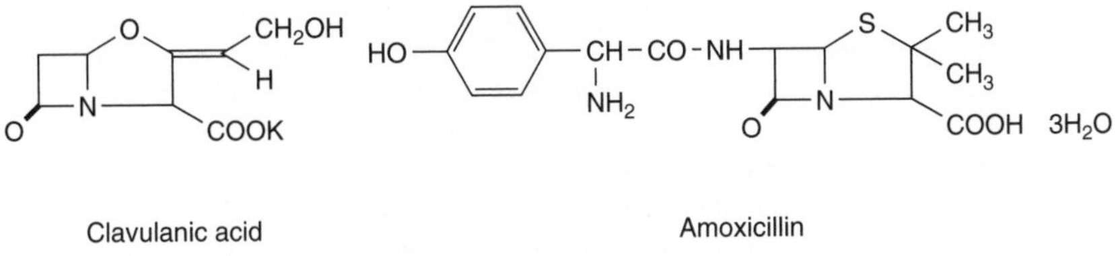

**Figure 20.2**   Structures of clavulanic acid and amoxicillin.

$\beta$-lactam antibiotics are bactericidal and act by inhibiting the synthesis of the peptidoglycan layer of bacterial cell walls. The peptidoglycan layer is important for cell-wall structural integrity, especially in Gram-positive organisms (Figure 20.3).

The cross-linking (transpeptidation) of the peptidoglycan chains is facilitated by transpeptidases known as penicillin-binding proteins. $\beta$-lactam antibiotics are analogues of D-alanyl-D-alanine, the terminal amino acid residues on the precursor N-acetylmuramic/N-acetylglucosamine (NAM/NAG)-peptide subunits of the nascent peptidoglycan layer. The structural similarity between $\beta$-lactam antibiotics and D-alanyl-D-alanine facilitates their

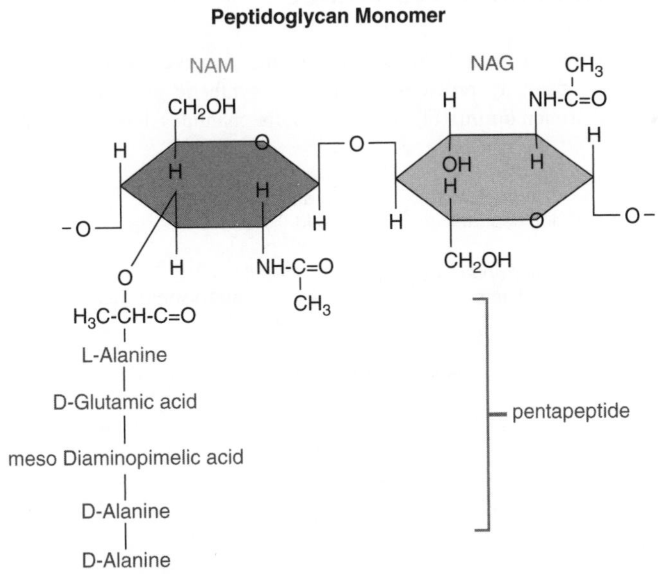

**Peptidoglycan Monomer**

**Figure 20.3**   The peptidoglycan monomer. Once the new peptidoglycan monomers are inserted, glycosidic bonds link these monomers into the growing chains of peptidoglycan. The long sugar chains are then joined to one another by means of peptide cross-links (transpeptidation) between the peptides coming off the NAMs. NAM = N-acetylmuramic acid, NAG = N-acetylglucosamine.

binding to the active site of PBPs. The $\beta$-lactam nucleus of the molecule irreversibly binds to (acylates) the $Ser_{403}$ residue of the PBP active site, preventing the final cross-linking of the nascent peptidoglycan layer and so disrupting cell-wall synthesis.

In the absence of antibiotic, peptidoglycan precursors signal a reorganisation of the bacterial cell wall, triggering the activation of autolytic cell-wall hydrolases. In the presence of antibiotic, a build-up of peptidoglycan precursors also triggers the digestion of existing peptidoglycan by autolytic hydrolases, but without the production of new peptidoglycan. As a result, the bactericidal action of $\beta$-lactam antibiotics is further enhanced.

A further mode of $\beta$-lactam resistance is due to an alteration in the PBP's structure, resulting in ineffective binding of the antibiotic. Notable examples of this mode of resistance include methicillin-resistant *Staphylococcus aureus* (MRSA) and penicillin-resistant *Streptococcus pneumoniae*.

---

Some six different PBPs are routinely detected in all strains of *E. coli*, with varying affinities for penicillin. They have been shown to catalyse a number of reactions involved in the process of synthesising cross-linked peptidoglycan from lipid intermediates and mediating the removal of D-alanine from the precursor of peptidoglycan; the enzyme has a penicillin-insensitive transglycosylase N-terminal domain (involved in the formation of linear glycan strands) and a penicillin-sensitive transpeptidase C-terminal domain (involved in the cross-linking of the peptide subunits).

---

## 20.3   Sulphonamide resistance

Humans cannot synthesise folic acid. Many bacteria, however, synthesise it from PABA; this bacteria-specific pathway provides a target for synthetic antimicrobial agents like the sulphonamides and trimethoprim (Figure 20.4). Sulphonamides inhibit dihydropteroate syn-

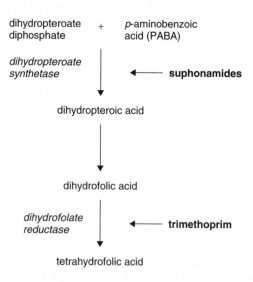

**Figure 20.4**   Biosynthesis of folic acid.

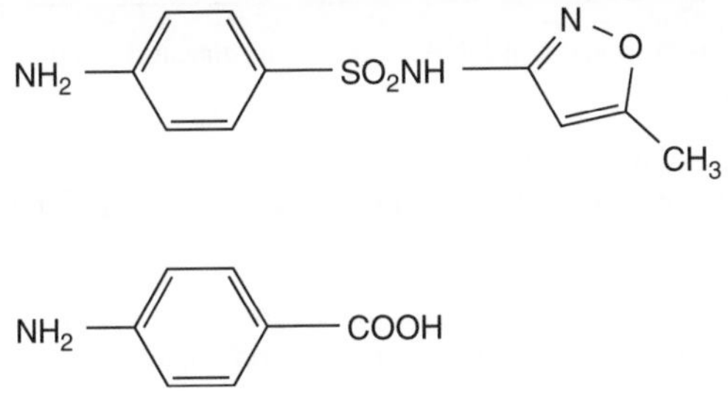

**Figure 20.5** Suphamethoxazole and PABA.

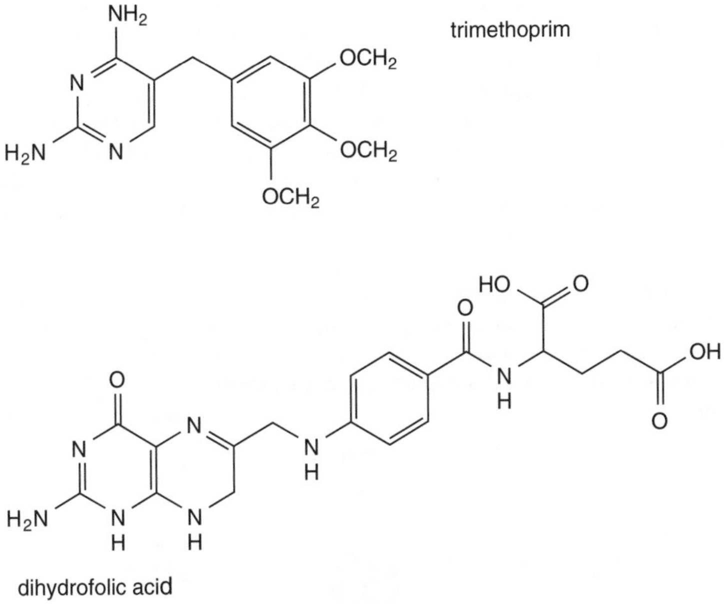

**Figure 20.6** Trimethoprim and dihdrofolic acid.

thetase; they are structural analogues of the normal substrate, PABA (Figure 20.5). Trimethoprim inhibits dihydrofolate reductase, the next step in the folic acid biosynthetic pathway (Figure 20.6).

Sulphonamides and trimethoprim have been used for many decades as efficient and inexpensive antibacterial agents, but resistance to both has spread extensively and rapidly, due to horizontal spread of resistance genes. Two genes, sul1 and sul2, mediated by transposons and plasmids, express dihydropteroate synthases that are highly resistant to sulphonamide. For trimethoprim, almost 20 phylogenetically different resistance genes, expressing drug-insensitive dihydrofolate reductases, have been characterised. They are efficiently spread as cassettes in integrons, and on transposons and plasmids.

- **Plasmids** are extra-chromosomal DNA, usually circular and double-stranded; they occur naturally in bacteria.

- **Transposons** are sequences of DNA that can move to different positions within the genome (transposition); once called 'jumping genes', they are examples of mobile genetic material.

- **Intergrons** are mobile DNA elements with the ability to capture genes, notably those encoding antibiotic resistance, by site-specific recombination.

## 20.4   Bacterial efflux pumps

Bacterial efflux pumps have evolved as survival mechanisms for bacteria to reduce the internal concentration of noxious chemicals from their environment. These same pumps can expel antibiotics and other drugs used in the therapy of infections.

Gram-negative bacteria, such as *E. coli* and *Pseudomonas*, are bound by two membranes, so the efflux pumps must traverse both in order to pump substances out (Figure 20.7).

Multidrug efflux pumps are tripartite export machines. A complex formed by an inner-membrane transporter and a periplasmic adaptor protein contacts an outer-membrane channel tunnel. Interaction with the adaptor protein leads to an opening of the periplasmic entrance of channel tunnel prerequisite for a successful export.

Efflux pumps belonging to the resistance-nodulation-division (RND) family are especially effective in generating resistance and often have a wide substrate specificity. An extreme case is the AcrB pump of *E. coli*, which pumps out the antibiotics tetracycline, chloramphenicol, ß-lactams, novobiocin, fusidic acid, nalidixic acid and fluoroquinolones, as well as detergents, bile salts, various cationic dyes, disinfectants and even solvents.

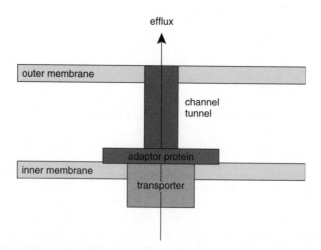

**Figure 20.7** Multidrug efflux pump. Model of a channel tunnel-dependent export apparatus. Interaction with the adaptor protein opens the entrance of the channel tunnel, allowing export of proteins or drugs. In contrast to the channel tunnel, the structure of the adaptor protein is unknown.

## 20.5 Pseudomonas aeruginosa

*Pseudomonas aeruginosa* is a highly prevalent opportunistic pathogen; it has a low antibiotic susceptibility, attributable to a concerted action of multidrug efflux pumps with chromosomally encoded antibiotic resistance genes and the low permeability of the bacterial cellular envelope. It also easily develops acquired resistance, either by mutation in chromosomally encoded genes, or by the horizontal gene transfer of antibiotic resistance determinants. Development of multidrug resistance by *P. aeruginosa* isolates requires several different genetic events, including the acquisition of different mutations and/or horizontal transfer of antibiotic resistance genes. Hypermutation favours the selection of mutation-driven antibiotic resistance in *P. aeruginosa* strains producing chronic infections, whereas the clustering of several different antibiotic resistance genes in integrons favours the concerted acquisition of antibiotic resistance determinants.

## 20.6 Vancomycin

Vancomycin is a glycopeptide antibiotic structure, used in the prophylaxis and treatment of infections caused by Gram-positive bacteria.

It is often reserved as the 'drug of last resort'. Vancomycin has increasingly become a first-line therapy in resistant *Staphylococcus aureus* infections. Vancomycin prevents NAM acid and NAG-peptide subunits from being incorporated into the peptidoglycan matrix; the large hydrophilic molecule forms hydrogen-bond interactions with the terminal D-alanyl-D-alanine moieties of the NAM/NAG-peptides (Figure 20.8).

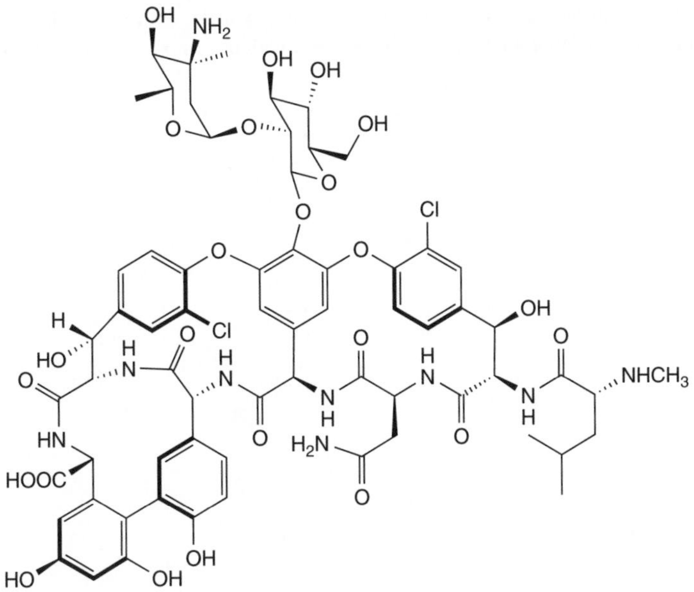

**Figure 20.8** Structure of vancomycin.

## 20.7 Staphylococcus aureus

*Staphylococcus aureus* is one of the major resistant pathogens. Found on the mucous membranes and the skin of around a third of the population, it is extremely adaptable to antibiotic pressure. MRSA was first detected in the early 1960s and is now 'quite common' in hospitals. Strains with intermediate (4–8 μg/ml) levels of resistance, termed glycopeptide intermediate *Staphylococcus aureus* (GISA) or vancomycin intermediate *Staphylococcus aureus* (VISA), began appearing in the late 1990s; the first documented strain with complete (>16 μg/ml) resistance to vancomycin, termed vancomycin-resistant *Staphylococcus aureus* (VRSA), appeared in 2002.

Community-acquired methicillin-resistant *Staphylococcus aureus* (CA-MRSA) has now emerged as an epidemic that is responsible for rapidly progressive and fatal diseases, including necrotising pneumonia, severe sepsis and necrotising fasciitis ('flesh-eating disease'). The epidemiology of infections caused by MRSA is rapidly changing. In the past 10 years, several infections caused by this organism have emerged in the community.

MRSA is acknowledged to be a human commensal and pathogen. MRSA in animals is generally believed to be derived from humans; pet owners can transfer the organism to their pets, and vice versa.

## 20.8 Clostridium difficile

*Clostridium difficile* is a commensal Gram-positive anaerobic bacterium of the human intestine, found in about 2–5% of the population. *C. difficile* is the most serious cause of antibiotic-associated diarrhoea and can lead to pseudomembranous colitis, a severe infection of the colon, often resulting from eradication of the normal gut flora by antibiotics. Discontinuation of causative antibiotic treatment is often curative. In more serious cases, oral administration of metronidazole or vancomycin is the treatment of choice. The bacterium produces several known toxins, including enterotoxin (toxin A) and cytotoxin (toxin B), both of which are responsible for the diarrhoea and inflammation seen in infected patients; another toxin, binary toxin, has also been described.

# Index

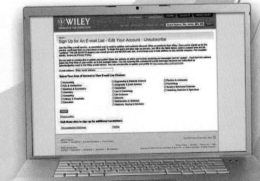